THE AUTHOR

Dr. Dhamo Kessowdas Butani, an Eminent Agricultural Entomologist, formerly Postgraduate Faculty at IARI, New Delhi. After his retirement, he worked as a Consultant at Tribhuvan University, Kathmandu, Nepal.

Insects and Fruits

Insects and Fruits

Dhamo K. Butani

Formerly Senior Entomologist (Fruits)
Indian Agricultural Research Institute
New Delhi - 110 012

2016

Daya Publishing House®

A Division of

Astral International Pvt. Ltd.
New Delhi - 110 002

© 2016, Publisher

Reprinted 1979 edition in arrangement with M/s Periodical Expert Book Agency, Delhi.

ISBN: 978-93-5130-673-3 (International Edition)

Published by   :   **Daya Publishing House®**
A Division of
Astral International Pvt. Ltd.
– ISO 9001:2008 Certified Company –
4760-61/23, Ansari Road
Darya Ganj, New Delhi - 110 002
Phone: 011-4354 9197, 2327 8134
Fax: +91-11-2324 3060
E-mail: info@astralint.com
Website: www.astralint.com

Digitally Printed at   :   Replika Press Pvt. Ltd.

FOREWORD

ADVANCES made in science and technology resulted in better living conditions. The world population rose from 1500 million in 1900 to nearly 4000 million in 1975. At the current rate of growth of 1.93% (2.4% in several tropical countries) the population will be more than 8000 million by 2000 AD. We are always faced with the problem of providing adequate balanced food supplemented with fruits of various kinds for the ever increasing mass of humanity. The problem is more serious in developing countries where about 1500 million individuals still suffer from malnutrition. To solve the problem, food production in these countries has been given top priority, and efforts are being made to intensify agriculture by using high yielding varieties and by preventing losses due to insects and diseases which destroy at least 33% of food man tries to produce.

Dr. Butani has realised the increasing emphasis now being placed on the greater consumption of fruits of good quality as a part of normal diet of general population. He is aware that quality and hence nutritive value depends on healthy fruits. He has compiled very methodically the vast information on insect pests associated with fruit trees in India, giving their nature of damage, bionomics, diagnostic characters and control. Through this book, the author has assiduously prepared and consolidated a solid base on which horticulturists and plant protection personnel can confidently build pest control policies and schedules on regional and national scale. Butani's vast experience is clearly reflected in the manner he has dealt with this difficult subject of apllied horticultural entomology. He has transgressed the limits of 'entomon' to deal with important pests belonging to other groups like mites and nematodes. This obviously, he had to do to forewarn the types of enemies, a fruit grower is likely to confront with, in different parts of the subcontinent.

I congratulate Dr. Dhamo K. Butani for bringing out this compendium which I am sure will serve a good reference book to orchardists, teachers and students of Horticultural Entomology.

Commonwealth Institute
of Entomology, London.

N.C. Pant Ph.D., D.I.C., F.N.A.
Director

with fruit trees in India, giving their

PROLOGUE

INDIA is a vast country having diversified climatic conditions and as such almost all kinds of fruits can be grown in one part or the other - tropical, subtropical and temperate. Over two dozen fruits are commercially grown in India for the past several decades, but the efforts made so far for their improvement both in quantity as well as quality have been negligible as compared to other agricultural crops. Fruits are rich in vitamins and many minerals essential for the proper development of human body. It is rather unfortunate that in India, fruits are not within the easy reach of an average man due to their limited production and consequent high prices. Opportunities for improvement of fruits and increasing area under their cultivation are almost unlimited. Efforts in this direction can lead to higher production of better and cheaper fruits. This has been convincingly shown by a number of countries specially in Europe and America. In our country quantum jumps in the productions of various agricultural crops, like wheat, etc., have shown that we too can achieve the same results, in case of fruits, by proper and concerted efforts. In this connexion a realistic approach will be to lay equal emphasis on both 'production' and 'protection' technologies. Evolving improved varieties and developing appropriate agronomic practices will not be sufficient, special attention will have to be paid to control pests and diseases as well as on the suitable post-harvest technology. For this purpose basic knowledge about the pests, their habits and habitats is an essential pre-requisite.

The earliest available record of Indian Entomology is the inclusion of 12 Indian insects by Carolus Linnaeus (1758) in 10th edition of *Systema Naturae*, followed by Joh. Christ Fabricius (1792-98) who included more than 1000 Indian species in his *Entomologia Sytematica* volumes. The beginning of agricultural entomology in India, can be easily traced to the end of last century. It started with publication of Indian Museum Notes (1889 to 1904). This was followed by monumental works of Maxwell H. Lefroy and F.W. Howlett (1909) and Bainbrigge T. Fletcher (1914) who wrote, *Indian Insect Life* and *Some South Indian Insects* respectively. During the next fifty five years, hardly any book on entomology was published, the only notable exception being, *Handbook of Economic Entomology for South India* by T. V. Ramakrishna Ayyar (1943 reprinted 1963). Quite a few books have been published during the last decade, including, *Insect Pests of Crops* by S. Pradhan (1969), *Textbook of Agricultural Entomology* by Hem Singh Pruthi (1969), *Insects and Mites of Crops in India* by M.R.G.K. Nair (1975), and *General and Applied Entomology* by K.K. Nayar, T.N. Ananthakrishnan and B.V. David (1976). Besides there are, *Agricultural Insect Pests of the Tropics and Their Control* by Dennis S. Hill (1975) and *Diseases, Pests and Weeds in Tropical Crops* edited by Jürgen Kranz, Heinz Schmutterer and Werner Koch (1977), published from Cambridge and Berlin respectively. All these books have dealt with crop pests in general giving very brief account of major pests only. As horticulture is gaining more and more importance and area under fruits is also increasing day by day, necessity has been felt to have a book giving detailed account of the pests of fruit trees. So far, the only publication on this topic is an ICAR bulletin, *Important Fruit Pests of North-West India* by Hem Singh Pruthi and H. N. Batra (1960). This bulletin too was written two decades back and since then rapid advances have been made in the field of entomology as well as in our knowledge of fruit pests. This information about insect pests of fruit trees and their control is scattered in numerous journals and magazines and is not easily available to the interested workers. An attempt has therefore been made to compile the available information on the various aspects of insect pests of fruit crops grown in India.

While writing this book, I have freely drawn upon the available literature. My own articles published in *Pesticides* (1973-78) have served as the basic frame work. The diagrams presented in this book are mostly drawn from the specimen available in the National Pusa Collection, New Delhi. A few sketches have also been redrawn from *Indian Insect Life* and *Some South Indian Insects.* The book, however, should not be taken as complete in all respects. The enormity and complexity of the subject due to large number of fruits involved coupled with the frequent changes in taxonomic status of various insect species, may have undoubtedly left some scope for improvement. This is, nevertheless, no excuse for any ommisions or errors. Suggestions, if any, for improvement of this book are welcome for incorporating in the future editions.

I am very grateful to Dr. N.C. Pant, Director, Commonwealth Institute of Entomology, London, for writing the foreword and to Mr. S. L. Katyal, Assistant Director General (Horticulture) and Dr. K. N. Mehrotra, Assistant Director General (Plant Protection) for their suggestions and encouragement. I express my sincere gratitude to Dr. M. G. Jotwani, Head, Division of Entomology, IARI, New Delhi, who assisted me from beginning to end in the compilation of this dissertation. His constructive critism and useful suggestions were extermely helpful in improving the manuscript. I am thankful to Mr. R. V. Raghavan, Editor-Publisher, *Pesticides*, Bombay, for permitting me to reproduce some of the illustrations, etc., published earlier in my articles in *Pesticides*. Help rendered by Mr. A.K. Roy, who has drawn most of the line drawings presented in this book, is gratefully acknowledged. I am also obliged to Dr. (Miss) Swaraj Ghai and Dr. (Miss) Ramney Fotidar for compiling and checking the list of mites; Dr. B.V. David for checking the names of whiteflies; Dr. S. Mohammad Ali for scrutinising the list of scale insects; Dr. (Miss) Prabha Grover for providing information on gall-midges; Dr. K. P. Srivastava for checking the index; Miss Shashi Sood for assistance in proof reading and to Mrs. Shabnum Raisinghani for typing the manuscript. Lastly, my sincere thanks are due to my wife Sushilarani who by her forebearance, patience and cooperation made it possible to produce this treatise.

Indian Agricultural Research
Institute, New Delhi - 110012
November 8th 1979

DHAMO K. BUTANI, D.Phil. (Agri.)
Senior Entomologist (Fruits)

Printed by Mr. MAHENDRA PAL SINGH GAHLOT

IN THIS BOOK

1
INTRODUCTION

THE FRUITS

FRUITS are the man's oldest food. Even the pre-historic man, who mainly hunted animals, had learnt to collect and enjoy eating wild berries and other fruits. The history of fruits is older than that of Adam and Eve who wore fig leaves and ate the forbidden apple. A reference to grape cultivation exists in Arthashastra (4th century B.C.) while Old Testament (1000 B.C.) gives directions on wine making. In India, the fruits are being cultivated since times immemorial. Nehru (1946) pointed out that even during pre-Budhist period horticulture was a prominent vocation in India. In earlier days fruit trees were cultivated only in backyards or as border trees along the fields, around Persian wheels and later along the roads. Then some aristocrats took orcharding as a hobby and collecting different fruit trees or a large number of varieties of same fruit became a status symbol. In 19th century several European settlers and missionaries did some pioneering work by introducing new varieties from England, France and East Indies and established commercial orchards.

Aonla, bael, banana, *falsa,* jack-fruit, a few varieties of citrus and mango are indigenous to India, whereas the rest have been introduced from different countries. Litchi, loquat, peach and sweet orange came from China; guava, custard apple, papaya, pine apple and sapota from tropical America and almond, apple, apricot, cherry, grapes, pistachio, pear, plum, pomegranate, walnut, etc. from central Asiatic region.

Importance of fruits in human diet is universally recognized. These are the chief sources of vitamins, without which the human body cannot maintain proper health and resistance to diseases. Apart from vitamins, minerals like calcium, phosphorous and iron are also found in a number of common fruits. 'Eat more fruits' and 'Grow more fruits' are gradually assuming an importance as popular national slogans in this country.

The practice of growing fruits on commercial lines has gained momentum during the last 30 years. The lead in this direction has been provided by the Indian Council of Agricultural Research, New

Delhi. Its effort created considerable awakening amongst the
cultivators and convinced them about the possibilities of fruit
growing as an economic proposition. Fruit research stations were
established in various States and now the Agricultural Universities
too have given Horticulture the status of a major subject. With the
result, the erstwhile haphazard system of fruit farming is rapidly
giving place to improved systematised and scientific methods of
fruit culture.

The statistics regarding the area covered by various fruits in
India are not very reliable. There are many scattered and roadside
trees which are hard to survey. It has been estimated that in India
there are about 1.5 million hectares under fruits (except cashew nut
and coconut). Of these nearly 78 percent area is under tropical
fruits, 13 per cent under subtropical and 9 per cent under temperate
fruits. Introduction of new varieties, improvement in various
cultural practices, application of adequate fertilizers and providing
timely irrigations will no doubt enhance the fruit production but in
order to harvest the potential bumper yields, it is essential to know
all about our enemies – the insect pests. For quality as well as
quantity, it is necessary to check their nefarious activities and safeguard our trees at all stages of their growth.

There are over 1000 species of insects found damaging fruit
trees all over the world; of these, as many as 800 have been intercepted from India so far. A classified list of these insects along
with their fruit host plants is given in appendix 7.1. It will be
observed that there is hardly any part of fruit trees, which does not
suffer insect damage at one or the other stage. There is not just
one insect species involved ; roots may be affected by some species,
stems by the others. There may be caterpillars feeding on tender
leaves and simultaneously the sucking insects depriving the trees of
the vital cell sap, and yet other insects may appear at the most
important fruiting stage. In addition to insect pests, a few species
of mites (appendix 7.2) and nematodes (appendix 7.3) also cause
substantial loss and so do some animals specially monkeys, squirrels,
rodents, etc. and birds like, finches, parakeet, crows, *bulbuls, mynas,*
sparrows, etc.

While we deplore the damage done by insect pests and worry
about the serious menace, we should not overlook to thank
braconid wasps, chalcid wasps, ichneumon wasps, scelionid and

tachanid flies because all these are parasites upon the insects that beset the fruit crops. We may also thank hover flies, antlions, lady bird beetles, etc. which are all predators and prey upon the harmful insects, often checking their population build up.

There is yet another important role played by insects – cross pollination. Cross pollination is necessary to give bigger, better and healthy fruits. If the world was suddenly deprived of insects there soon would be serious dearth of vegetation. Most of the fruits would disappear from the earth. Without the activities of bees, our tables will starve of plums and figs; there would be no melon, orange or apple, as all these fruits need services of insects for pollination.

Thus insects are a curse as well as a blessing, for agriculturist in general and orchardist in particular. This book however, presents only the destrutcive role played by the insects as also the ways and means to prevent the avoidable losses caused by the insect pests.

For bringing up trees beautiful and healthy
Rich soil, improved seed, copious irrigation
Use of fertilizers, alone are not adequate
It is orchardist's personal care and affection
That makes the trees vigourous and wealthy
So, start from today, paying more attention
and learn all about plant protection.

2
PESTS OF TROPICAL FRUITS

2.1
MANGO

MANGO, *Mangifera indica* Linnaeus – the King of fruits, is one of the ancient fruits of Indian origin. Its cultivation is estimated to have started more than 6000 years back. Forests and gardens of mangoes have been mentioned even in Ramayana and Mahabharta. Portuguese carried the mango seeds from India to East Africa in sixteenth century. By now, besides India, mangoes are grown in Sri Lanka, Bangladesh, Burma, Thialand, Malaysia, Philippines, Indonesia, Hawaii, tropical Australia (Queensland), USA (Florida), West Indies, Brazil, Mexico, South-east Africa, Egypt, Israel, Pakistan, etc. (Singh, 1960). In India, mango orchards cover an area of more than 750,000 hectares (Vevai, 1969 c) – Uttar Pradesh (120,000 hectares), Andhra Pradesh (100,000 hectares), Bihar (90,000 hectares), Bengal (58,000 hectares), Kerala (68,000 hectares).

India accounts for nearly 80 per cent of the world's mango production and exports substantial quantities. It earned foreign exchange of Rs. 90 millions in 1975-76. There is a growing demand for fresh mango, tinned mango slices, juice, jam, pickles, etc., from all over the world. The fruits besides being sweet and succulent are a rich source of vitamin A and also contain vitamin C (Singh, 1969). Mango also has medicinal value – bark of the tree serves as an astringent and cures dysentery and bleeding piles; juice of the bark with lime water is given for gonorrhoea. Powder of tender leaves is good remedy for diarrhoea and diabetes. The smoke from burning leaves is good for throat diseases. Dried mango flowers are useful in treating dysentery. Ripe fruit is diuretic. The seed contains large quantity of gallic acid and is used as a cure for round worms. Ointment made of the resinous gum from the tree is a good dressing for certain skin diseases.

Over 175 species of insects have been reported damaging mango trees (Fletcher, 1917; Vevai, 1969c; Nayar *et al.*, 1976). Of these only a handful are of major importance, namely, mango hoppers, mango mealy bugs, bark eating caterpillars, stem borers, scale insects, fruit flies, stone weevil, gall midges and termites. The status of red ant is still doubtful as it has been recorded preying upon a number of other insects. The insect pests of minor importance include, field crickets, mango psyllids, leaf webbers, leaf miner, leaf eating weevils, beetles and eaterpillars, thrips, whiteflies and other sucking bugs, shoot borers, fruit borers and fruit sucking moths.

The importance of these pests can be well judged from the fact that the U.S. Government has placed an embargo on the import of mango fruit from India because of the risk of introducing Oriental fruit fly and mango stone weevil in that country (Wadhi, 1964). The quarantine significance has been high lighted by Pradhan and Wadhi (1962). These authors have also listed 87 species of insects and 5 species of mites reported on mango from countries other than India, along with brief notes on their biology, nature of damage, etc.

MANGO HOPPERS

These are considered to be the most destructive pests of mango. There are three species of these hoppers. These were first recorded from Saharanpur (Uttar Pradesh) and described by Lethierry (1889) as *Idiocerus atkinsoni, I. clypealis* and *I. niveosparsus*. Baker (1915) errected a new genus *Idioscopus* and placed *clypealis* under this genus. Later, Maldonada-Copriles (1964) transferred *atkinsoni* and *niveosparsus* also under *Idioscopus*. Anufriev (1970) shifted *atkinsoni* and placed it under a new genus *Amaritodus*. So now there are *Amaritodus atkinsoni* (Lethierry), *Idioscopus clypealis* (Lethierry) and *I. niveosparsus* (Lethierry).

These hoppers have also been recorded from Philippines, Taiwan, Vietnam, Indonesia, Sri Lanka, Burma, Bangladesh and Pakistan. In India, these hoppers are widely distributed in all the mango growing regions. *A. atkinsoni* is comparatively more common specially in North India. *I. clypealis* though found all over India is more predominent in South Gujarat, Maharashtra and Karnataka, while *I. niveosparsus* has been recorded from peninsular India. These hoppers have also been found on *Citrus* spp. and *Calophyllus inophythum* but these plants do not serve as alternate hosts (Uppal and Wagle, 1944). Nayar *et al.* (1976) has reported these hoppers on sapota.

Enormous number of nymphs may be found clustering on the inflorescences during Spring, sucking the sap. The infested flowers shrivel, turn brown and ultimately fall off (Husain and Pruthi, 1921). On attaining maturity the hoppers leave the blossoms and move on leaves and trunks of the trees. Swarms of adults are commonly seen hovering in mango groves and sitting on all plant parts. Both nymphs and adults suck the sap usually from the ventral surface of leaves. As a result, growth of the trees is stunted. The hoppers also excrete honeydew which encourages the development of fungi, *Capnodium mangiferum* Cooke and Brown and *Meliola mangiferae* Earle, resulting in growth of sooty mould on dorsal surface of leaves, branches, and even fruits. This black coating interferes with the photosynthetic activity of the plant, ultimately resulting in non-setting of flowers and dropping of immature fruits. Besides, heavy egg-laying within florets and stalklets cause physical injury, resulting in withering of affected parts.

There are no authentic figures available regarding the extent of loss caused by these hoppers. According to Rao (1930) these hoppers cause a loss of 20 to 100 per cent of inflorescences, whereas Cheema *et al.* (1954) and Gangolly *et al.* (1957) reported the loss to be 25 to 60 per cent.

Eggs and nymphs of the three species are rather difficult to distinguish from each other but the adults though looking alike vary in size and colour shade. Freshly hatched nymphs are wedge shaped and whitish in colour with two small red eyes ; gradually with each moulting the colour changes to yellow, yellowish-green, green and ultimately to greenish-brown. Adults are also wedge shaped having greenish brown body and pale yellow vertex. Fore-wings are thicker than hind wings, bronzy sub-hyaline with veins pale

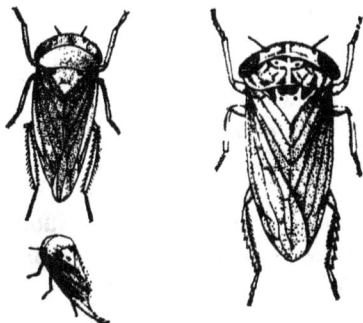

Idioscopus clypealis
(Lethierry)

I.niveosparsus
(Lethierry)

yellow and a white line along the costal margin forming distinct mid longitudinal line when the insect is at rest. Adults of *A. atkinsoni* are darker in colour and bigger in size, males being 4.2 to 4.8 mm long and females 4.7 to 5.1 mm; those of *I. clypealis* are smaller and narrower, males being 3.4 to 3.7 mm long and females 3.6 to 3.9 mm ; while *I. niveosparsus* are medium sized, males being 3.9 to 4.3 mm long and females 4.9 to 5.3 mm. Srivastava (1958) has described in detail the external morphology of *I. clypealis*.

A. *atkinsoni* is a shy breeder whereas *I. clypealis* is the most prolific breeder. Egg-laying starts around end of January or early February and continues till March. The females make tiny slits in the tissues of flowering shoots, flower buds or tender leaves with their ovipositor and lay the eggs singly therein. A single female lays about 100 to 200 eggs. The eggs hatch in 4 to 7 days. The nymphs undergo 4 to 5 moultings and the total nymphal period ranges from 10 to 13 days in South India and 18 to 20 days in North India. In North India, there are two distinct generations in a year ; spring generation in February to April and summer generation during June to August ; the former being definitely more destructive than the later, as during that period the hoppers feed on inflorescences. Overwintering is in the adult stage by hiding in cracks and crevices or in the bark of the trees. The hoppers are active throughout the Summer and Spring. They love damp and shady places and their population rises to phenomenal proportions in neglected and water-logged orchards. The activity declines with the onset of monsoons and the adult hoppers remain inactive from August to January. A spell of cold weather, with temperature dropping to 10–15° C results in increase in egg-laying.

Keeping the orchards clean and avoiding over-crowding of trees and water-logging, keeps the pest at bay. As regards chemical control measures, Chari *et al.* (1969) observed 95.3% reduction by spraying 0.1% endosulfan and 87.3% with 0.1% carbaryl. Sathianandan *et al.* (1972) and Singh *et al.* (1974) found spraying with 0.1% carbaryl to be effective. Butani (1974d, 1975c) recommended spraying with 0.03% phosphamidon, diazinon or monocrotophos. Four to six sprayings may be necessary to check the pest effectively – first at the time of bud swelling (January), one to two soon after flowering at an interval of 13 to 18 days and another two to three for summer generation during June-July.

MANGO MEALY BUGS

Drosicha mangiferae (Green) is another major pest of mango. It was originally described by Green (1903) as *Monophlebus mangiferae* sp. nova. It has been reported from India, Bangladesh, China and Pakistan. In India, it is widely distributed along the Indo-Gangetic plain and has a very wide range of host plants. Rahman and Latif (1944) have listed 62 host plants of this pest. Among the fruits mention may be made of apple, apricot, *ber*, cherry, citrus, *falsa*, fig, grapevine, guava, *gular*, jack-fruit, *jamun*, litchi, mango, mulberry, papaya, peach, plum, pomegranate, etc.

Soon after hatching, the majority of nymphs start crawling up the tree trunks and clusters of these may be seen on young shoots and panicles, sucking the cell sap therefrom. A few nymphs crawl away to neighbouring trees as well. These are more active on bright sunny day.

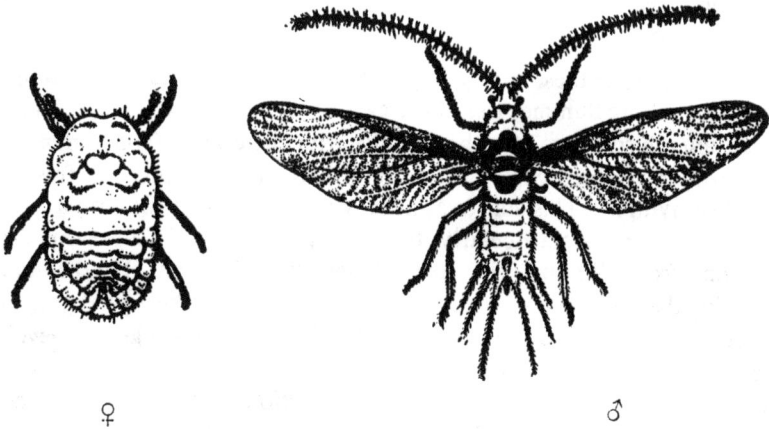

♀   ♂

Drosicha mangiferae (Green)

The nymphs and adult females are flat, oval, waxy-whitish insects, sometimes mistaken for fungal growth. Adult females are wingless while males are crimson coloured bugs with two dark brownish black wings and cause no damage, except fertilizing the females. The gravid females crawl down the trees during end March to end May and enter the soil (80 to 150 mm deep) wherein they excrete whitish foam. This forms a pouch in which the females go on depositing the eggs for 7 to 16 days and soon after completing the oviposition, the females die. The dead bodies are invariably seen sticking near the eggs. Each female gives 400 to 500 eggs and these remain in the soil in a state of diapause. Winter chilling terminates this diapause (Atwal, 1961). The hatching of these eggs also depends upon the humidity of the soil (Singh, 1946) and normally starts from end of November and continues till January. There are three nymphal stages ; first instar, second instar, third instar female and third instar male live for 45 to 71, 18 to 38, 15 to 26 and 5 to 10 days respectively (Rahman and Latif, 1944). The total life-cycle occupies 67 to 119 days in case of males and 77 to 135 days in case of females.

Another species, *Drosicha stebbingi* (Green) was originally recorded on *Shorea robusta* (*Shal* - a forest tree) at Dehra Dun (Uttar Pradesh) and described by Green (1903). Subsequently this species got completely confused with *D. mangiferae* and Latif (1949) even considered the two to be synonymous. The two are distinct species.

Drosicha contrahens (Walker) and *D. dahlbergiae* (Green) have also been recorded on mango trees but the damage caused by these is negligible.

To control these mealy bugs, dig or plough repeatedly around the trees during Summer to kill the eggs by exposing the same to Sun's heat, ants, birds, etc. Mix thoroughly, with the upper 400-500 mm of soil around the tree, 5% aldrin or chlordane dust, to kill the freshly hatched nymphs. This soil treatment will also control the other soil borne insect pests particularly the termites, if any. To prevent the nymphs from climbing the trees, Srivastava and Butani (1972) found banding the tree trunks with 300 mm wide alkathene sheet (400 gauge) to be effective. The band is fixed about 300 mm above the ground level. Before tying the band, paste a layer (400-500 mm wide) of mud or wet soil around the tree trunk, to make that portion smooth. Then wrap the alkathene sheet around the tree on this mud plaster and tie the sheet on both sides with *sutli* (jute or sunnhemp thread). Clean the alkathene sheet occasionally with a piece of moist cloth, to keep it smooth and slippery. This will prevent the nymphs from going up the tree. Such of the nymphs as may be found congregating at the lower edge of alkathene band must be killed mechanically or by spraying diazinon or monocrotophos. If nymphs be observed on trees, spray with diazinon or monocrotophos. Birat (1963) got 75.3 per cent reduction in pest population by spraying 0.06% diazinon whereas Srivastava *et al.* (1973) reported 88.4 per cent reduction with 0.03% monocrotophos. Butani (1974 d) recommended spraying with

0.04% diazinon or monocrotophos. These pesticides are quite effective in controlling the young nymphs but not full grown nymphs or adults.

BARK EATING CATERPILLARS

Indarbela spp. are found all over India, damaging *aonla, ber,* citrus, guava, *jamun,* litchi, loquat, mango, mulberry pomegranate and a number of forest and ornamental trees, These occur throughout the Indian sub-continent including Bangladesh, Burma, Sri Lanka and Pakistan. The freshly hatched larvae nibble the tree trunk and after 2 to 3 days bore into the same and feed within. This interrupts with translocation of cell sap, with the result the growth of the tree is arrested and fruiting capacity adversely affected (Butani, 1977 a). The caterpillars spin silken webs consisting of their excreta and chewed wood particles which are seen hanging loosely on the bark of affected trees, more commonly at the junction of two stems or main branches. The older trees are more prone to attack by these pests than the young ones. Similarly orchards that are neglected and not kept clean show higher percentage of pest incidence than the clean and well kept orchards. Though these are major pests of some fruit trees, no systematic studies have been carried out so far to study the extent of loss caused by their nefarious activities.

The species most commonly found on mango in India is, *Indarbela quadrinotata* (Walker). *I. tetraonis* (Moore) – a serious pest of guava, is also found attacking mango specially when guava trees are around. *I. dea* (Swinhoe) and *I. theivora* (Hampson) are other species reported from India. Caterpillars of *quadrinotata* are 50 to 60 mm long with dark brown heads and dirty brown bodies. Pupae are 16 to 20 mm long, stout, reddish-brown with 2 rows of spines on each abdominal segment arranged transversely on anterior and posterior margins. Adults are pale brown moths with head and thorax dark rufous; abdomen fuscous; forewings pale rufous with numerous dark rufous bands of strigae and hind wings fuscous. Wing expanse is 34 to 38 mm and 38 to 42 mm in case of male and female moths respectively. A single female lays about 2000 eggs in clusters of 15 to 20 each on bark of mango trees during April to June. These hatch in 8 to 10 days. Larvae are fully grown by December but continue to feed slowly till April when pupation takes place. Pupal period varies between 21 and 31 days. Moth emergence continues till June and their longevity is hardly 3 days. There is only one generation in a year.

In addition to *Indarbela* spp., Singh (1954 a) recorded *Lymantria mathra* Moore from Uttar Pradesh causing damage to the bark and inflorescences of mango and litchi trees.

To avoid the occurrence of these pests, keep your orchards clean and aviod over-crowding of trees. In case of infestation, Srivastava (1964) suggested inserting an iron spike into the hole to kill the larvae. This is quite effective and may be practiced in small orchards or when the infestation is low. The other method suggested is to clean the tree trunks by removing all the webs and inserting into the holes cotton wool soaked in a good fumigant like carbon bisulphide, chloroform, petrol, etc. Patel *et al.* (1964) observed complete mortality of the caterpillars by injecting into the holes kerosene oil whereas Khurana and Gupta (1972) suggested injecting the holes with 0.013% dichlorvos (DDVP), 0.05% trichlorfon or 0.05% endosulfan. Immediately after treating the holes, seal the same with mud. Such of the holes as may reopen, need be retreated.

MANGO STEM BORERS

Batocera spp. are widely distributed all over India, and Bangladesh attacking fig, jack-fruit, mango, mulberry, papaya, apple, etc. Eggs are laid singly either in the slits of tree trunks or in the cavities (niche) in main branches and stems which are then covered with a viscous fluid. On hatching, the grubs make zig zag burrows beneath the bark and tunnel into the trunks or main stems, moving upwards, feeding on the internal tissues. When the grubs reach sapwood, the attacked stems die and wither away, Normally the attack by this pest goes unnoticed till a branch or two start shedding leaves and drying up. Sometimes however, sap and masses of fras may be seen exuding from the bored holes. Adult beetles feed on the bark of young twigs and petioles. Pupation takes place within the affected trunks or stems, usually during March-April in northern India. Adults emerge with the onset of monsoon (May-June) and the emergence continues during the monsoon season.

The species recorded from India include, *B. rufomaculata* (de Geer), *B. rubus* (Linnaeus), *B. roylei* (Hope), *B. numetor* (Newmann) and *B. titana* (Thomson). Of these first one is more common and destructive while the last three are of minor importance. Eggs of *B. rufomaculata* are 5.0 to 6.5 mm long, white, shining and oval in shape. Full grown grubs are 85 to 95 mm long, fleshy, stout, yellowish-ivory in colour with well defined segmentation. Pupae are 50 to 55 mm long and yellowish-brown to dark brown in colour. Adults are stout, dark brown longicorn beetles, 50 to 55 mm long (females larger than males) with yellowish-green pubescence; prothorax with two large kidney shaped orange spots and a short thick spine-like projection on either side. Elytra irrorated with small light orange spots.

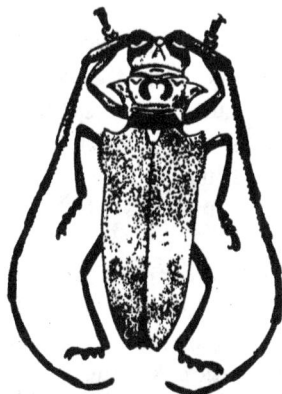

Batocera rufomaculata (de Geer)   *B. titana* (Thomson)

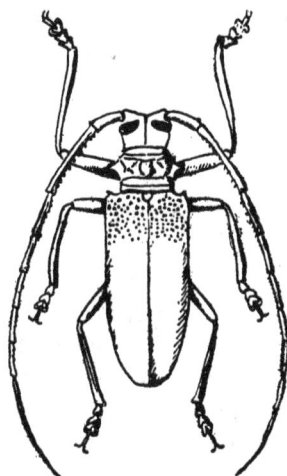

The beetles mate after 1 to 2 days of emergence and oviposit for 22 to 25 days, laying on an average one egg per day. Incubation period is 7 to 13 days; grubs remain active for 140 to 160 days while pupal period lasts for 20 to 25 days. Total life-cycle occupies 170 to 190 days and adult longevity is 60 to 100 days.

Another species commonly found attacking mango in India is *B. rubus*. These beetles are 30 to 45 mm long, brownish-grey with elytra having shining granulations at base and 4 or more dirty yellowish spots, of variable size, on each elytron.

To control these pests, cut and destory the infested branches with grubs and pupae within. If the holes are located on main stems or trunks, these may be treated in the same manner as suggested in the case of bark eating caterpillars.

Other longicorn beetles attacking mango trees in India, include, *Acanthophorus serraticornis* Olivier, *Anoplophora versteegi* (Ritseema), *Epepeotes luscus* (Fabricius), *Rhytidodera bowringi*, *R. simulans* White and *Stromatium barbatum* (Fabricius). In addition to these cerambycids, a few scolytids (shot-hole beetles) have also been reported boring the mango stems and tender shoots of mango trees. These include, *Hypocryphalus mangiferae* (Stebbing), *Xyleborus kraatzi* Eichhoff and *X. semigranosus* Blandford, but none of these is of any major importance.

COCCOIDS

A number of scale insects – fluted (Monophlebids), armoured (Diaspids)and soft (Lecanids) as also a few mealy bugs(Pseudococcids), have been reported damaging mango trees all over India. These insects, on hatching from eggs, crawl a bit in search of a succulent spot, get affixed there and become immobile. They suck the plant sap and though each insect by itself can hardly cause any loss, their enormous number collectively, causes substantial loss of vitality. In case of servere infestation, the growth and fruit setting capacity of the tree is adversely affected. The life-cycle and bionomics of these insects have not yet been studied in India.

The presence of an impervious coating or shield over the bodies of these insects makes it difficult to control them effectively by use of insecticides. Pruning the affected leaves, twigs and branches and their prompt destruction prevents the increase in population of the pest. Spraying with 0.04% diazinon or monocrotophos also helps in minimising the nymphal population.

The various species recorded on mango, include *Icerya minor* Green, *I. pulcher* (Leonardi), *I. purchasi* Maskell, *I. seychellarum* (Westwood), *Aspidiotus destructor* Signoret (*transparens* Green), *Aulacaspis rosae* (Bouché), *A.tubercularis* Newstead, *Chrysomphalus ficus* Ashmead (*aonidum* Linnaeus), *C. dictyospermi* (Morgan), *Genaparlatoria pseudaspidiotus* (Lindinger), *Insulaspis pallidula* Williams, *I. gloverii* (Packard), *Lindingaspis floridana* Ferris, *L. greeni* (Brain and Kelly), *L. rossi* (Maskell), *Parlatoria camelliae* Comstock, *P. cinerea* Hadden, *P. crypta* Mc Kenzie, *P. oleae* (Colvée), *P. pergandii* Comstock, *Phenacaspis cockerelli* (Cooley) *P. vitis* (Green), *Pinnaspis aspidistrae* (Signoret), *Pseudaonidia trilobitiformis* (Green), *Pseudaulacaspis barberi* (Green), *Radionaspis indica* (Marlatt), *Ceroplastes actiniformis* Green,

C. ceriferus (Anderson), *C. floridensis* (Comstock), *C. pseudoceriferus* Green, *C. rubens* Maskell, *Chloropulvinaria polygonata* (Cockerell), *C. psidii* (Maskell), *Coccus acutissimus* (Green), *C. adersi* (Newstead) *C. bicruciatus* (Green), *C. discrepans* (Green), *C. hespridum* Linnaeus, *C. mangiferae* (Green), *Lecanium latioperculatum* Green, *Pulvinaria cellulosa* Green, *Saissetia coffeae* (Walker), *Vinsonia stellifera* (Westwood), *Nipaecoccus viridis* (Newstead), *Rastrococcus iceryoides* (Green), *R, mangiferae* (Green), *Pseudococcus adonidum* (Linnaeus), *Kerria lacca* (Kerr), etc. (Ayyar, 1924; Misra, 1924; Singh, 1960; Ali, 1961 and 1964).

Of these, the one comparatively harmful is *Aspidiotus destructor*. Besides India, this armoured scale insect has been reported from Sri Lanka, China, Taiwan, Fiji Island, Mexico, West Indies, British Guiana, Africa (Uganda and Gold Coast), Mauritius, etc. In India, it is found throughout the plains and foot of the hills. It is a major pest of various plamae and sporadic pest of citrus and mango. It is also found on banana, guava, *jamun*, papaya, etc. (Rahman and Ansari, 1941). Fotidar and Kapur (1941) observed this scale as a serious pest of mango seedlings in the nurseries. It prefers grafted varieties of mango. The pest is active during Summer and affects the fruit setting adversely (Ansari, 1942). Small (2 mm long), oval, translucent, yellowish-brown crawlers (nymphs) are found on underside of leaves and tender shoots. These get covered with waxy material and become immobile. Female scale is circular, semi-transparent and pale brown. It reproduces oviparously. The eggs hatch under the scale giving rise to young ones (Taylor, 1935). Total life-cycle lasts for 32 to 34½ days.

FRUIT FLIES

Mango fruit fly, *Dacus dorsalis* (Hendel) is the most destructive fruit fly. It is widely distributed in the Orient region (CIE map No. A-109) from Australia and Hawaii to Pakistan, hence it is also called Oriental fruit fly. Of late it has also been reported from California (USA). The pest is active throughout the year in South India whereas in northern parts the pest hibernates during Winter (November to March) in pupal stage. The flies appear late in Spring on such of the fruits that are about to ripe and the population increases rapidly during Summer. The pest being polyphagous breeds profusely on guava during March, shifts to loquat, apricot and plum during April-May, thereon to peach and fig (June) and finally on mango (June to August). After August the flies breed mostly on guava but are also found on pear, fig, apple, citrus and even ripe banana (Narayanan and Batra, 1960). The flies are strong fliers and can fly upto two kilometers in search of food.

The female flies lay eggs just below the fruit epidermis (1 to 4 mm deep). On hatching the maggots feed on pulp of those fruits. As a result a brown patch appears around the place of oviposition and the infested fruits start rotting. These affected fruits drop down prematurely and the maggots come out from these fallen fruits to pupate in the soil. Adult flies are very conspicuous. These are about 7 mm long, with hyaline wings (expanse: 13 to 15 mm), thorax ferrugineous without yellow middle stripe, legs yellow, abdomen conical in shape and dark brown in colour.

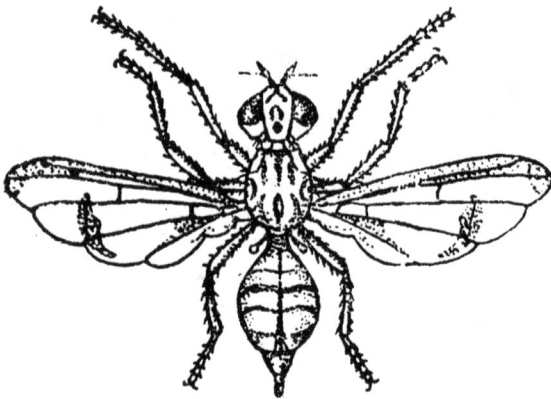

Dacus dorsalis (Hendel)

Bionomics have been studied by Janjua (1948) and Shah *et al.* (1948). Preoviposition period is 2 to 5 days. A single famale can lay 150 to 200 eggs (average 50) in about a month. The eggs are laid in clusters of 2 to 15 eggs and these hatch in 2 to 3 days during March and 1 to $1\frac{1}{2}$ days during April. Maggot duration is 6 days in Summer and extends upto 19 days with the fall in temperature. Pupation usually takes place 80 to 160 mm below the soil surface and pupal period ranges from 6 days (Summer) to 44 days (Winter). Bess and Haramoto (1977) reported that the lowest average temperature at which the immature stages can develop is about 14° C and temperature above 21° C is necessary for the flies to attain sexual maturity without undue prolongation in preoviposition period.

Besides *Dacus dorsalis,* other species found on mango include, *D. correctus* (Bezzi) (*Bactrocera zonata* Bezzi), *D. diversus* (Coquillett,) *D. incisus* Walker, *D. tau* (Walker) (*hageni* de Meijere, *caudatus* Fabricius) and *D. zonatus* (Saunders) (Ayyar, 1963; Vevai, 1969 c). The immature stages of these species are similar to each other and difficult to distinguish from one another but the adults however, can be distinguished through microscopic inspection by the taxonomists (Kapoor, 1972).

The best way to avoid infestation of fruit flies is to harvest the fruits before ripening, To check the carry over of the pest collect and destroy all fallen and attacked fruits; plough around the trees during Winter to expose and kill the pupae. The adult flies may be trapped and killed by poison-baiting or bait-spray (20 gm malathion 50% wettable powder or 50 ml diazinon E.C.+200 gm molasses in 2 litres of water for baiting and 20 litres of water for spraying). Talgeri (1967) suggested spraying with 0.03% oxydemeton methyl, 0.03% phosphamidon or 0.06% dimethoate while Tandon *et al.* (1974) recommended spraying with 0.2% carbaryl or 0.06% dimethoate.

MANGO STONE WEEVIL

Sternochetus mangiferae (Fabricius) is widely distributed in the tropics (CIE map No. A-180). It is an specific pest of mango; sweet varieties like, *Alfanso, Bangalora, Neelum,* etc. being preferred (Gandhi, 1955; Atwal, 1963). According to Sundra Babu (1969) these varieties had 73, 93 and 100 per cent infestation respectively. The pest is more common in South India where late varieties suffer the most.

Eggs are laid singly on the epicarp of partially developed fruits or under the rind of ripening fruits. Freshly formed grubs bore through the pulp, feed on seed coat and later damage the cotyledons. Pupation takes place inside the seed along the concave side (Singh, 1960). Later, the adults eat their way out of the ripe fruits. Since the grub passes its entire life inside the seed there are no

external symptoms of injury on the fruits. On cutting open the fruits,
the pulp adjacent to the affected stone is seen discoloured due to
excretion of the grub.

Life-history has been studied by Subramanyam (1926). Eggs
are minute in size and white in colour; grubs are also white in colour.
Adult weevils are 5 to 8 mm long, stout, and dark brown in colour.
Life-cycle is completed in 40 to 50 days during June-July; adult
weevils hibernate from July-August till next fruiting season – thus there
is only one generation in a year.

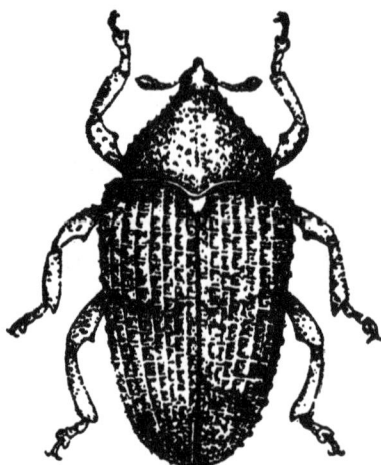

Sternochetus mangiferae (Fabricius)

Mc Bride and Mason (1934) studied the effect of sub-freezing
temperature on mango seeds containing grubs and pupae of this
weevil and observed that exposure of the stones for 48 hours at 12° C
killed all grubs and pupae as also 77.7 per cent adults. Five days
exposure gave cent per cent mortality of adults. In India, Cheema
et al. (1954) suggested destruction of affected fruits and digging of
soil to expose the hibernating weevils. David and Sundra Babu (1962)
found spraying with 0.01% fenthion to be effective. Wadhi (1972)
advocated dipping the hard fruits in ethylene dibromide emulsion at
50° C for two hours.   The treatment will not affect the taste, flavour,
etc. of the fruits specially of *Alfanso* and *Dashahri* varieties; *Langra*
variety, however, does not tolerate this treatment. Wadhi and
Sharma (1972) estimated the bromide level in treated fruits and
observed that level of total bromide in the pulp of treated fruits falls

below 5 ppm and 2 ppm in 24 and 72 hours respectively which is far below the tolerance level of 10 ppm.

Another species commonly met with in Assam and Bengal is mango pulp weevil, *S. frigidus* (Fabricius) (*gravis* Fabricius) (Sen 1923; Gangoly *et al.*,1957).Voute(1935a) reported this species from Indonesia where it has egg, larval and pupal periods of 5½ to 7, 30 to 36 and 5 days respectively. In China and Indonesia, the red ant, *Oecophylla smaragdina* (Fabricius) preys upon this weevil.

MANGO GALL MIDGES

A few species of gall midges have been reported causing slight to severe damage to mango trees in India. These include midges causing leaf galls, namely *Allassomyia tenuispatha* Kieffer, *Amradiplosis allhabadensis* Grover, *A. echinogulliperda* Mani [*A.keshopurensis* (Rao), *Amraemyia amraemyia* Rao, *A. brunneigallicola* (Rao), *A. viridgallicola* (Rao). *Bungomyia brachyrhyncha* Nayar], *Procontarinia mangiferae* (Grover) and *P. matteriania* Kieffer and Cocconi. These cecidomyids lay their eggs on ventral surface of leaves. On hatching, the maggots bore inside the leaf tissues and feed within, resulting in formation of small raised wrat-like galls on the leaves. The affected leaves get badly deformed and drop prematurely. The most common one is *P. matteriana* (Rao, 1956; Sengupta and Behura, 1957) which is found in India, Indonesia, Mauritius and South Africa.

In addition to the above mentioned species, *Rhabdophaga mangiferae* Mani forms galls on leaves and twigs; whereas mango blister midge *Erosomyia indica* Grover feeds on different parts of tree including inflorescence buds, axis of inflorescences, panicles and minute

Erosomyia indica Grover

fruits. *E. indica* is found all over India. These adults are yellow coloured minute flies having two tiny transparent wings; males are shorter than females. Life expectancy is very short, males die soon after mating and females after ovipostion.

Mango blossom feeders are, *Dasyneura amaramanjarae* Grover *D. citrt* Grover, *Procystiphora mangiferae* (Felt) and *P. indica* Grover.

Dasyneura amaramanjarae has been recorded only from northern India causing on an average a loss of 9 to 12 percent (Prasad, 1974). Eggs are laid in tender. buds; a single female lays 40 to 50 eggs, depositing 1 to 6 eggs per bud. On hatching, the grubs feed on reproductive and other internal parts of the bud and drop down to pupate in the soil. The affected buds fail to develop. Grover (1965) observed that this midge perfers *Alfanso* variety of mango.

Procystiphora mangiferae is a minute orange coloured midge, males being smaller than females. Eggs are laid inside the tender buds generally one egg per bud. On hatching the grubs feed within these buds. The affected buds swell into small cone-like pointed galls and fail to open. Pupation takes place in the buds. Grubs are spindle-shaped, 0.2 to 0.25 mm long and creamy yellow in colour. Pupae are yellowish-brown to dark brown and the adults are small flies having wing expanse of 4 to 5 mm. Larval and pupal durations occupy 6 to 9 and 2 to 3 days respectively while the entire life-cycle is completed in 8 to 11 days (Kulkarny, 1955).

To check the loss caused by the leaf gall forming midges, collect and destory the affected leaves, this will prevent the population build-up of the pest. Spraying with 0.02% phosphamidon+0.03% diazinon is also effective in controlling the gall midges injurious to mango inflorescences (Prasad, 1972).

TERMITES

Termites or white ants are social insects that live in colonies – the termitaria. In each termitarium, several castes can be distinguished; of which only the worker caste is of direct economic importance. Kings and Queens function as sexual forms, while the soldier caste defend the termitaria against invasions by enemies. Winged sexual forms appear with the first shower of monsoon.

Males are attracted to females by pheromones (Schmutterer, 1977). The mating takes place either up in the air or on the ground and soon after mating the insects lose their wings, re-enter the soil and tart a new colony. The Queen lays one egg every 2-3 seconds continuously during her entire life span of 7 to 10 years. Termites are warmth loving insects and inhabit the entire tropical and subtropical regions of the world except higher altitudes where temperatures are too low and the deserts where there is no food. They are polyphagous and have the widest range of host plants. They devour not only the live plant material but also the dead wood including fencing, fittings and furniture.

Termites abound in sandy and sandy-loam soils and cannot thrive under conditions of bad aeration and poor drainage. Theyalso appear and cause havoc in orchards that have been raised on virgin lands. The insects are heliphobic and either remain underground and feed on roots and then move upwards making the trunks completely hollow or they construct mud galleries on tree trunks. These galleries are built mostly during night and under the protection of these galleries, termites feed on the bark of the trunks. They are active all the year round, though the incidence during monsoon months is rather low. The damage is more severe in nurseries and young newly planted orchards where the entire seedlings or saplings may dry and die away. Formation of earthen galleries on tree trunks decreases with rise in temprature from March onwards and it again increases from September onwards.

To some extent termites are also beneficial as they help in decomposition of organic matter in the soil. In tropical Africa, some tribes eat the sexual forms roasted in oil as these are said to be rich in fats and proteins.

The species recorded damaging mango trees in India, include, *Coptotermes heimi* Wasmann, *Neotermes bosei* Snyder (*gardneri* Snyder), *N.mangiferae* Roonwal and Sen-Sarma, *Heterotermes indicola* Wasmann, *Microtermes obesi* (Holmgren) (*anandi* Holmgren), *Odontotermes assmuthi* Holmgren, *O. feae* (Wasmann), *O. obesus* (Rambur) (*assamensis* Holmgren, *bangalorensis* Holmgren), *Trinervitermes biformis* (Wasmann) (*heimi* Wasmann), *T. rubidus* (Hagen) and *Stylotermes fletcheri* Holmgren. Of these, the most destructive one, found all over India. is *Odontotermes obesus*.

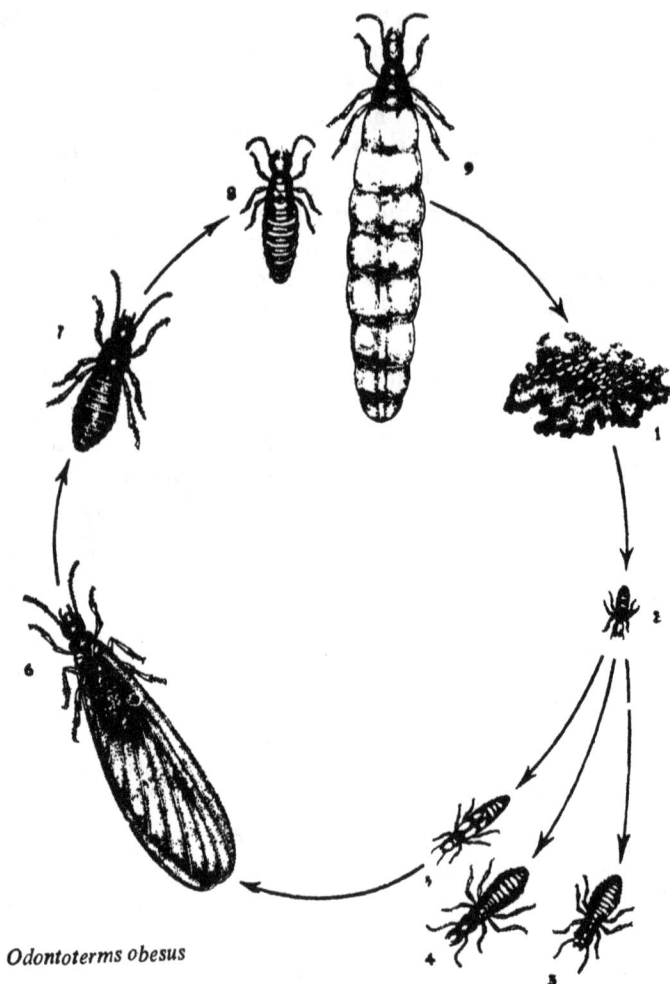

Odontoterms obesus

1 EGGS   2 NYMPH   3. WORKER   4 SOLDIER
5 REPRODUCTIVE NYMPH   6   WINGED   P PRODUCTIVE
7. REPRODUCTIVE AFTER LOSS OF WINGS 8 KING 9. QUEEN

To control the termites, deep ploughing may be done around
the trees followed by copious irrigations. Soil treatment with 5%
aldirn, chlordane or heptachor dust @ 200 to 300 gm per tree is also
effective. If the mud galleries appear on tree trunks, remove the
same and sponge the trunks with kerosene oil.

RED ANT

Oecophylla smaragdina (Fabricius) has been reported from the entire Oriental region extending from Australia to Africa (Atwal, 1963). The ants web and stitch together a few leaves usually at the top of the branches and build their nests on citrus, jack-fruit, *jamun*, litchi, mango, sapota trees, etc. These nests though not airtight are certainly waterproof and the leaves also remain green as the same are not detached from the trees. The ants are carnivorous and are considered beneficial in several countries, as they prey upon some hoppers, moths, caterpillars, beetles, bees, etc. Voute (1935a) reported that in China and Indonesia these ants feed on the mango pulp weevil, *Sternochetus frigidus* (Fabricius). Garcia (1935) reported these ants preying on nymphs of citrus green bug, *Phynchocoris serratus* in Philippines. The ants also feed on honeydew excreted by aphids and coccids. For this purpose mango aphid, *Toxoptera odinae* van der Goot and coccid, *Rastrococcus iceryoides* (Green) are commonly found in their nests (David, 1961).

The ants are active all the year round, though their activity slakens during monsoon months and gets a flip on sunny days. The ants do not cause any direct injury or loss to the trees. Indirectly the damage is caused by protecting aphids and scale insects from being preyed upon by their parasites and predators and also carry the nymphs of aphids, mealy bugs and scale insects from tree to tree thus spreading the infection of these noxious pests. Besides, being very ferocious, they also prove to be a nuisance to the persons who climb the trees and other workers around, who often get badly bitten by these ants.

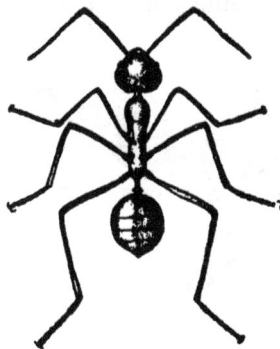

Oecophylla smaragdina (Fabricius) – worker

Eggs are oval in shape and whitish in colour. Larvae are also whitish, 1.2 to 1.4 mm long when freshly formed while full grown ones are 9 to 11 mm long. Pupae are also pure white in colour. Adults are light orange red in colour ; the workers are 14 to 18 mm long, wingless and infertile. Sexually functional males and females are winged and usually mate outside the nests in the course of their nuptial flights. The fertilized females, also known as Queens, shed their wings at the time of nest formation. David (1961) has studied their bionomics ; egg, larval and pupal periods occupy 4 to 8, 10 to 17 and 5 to 7 days respectively.

To control these ants, it is recommended that their nests be removed and destroyed mechanically or the same should be sprayed to run-off point with 0.1% BHC (Butani and Tahiliani, 1974).

FIELD CRICKETS

These are minor pests, which cause occasionally some damage. The species commonly found is *Brachytrypes porteniosus* (Lichtenstein) popularly known as large brown cricket. Both nymphs and adults may be found feeding at night on young shoots. Eggs are laid soon after the rainy season in the soil in batches of 40 to 50 eggs. On hatching the nymphs make fresh burrows in the soil below the trees. It is the appearance of very small soil heaps near the holes that indicate the presence of these crickets. Eggs are white and cylindrical. Nymphs and adults are greyish-brown and more or less similar in shape except that the nymphs have wing pads and the adults fully developed wings. Adults are about 5 mm long; males have a pair of anal styles and females have a long ovipositor. Egg and nymphal durations are 2 to 4 and 28 to 32 weeks respectively and adults live for 16 to 20 weeks.

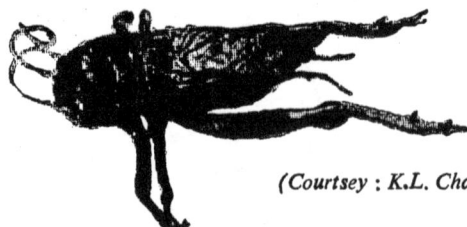

(Courtsey : K.L. Chadha)

Brachytrypes porteniosus (Lichtenstein)

Platygryllus melanocephalus (Serville) is another species found feeding on mango leaves. This is a surface burrowing species and does not make deep burrows in the soil.

Dusting the mango seedlings with 10% BHC is quite effective for protecting from damage by these crickets.

MANGO PSYLLIDS

Apsylla cistellata (Buckton) is an sporadic pest of mango that often causes serious damage in North Bihar and submontane region of Uttar Pradesh. Eggs are laid in midribs as well as lateral veins of new leaves. The nymphs burst open the tissues and crawl to the adjacent buds to suck the cell-sap therefrom. The irritation caused by feeding of these nymphs makes the buds develop scaly leaves which imbricate the central axis and form hard conical green galls in the growing buds (Butani, 1962). The galls appear usually during September-October. As a result of gall-formation there is no fruit setting. Eggs are white and nymphs flat round and pale yellow in colour. Adults are 3 to 4 mm long with black head and thorax and light brown abdomen.

♂   ♀

Apsylla cistellata (Buckton)

Biology and bionomics have been studied by Singh (1954 b) and Prasad (1957). A single female lays on an average 150 eggs during the end of March or early April. These hatch in August-September. Nymphal period is 5 to 6 months. The nymphs overwinter inside these galls and the adults emerge from the galls in February-March. There is only one generation in a year.

Pauropsylla brevicornis Crawford and *Arytania obscura* Crawford are other psyllids found on mango. But none of these cause any serious damage. *A. obscura* has been reported from Bengal. Its eggs are laid on tender tissues of top leaves. Small flat pale yellowish nymphs and light brown adults with spotted wings are seen in large number attacking tender shoots and sucking the plant sap. As a result, the leaves around the growing point curl and twist and ultimately the growing point dries up.

There is no effective control of psyllids. Collect and burn the galls with nymphs within, to prevent the carry over of the pest. Repeated application of BHC at the time of adult emergence (mid-February to end April) affords some relief.

MANGO LEAF WEBBERS

Macalla moncusalis Walker and *Orthaga exvinacea* Hampson are commonly found throughout the plains of southern India; both are minor pests of mango. Eggs are laid singly or in clusters within silken webbings or on leaves. On hatching the caterpillars feed gregariously by scrapping the leaf surface. Soon they web together tender shoots and leaves and feed within. Several caterpillars may be found in a single webbed up cluster of leaves and pupation also takes place inside these webs in silken cocoons. Caterpillars of *M. moncusalis* are stout, about 25 mm long, brown in colour with paler stripe down the back while those of *O. exvinacea* are slender, pale green with dark bands. Moths of *M. moncusalis* are fuscous, with dark patches on forewings; wing expanse is 26 to 32 mm.

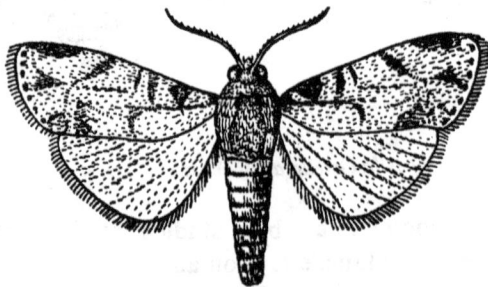

Macalla moncusalis Walker

Egg, larval and pupal periods of *O. exvinacea* are 4, 28 to 33 and 11 to 14 days respectively while the adult longevity is 4 to 5 days.

Lamida sordidalis (Hampson) has been reported from Bihar (Sinha and Sinha, 1961) feeding on shoots and young leaves, webbing the same together during November-December.

To control these leaf webbers, remove and destroy immediately the webs with larvae within.

MANGO LEAF MINER

Acrocercops syngramma Meyrick, a microlepidopterous species has been reported from Bihar, Bengal, Maharashtra, Karnataka and Tamil Nadu. It is a minor pest, active usually from October-November to April-May. The tiny caterpillars mine under the dorsal epiderms of tender leaves and feed within. As a result, greyish-white round blisters appear on the leaves. Pupation takes place within the tunnels. The caterpillars are apodous and pale greenish-yellow in colour while the adults are very small, delicate moths with narrow long fringed wings.

Acrocercops syngramma Meyrick

Remove and destroy promptly the affected leaves in the initial stage of attack. This will prevent the pest infestation from spreading.

LEAF EATING WEEVILS

Mango leaf cutting weevil, *Deporaus marginatus* (Pascal) is found all over India, Bangladesh and Burma. It is specific pest of mango. In India the pest is active from August to October (Fletcher, 1914), when it attacks the new flushes of leaves and destroys the same completely leaving only the stems. Young trees suffer comparatively more than the older ones. The female excavate small cavities on either side of the midribs on lower surfaces of tender leaves. Eggs are laid singly in these cavities. About 10 to 20 eggs are laid in a single leaf which is then cut by the weevils, near the base. On hatching the grubs mine between the two epidermal layers of the leaf and feed within. When full grown, the grubs come out of the mines to pupate in the soil.

Eggs are oval and transluscent. Grubs are dirty green and about 5 mm long. Adults are small, greyish-brown weevils having long brown snout and shining black elytra. Egg, grub and pupal durations are, 2 to 3, 6 to 8 and 7 to 8 days respectively (Khanna, 1952). There are three generations in a year.

Deporaus marginatus (Pascal)   *Apoderus transquebaricus* (Fabricius)

Mango leaf twisting weevil, *Apoderus transquebaricus* (Fabricius) is another minor pest of mango reported from South India. Its other main host is country almond, *Terminalia catappa* Linnaeus. Eggs are laid singly on leaf tips. These leaves are then rolled tip downwards into neat thimble-shaped structures and the earlier stages of the pest are passed within these rolled leaves. The adults come out by making a hole at the side of these rolled leaves.

Eggs are oval in shape and yellowish in colour. Grubs are legless and pale yellow while pupae are bright yellow in colour. Adult are medium sized weevils, reddish-brown in colour with long neck and snout.

Mango leaf mining weevil, *Rhynchaenus mangiferae* Marshall is widely distributed in South India (Ayyar, 1922) where it is active during March to July. Like *D. marginatus* these weevils also bite and make small oval pouches on the ventral side of leaves and deposit therein their eggs but thereafter unlike *D. marginatus*, these weevils do not cut the leaves. On hatching the grubs mine the leaves and feed within on mesophyll. As many as 20 to 30 grubs may be found in a single leaf. Affected leaves turn reddish-brown, crumple, dry and fall off. Singh (1954c) reported this weevil from Uttar Pradesh where it is found during February to May damaging inflorescences. Eggs are laid in the flowers; grubs mine the thalamus of flowers and buds and pupate also therein. Affected flowers and buds wither and dry away. After the bloom is over, the weevils feed on malformed inflorescences and young tender leaves. *Langra* variety of mango is preferred by this weevil.

Eggs are minute, flattend, oval and transluscent. Grubs when full grown, are about 3 mm long, legless, flattend and transluscent whitish in colour. Adult weevils are 1.5 to 2.0 mm long, pale reddish-brown in colour with dark brown elytra. They possess elongated hind femora which help the weevils to jump like flea beetles. Egg, grub and pupal periods are 2 to 3, 4 to 5 and 3 to 4 days respectively with total life occupying 10 to 12 days There are more than eight overlapping generations in a year.

Other curculionid weevils that feed on the mango leaves include, *Alcidodes frenatus* (Faust), *Amblyrrhinus poricollis* Boheman, *Atmetonychus peregrinus* Olivier, *Camptorrhinus mangiferae* (Marshall), *Myllocerus discolor* Boheman. *M. laetivirens* Marshall, *M. undecimpustulatus maculosus* Desbrocher, *M. sabulosus* Marshall, *Oncideres repandator* Faust, *Peltotrachelus pubes* Faust and *Platymycterus sjostedti* Marshall, These are all pests of minor importance reported occasionally from various parts of India. While the adult weevils feed on leaves, the grubs of *Myllocerus* spp. feed on rootlets and those of *Alcidodes frenatus* bore into the top shoots.

To control these weevils, spray with 0.03% monocrotophos or endosulfan (Butani, 1975c). In case of young trees dusting may be done with 5% BHC. Collection of affected and fallen leaves and their prompt destruction by burning also help in checking the pest population especially that of *Deporaus marginatus*.

LEAF DEFOLIATING BEETLES

Adoretus lasiopygus Burmeister (*ovalis* Blanchard) is found in South India (Fletcher, 1914) while this and *Anomala varicolor* Gyllenhal have been reported from North-West India (Pruthi and Batra, 1960). The adult beetles appear with the onset of monsoon and feed upon leaves of mango trees during night and are active for about two months. These are polyphagous and besides mango, *A. lasiopygus* has also been recorded feeding on leaves of apple, fig, grapevine, guava. loquat, peach, pear, plum, etc. and *A. varicolor* on peach, pear and plum leaves. Other chafer beetles recorded on mango include, *Anomala dussumieri* Blanchard, *Holotrichia consanguinea* (Blanchard) and *H. insularis* Brenske.

In the initial stage of attack, collect and destroy the beetles when these congregate at night for feeding and mating on the mango as well as shade trees around. If attack is severe, spray with 0.2% carbaryl twice at 7 days' interval.

LEAF EATING CATERPILLARS

A number of lepidopterous larvae have been recorded feeding on mango leaves. The most common one is castor slug caterpillar, *Latoia lepida* (Cramer), which occurs throughout India. Though castor and coconut are its preferred hosts it is found on banana, citrus, country almond, fig, mango, pomegranate, wood apple, etc. Eggs are laid in clusters on ventral side of leaves. On hatching the caterpillars move gregariously but sluggishly devouring the leaves completely, leaving only the midribs. Eggs are flat and shining. Larvae are fleshy, soft bodied, apple green, hairy with greenish-blue stripes and two yellowish-green lateral stripes (one on each side). They also have a series of tufts of spines on their body and these spines are highly irritant to touch. Full grown caterpillars (about 25 mm long) pupate in shell-like compact, elliptical, chocolate brown cocoons, attached to tree trunks. Moths are short, stout with forewings predominatly green fringed with big brown patches.

Incubation, larval, pupal and total life-cycle durations occupy on an average one, 6, 3 and 10 weeks respectively (Ananthakrishnan and Abraham, 1965).

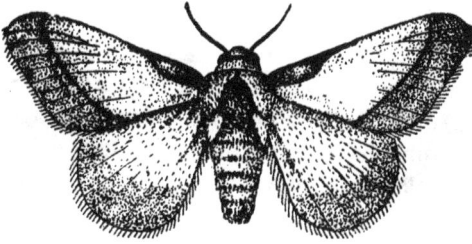

Latoia lepida (Cramer)

Normally, no control measures are warranted against this pest. It and when the caterpillars appear, clip off and destroy the leaves with gregarious larvae thereon. Dustrng with 5% BHC is also effective.

Natada velutina Kollar is another slug caterpillar found feeding on mango leaves but does not cause any severe damage. Other lepidopterous caterpillars occasionally recorded on mango leaves in various parts af India, include, *Perina nuda* Fabricius, *Dasychira mendosa* (Hübner). *Euproctis flava* (Bermer). *E. fraterna* (Moore) *E. lunata* (Walker), *E. xanthosticha* (Hampson), *Lymantria beatrix* Stoll, *Cricula trifenestrata* Helfer, *Bombotelia jocosatrix* (Guenée),

Bombotelia jocosatrix (Guenee)

Oenospila quadraria Guenée, *O. veraria* (Guenée), *Euthalia garuda* Moore, *Rapala melampus* (Cramer), *Agrius convolvuli* (Linnaeus) and *Acherontia styx* Westwood, *Selepa celtis* Moore, *Spilonota rhothia* Meyrick, *Stauropus alternus* Walker. None of these are of any major importance though some of these, specially *P. nuda* occurs quite regularly.

Perina nuda is a minor pest of fig, jack-fruit and mango. Its eggs are laid in clusters on leaves. On hatching the caterpillars feed on the leaves and pupate in leaf-folds. Pupae are hairy and brightly coloured dorsally. Male moths have half ochreous and half transparent forewings while females have forewings, dull ochreous in colour.

THRIPS

These are minute, about one mm long, slender and soft bodied fragile insects with the wings heavily fringed. They have asymmetrical, rasping and sucking type of mouth-parts with which these insects lacerate the plant tissues and suck the oozing cell-sap. Metamorphosis being of hemimetabolic type, the nymphs and adults have similar feeding habits. However, in-between these there are quiescent pre-pupal and pupal stages. Adult females have serrated (saw-like) ovipositor, which causes injury to the leaves, and other plant parts, at the time of egg-laying.

The species reported from India, on mango, include, *Aeolothrips collaris* Priesner, *Anaphothrips sudanensis* Trybom, *Caliothrips indicus* (Bagnall), *Rhipiphorothrips cruentatus* Hood, *Selenothrips rubrocinctus* (Giard), *Haplothrips ganglbaueri* Schmutz, *Neoheegeria mangiferae* (Priesner), *Ramaswamiahiella subnudula* Karny and *Scirtothrips dorsalis* Hood. Of these, first four species feed on the leaves and last four infest the inflorescences. These thrips seldom cause any economic loss and as such are of minor importance. In case of leaf feeding species, the nymphs and adults congregate on ventral surfaces, more towards the leaf tips and feed on mesophyll. The affected leaves show silvery sheen and bear small spots of faecal matter. In case of severe infestation, the leaf tips turn brown and get curled. When inflorescences are infested there is discolouration at the feeding sites which subsequently turn brown and die off.

Other sucking pests found feeding on mango leaves are, mango blackfly, *Aleurocanthus mangiferae* Quaintance and Baker, citrus blackfly, *A. woglumi* Ashby, leaf hopper, *Amrasca splendens* Ghauri, aphid, *Toxoptera odinae* van der Goot and cow-bug (membracid),

Otinotus oneratus (Walker). Besides, painted bug, *Coptosoma nazirae* Atkinson has been reported sucking the sap from tender shoots; *Antestia cruciata* (Fabricius) infests the inflorescences and *Spilostethus pandurus* (Scopoli) attacks the fruits. But none of these are of any major importance.

To control these sucking pests, if and when necessary, spray with 0.03% phosphamidon or monocrotophos or 0.04% diazinon or dimethoate.

MANGO SHOOT BORERS

Chlumetia transversa Walker and *C. alternans* Moore have been reported boring the mango shoots (Beeson, 1941); the former is comparatively more common and destructive. *C. transversa* was first recorded from Uttar Pradesh in 1900 (Anonymous, 1903) and is now found all over India. It has also been reported from Sri Lanka, Malaysia, East Indies (Hampson, 1912), Philippines (Palo, 1932) and Indonesia (Voute, 1935b). Besides mango, it has been reared on litchi leaves (Lefroy and Howlett 1909). Eggs are laid on tender leaves. Freshly hatched caterpillars bore into midribs of tender leaves and come out after a couple of days to bore into tender shoots near the growing point (Singh, 1957) tunnelling downwards (100 to 150 mm), throwing their excreta etc. out of the entrance hole. When full grown these caterpillars come out and enter into the slits and cracks in the bark of the tree, dried malformed inflorescences or cracks and crevices in the soil, for pupation. Leaves of affected shoots wither and droop down. Young grafted seedlings are severely affected and may even be killed. The pest is usually active from August to October.

Young caterpillars are yellowish-orange in colour with characteristic dark brown prothoracic shield. Full grown caterpillars are dark pink with dirty spots and measure 20 to 24 mm in length. Pupae are obtect type and 10 to 12 mm long. Adults have thorax and abdomen clothed with rufous, fuscous and grey scales. Forewings are dark grey beautifully patterned with wavy design. Hind wings are fuscous, apical side being darker than proximal side. Wing expanse is 15 to 20 mm. Moths of *C. alternans* differ from *C. transversa* in being bigger (wing expanse 26 mm) and having thorax and abdomen darker; forewings are green with basal area dark rufous and hind wings with a marginal series of dark striae (Hampson 1894).

Clumetia transversa Walker

Incubation, larval and pupal periods range from 2 to 3, 10 to 12 and 15 to 18 days respectively (Singh 1957). The pest overwinters from October to March in pupal stage and there are four overlapping generations in a year.

Anarsia melanoplecta Meyrick has been recorded from Bihar and Maharashtra (Fletcher, 1921). Its caterpillars bore into the buds and young shoots and feed on the inflorescences of mango (Beeson, 1941.) *Argyroploce erotias* Meyrick is found in North India rolling the tender leaves and boring into the shoots. *Platypelpa (Argyroploce aprobola* (Mayrick) occurs in Kerala; its larvae web together the young leaves and bore into the shoots (Nair, 1975).

To control these shoot borers, clip off and destroy promptly the affected shoots in the inital stage of attack. In case of severe attack, spray 2 to 3 times at three weeks' interval with 0.2% carbaryl, commencing from initiation of new flush of leaves (Tandon *et al.*, 1975).

FLOWER FEEDING CATERPILLARS

A few species of lepidopterous larvae have been reported from South India feeding on mango inflorescences (Nair, 1975). These include, *Asura ruptofascia* Hampson, *Celama analis* Will. and Westwood, *C. fasciatus* Walker ; *Chlorochystis* species, *Cosmostola laesaria* Walker, *Gymnoscelis imparatalis* Walker ; *Enarmonia anticipans* Walker, *Eublemma abrupta* Walker, *E. angulifera* Moore, *E. brachygonia* Hampson, *E. silicula* Swinhoe, *E. versicolor* Walker, *Nanaguna breviuscula* Walker, *Phycita umbratelis* Hampson. In addition Nayar *et al.* (1976) reported *Anatrachyntis simplex* Walsingham. However, none of these have been reported as pests of economic importance.

MANGO FRUIT BORERS

Noorda albizonalis Hampson has been reported as a major pest form Orissa, specially on grafted varieties like *Baiganpalli, Latsundri* and *Bangaloria.* The young caterpillars attack the freshly formed fruits at the distal end. The typical symptom of a damage by this pest, is appearance of a small dot (which also serves as exit hole) with dark pale brown ring encircling it. The caterpillars feed within the fruit and make series of tunnels in the kernel. Full grown caterpillars are about 25mm long, red in colour, slightly hairy and usually pupate inside the fruits from where the moths emerge through the exit holes; sometimes even the full fed caterpillars come out from these holes and pupate outside (Sengupta and Behura 1955).

Ctenomereistis ebriola Meyrick is another fruit borer reported from Orissa, where it is active during March-April and sometimes continues till June. The caterpillars bore into the developing fruits in the same way as *Noorda albizonalis.*

Castor capsuler borer, *Dichocrocis punctiferalis* (Guenée) and *Hyapsila leuconeurella* Ragonot have also been occasionally found boring mango fruits but none of these are of any economic importance.

To control these fruit borers, if and when these occurs, remove and destroy immediately all the infested fruits. In the case of severe infestiation, which is rather rare, spray with 0.1% BHC+0.1% DDT or if the infested trees are young, dust with 10% BHC.

FRUIT SUCKING MOTHS

Othreis fullonia (Clerck) and *O. materna* (Linnaeus) are the two common species reported from India. Their main host is *Citrus* Spp., but apple, banana, mango, peach, pear, plum, pomegranate and tomato fruits are also attacked (Pruthi and Batra, 1960). The caterpillars feed on various grasses growing around the fruit trees and the moths, which are nocturnal in habit, puncture the fruits, after dusk and feed on the juice thereof. The affected fruits rot and fall off prematurely. Vevai (1969 c) also reported *Achaea janata* (Linnaeus) and *Calpe emarginata* (Fabricius) as fruit sucking moths of mango.

If and when these moths appear in mango orchards, hand-picking and prompt destruction of caterpillars and moths is suggested. This will prevent the population built-up of the pest and reduce further damage.

2.2
BANANA

BANANA or plantain. *Musa* spp., is one of the oldest fruits known to mankind. It is indigenous to South-East Asia and its numerous species and varieties have originated from India (Assam), Burma, Thialand or Indo-China. These have by now spread to other tropical regions of America, Africa, Australia, Philippines and Hawaii (Singh, 1969). In India, cultivation of banana was started as early as 600 - 500 B.C. At present there are about 210,000 hectares under this crop; of which 48,000, 38,000, 31,000, 20,000 and 20,000 hectares are in Kerala, Tamil Nadu, Maharashtra, Gujarat and Assam respectively. Of late this crop has become an impotant foreign exchange earner for our country, Japan and USSR being our main customers. Banana is essentially a tropical crop that grows best on well drained fertile soils. It cannot tolerate frost and is very susceptible to wind damage. The fruit is one of the most important energy-producing foods. It is a good source of minerals and vitamins and some of the varieties contain as high as 20% sugar. The varieties with high sugar content are eaten as dessert while those with high starch are used for cooking and beer-brewing. Scott (1957) has given an illustrated account of preperation of banana beer (*mwenge*) in Uganda.

Simmonds (1966) has given an annotated list of 182 insect pests of bananas along with their world wide distribution. In India, banana trees do not suffer much from the ravages of these pests. The major pests in India, include, banana weevil, banana stem borer and flea beetles. Besides, there are about 36 insect species reported attacking banana leaves and fruits but most of these are mere records and only a few are sporadic pests causing occasional damage. Fortunately, banana scab moth *Nacoleia octasema* (Meyrick) – a major and specific pest of banana in South-East Asia and Australia, has not yet been recorded from India.

BANANA WEEVIL

Cosmopolites sordidus (Germar) is one of the most destructive pest of banana. Though a native of South-East Asia (Malaysia), it is found in all the banana growing countries of the world except Egypt, Israel and Hawaii (CIE map No. A - 41). In India, it is widely distributed and has been reported as a major pest from Assam, Kerala, Tamil Nadu, Karnataka and Maharashtra (Vevai, 1971a). A bibliography of the earlier work done on this pest has been published by Leonard (1931).

C. *sordidus* is an specific pest of *Musa* spp., found on good many varieties, *Malbhog* and *Champa* being the preferred ones. Eggs are laid singly in collar (above ground) or rhizomes (under ground). Often a hole is made in the decaying leaf sheath to lodge an egg. Soon after hatching, the grubs bore into the stem of same stool and feed within. Pupation is usually in the soil. Adults are sluggish and avoid day light, hiding in leaf sheaths or rotting pseudostems where the humidity is very low. They feed during night on the pseudostems and bore into the suckers. The attacked pseudostems get riddled with holes and the root origins are weakened. Secondarily. the tunnels made by these weevils are occupied by fungi and bacteria which help to accelerate the process of rotting. As a result, with the strong blast of wind the trees break down from the point of infestation. Even if the fruits are formed these are few in number and inferior in quality (Sen and Prasad, 1953).

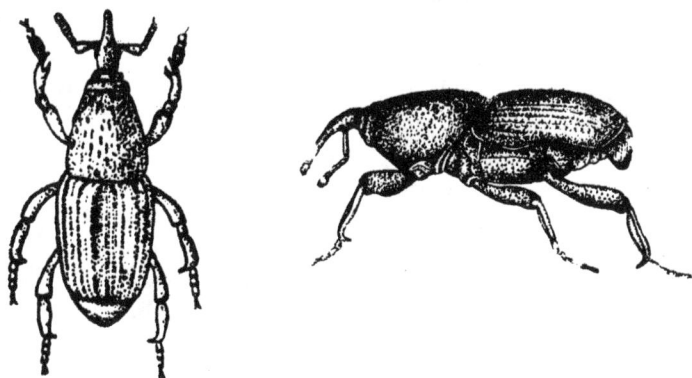

Cosmopolites sordidus (Germar)

Eggs are elongated oval in shape, about one mm long and pure white, in colour. Grubs are creamy white, stout, fleshy, highly wrinkled and legless with spindle-shaped, 8 to 12 mm long body.

Adult weevils are 10 to 13 mm long shiny black in colour, having fairly long and curved snout and short elytra striated longitudinally. Though these wings are functional the weevils seldom fly. Oviposition occurs throughout the year (Frohlich and Rodewald, 1970) and is more during rainy season (Smith, 1977). Incubation period is normally about a week and may extend upto 4 weeks depending upon the climatic conditions. Duration of grub stage is 2 to 6 weeks while pupatory stage is accomplished in 10 to 12 days. Adults live upto two years and can remain without food for 6 months.

To avoid the attack, healthy suckers or rhizomes should be used; areas previously infested should be avoided. At the time of planting, the pits should be treated with 5% aldrin, dieldrin or chlordane dust @ 60 to 80 gm per pit. Alternatively, before planting, the suckers should be cleaned, trimmed and soaked in 0.1% solution of aldrin, dieldrin or chlordane. This coupled with clean cultivation and destruction of sheltering and feeding places of adult weevils keeps the pest under check. Clean cultivation includes keeping the area around the base of pseudostems clear of weeds, good husbandry, practices such as prevention of soil erosion, regular removal of water suckers and dead leaves besides mulching and manuring; removing immediately pseudostems from which banana bunches have been cut – these pseudostems should be cut below the ground level and should not be left in the field as such. These cultural practices help to produce vigourous banana plants which are able to tolerate the weevil damage. In cases of severe attack, spray 0.05% dieldrin around the collar of affected plants (Saraiva, 1964) or 0.03% phosphamidon or dimethoate or 0.04% oxydemeton methyl or 0.08% fenitrothion (Butani, 1973b). The treatment may be repeated once a month or earlier. This will also kill other insect pests which may be present during that period. Biological control has also received a lot of attention in various countries (Cuille, 1950) but with little success except in Fiji and Tahiti Islands where the hysterid beetle, *Plaesius javanicus* has kept this pest under check.

BANANA STEM BORER

Odoiporus longicollis (Olivier) is another specific pest of banana reported as a major pest from India, Bangladesh, Burma, Sri Lanka and Indonesia. According to Simmonds (1966) this weevil is troublesome in various parts of South-East Asia and is likely to spread to other countries as well. In India, it is a serious pest in Assam, Bengal, Bihar and has also been reported from Delhi (Batra, 1952a) and Uttar Pradesh (Shukla and Kumar, 1969). The pest breeds throughout the year but is more active during Summer and monsoon months. Eggs are laid in air chambers through slits cut on leaf sheaths. On hatching, the grubs start feeding on tissues around the air chambers of leaf sheath and then bore inside the pseudostems. A number of grubs may be found boring a single plant. The pseudostems thus riddled become weak and start rotting. Ultimately, as in the case of *Cosmopolites sordidus* these trees also break down with strong blast of wind.

Eggs are yellowish-white in colour and cylindrical (2 x 1 mm) in shape. Grubs are apodous, soft, fleshy, wrinkled and sparsely covered with brown hairs of varying length. Pupae are exarate and pale yellow in colour. Adult weevils are chocolate to black in colour, 23 to 28 mm long with pointed head. Dutt and Maiti (1972 a) studied its biology in detail and observed egg, grub and pupal periods to occupy respectively, 3 to 5, 26 and 20 to 24 days during Summer and 5 to 8, 68 and 37 to 44 days in Winter.

Another curculionid weevil, found boring in stems and roots of banana in India, is *Polytus mullerborgi* Boheman (Fletcher, 1917) This is a common pest in Fiji and Philippines Islands and does little damage in India.

To check the population of these weevils, uproot and burn infested plants. Spraying with 0.3% BHC (Vevai, 1971a) or 0.05% endosulfan or 0.1% carbaryl is also effective. Dutt and Maiti (1972 b) suggested inserting aluminium phosphate tablets into the thick basal region of pseudostems @ three tablets of 0.5 gm each, per pseudostem.

FLEA BEETLES

Nodostoma subcastatum Jacoby and *N. viridipennis* Motschulsky are also major pests of banana reported from India, Bangladesh and Burma. In India, *N. subcastatum* is more common and has been reported from Assam, Bihar, Uttar Pradesh and Delhi (Nair, 1975). Besides banana, it also damages grapevines. *N. viridipennis* has been reported as pest of banana in Bihar, where it prefers *Alpan, Champa* and *Malbhog* varieties (Sen and Prasad, 1953). The pest appears around April-May and continues its activities till the end of rainy season (September-October); the maximum infestation is during August-September. Grubs are found underground near the roots while the beetles are found feeding on leaves and fruits. The central leaves of the plant forming the top whorl are the worst affected. In case of fruits, the beetles scratch the skin of newly formed fruits; the scratches become dilated with the development of fruits – thus the fruits are blemished and the flavour spoiled, reducing thereby the market value of such fruits. The mature fruits are not attacked as beetles probably are unable to scratch the hard peel. Roy and Sharma (1952) observed nearly 80% of banana bunches attacked by this pest during rainy season at Sabour (Bihar).

Clean cultivation and phyto-sanitation prevents the occurrence of this pest. The only chemical control measure suggested is spraying with 0.2% dieldrin (Sarma and Saikia, 1967). However, since the use of dieldrin as foliar spray is being discouraged, spray with 0.05% endosulfan or 0.1% carbaryl. To check the pest effectively, spray immediately after the emergence of inflorescneces and again after a fortnight.

LEAF EATING CATERPILLARS

Bag worms, hairy caterpillars, leaf rollers, slug caterpillars and tobacco caterpillar are some of the lepidopterous pests feeding on banana leaves. These are mostly of minor importance.

Bagworms, *Acanthopsyche minima* Hampson is reported from Kerala. The famale moths remain inside the pupal cases and also lay the eggs there. A single female lays from 150 to 200 eggs before extricating itself from the pupal case. Almost all the eggs hatch at the same time and a swarm of caterpillars emerges from the pupal case. These caterpillars scrap the leaves for a day or two, construct very small tubular bags or coverings using small pieces of leaf lamina They keep their body inside the bag and with head and thorax protruding continue feeding on the leaves, making irregular holes in the leaf lamina. Eggs are small (less than one mm long), cylindrical and yellowish-white in colour. Full grown larvae are 9 to 13 mm long and purple-brown in colour. Pupae are obect type, males light brown and females, yellowish-brown. Male moths are slender, blackish in colour; forewings hyaline with outer margins transparent and hind wings uniformily hyaline bearing long white setae. Female moths are very degenerate and wingless, devoid of antennae, mouth-parts and legs. Eggs hatch in 7 to 10 days. Larval period of males is 45 to 18 and 56 to 67 days during Summer and Winter respectively whereas that of females is 55 to 58 days in Summer and 67 to 78 days during Winter. Pupal period is 9 to 11 and 7 days in case of males and females respectively (Johnson, 1975).

Kophene cuprea Moore is another bagworm found on banana leaves in Kerala. Its incubation, larval and pupal durations are about a week, 2½ months and 7 to 8 days respectively (Brahmachari, 1938).

Black hairy caterpillar, *Pericallia ricini* Fabricius also known as wooly bear, is a polyphagous pest found all over India. It is an sporadic pest of banana, causing considerable injury to the leaves Eggs are laid in clusters on ventral leaf surface. On hatching the larvae feed on epidermis of the leaves for a couple of days then they eat the whole leaves. Larvae have red head and blackish-brown body with long thick hairs arising on bluish wrats. Pupation takes place in cocoons made of hair and silk. Adults are stout moths having fuscous forewings bearing pale ringed black spots and crimson hind wings with black patches.

Pericallia ricini Fabricius

Incubation, larval and pupal periods extend to 4, 26 to 32 and, 10 to 12 days respectively and the total life-cycle is completed in about 40 days during April-May (Nair, 1975).

Other hairy caterpillars reported feeding occasionally on banana leaves, are, *Asura ila* Moore, *Creatonotus gangis* Linnaeus *Diacrisia obliqua* Walker, *Dasychira mei dosa* Hübner, *Euproctis fraterna* (Moore) and *Eupterote geminata* Walker.

Diacrisia obliqua Walker

Erionota thrax Linnaeus, popularly known as banana skipper or leaf roller has been reported from India, China, Vietnam, Malaysia, Indonesia, Philippines and Mauritius (Feakin, 1977). Eggs are laid in clusters on young leaves and sometimes even on fruits. On hatching the caterpillars cut the leaf lamina at the edge and roll a strip of leaf lamina towards the midrib and feed within. A number of such leaf rolls may be seen hanging from a single leaf. The larvae spent their entire life in the same leaf roll and pupate *in situ*. Eggs are bright yellow, which turn red with age and finally become white, just before hatching. Caterpillars are greenish in colour. covered with white powdery secretion. Pupae are brown. Adults are brownish butterflies with yellow spots on forewings; wing expanse is 70 to 80 mm. Egg, larval and pupal durations last for 5, 25 and 10 days respectively (Simmonds, 1966).

Slug caterpillars, *Belippa laleana* Moore, *Contheyla rotunda* Hampson, *Latoia lepida* (Cramer), *Nemeta lohor* (Moore), *Microlimax* species and *Miresa decedens* Walker have been reported feeding occasionally on banana leaves, specially during the rainy season. Their larvae are lethargic and remain feeding at one place cutting out large patches of leaf blade. Being nocturnal in habit, the larvae remain hidden within leaf axils during day and feed at night. The species commonly found in Assam and Nilgiris is *M. decedens*. Its caterpillars are pale greenish-white with a median dorsal bluish-black band and 4 rows of tubercles bearing setae. Pupation takes place in hard shell-like cocoons. Adults are stout greyish-brown moths with velvety golden bloom all over the dorsal side and a pale streak on the forewings; wing expanse is 35 to 40 mm. There is only one generation in a year, as pupal period lasts for about 10 months, from September-October to June-July of the following year (Ayyar, 1924).

Spodopiera litura (Fabricius) [*Prodenia litura* (Fabricius) *auctt.*] commonly known as tobacco caterpillar has been often confused with and considered as synonym of *S. littoralis* (Boisduval). However, the two are distinct species and quite allopatric in distribution. *S. litura* has been reported from Pakistan, India, Bangladesh, Sri Lanka, S. E. Asia, China, Korea, Japan, Philippines, Indonesia, Australia, Pacific Isands, Hawaii and Fiji (CIE map No. A–61); whereas *S. littoralis* is confined to Africa and Mediterranean region (CIE map No. A–232). *S. litura* is a polyphagous pest having a very wide range of host plants including, banana, citrus, melons, mulberry papaya and though tobacco and cauliflower are its preferred hosts. The freshly hatched caterpillars feed gregariously on the same leaf and later move to other leaves become solitary and feed voraciously at night. Pupation takes place in earthen cocoons, 50 to 80 mm deep down in the soil.

Eggs are pale yellow to light green in colour and round in shape, diameter about 0.5 mm. Full grown caterpillars are plump, cylindrical, 35 to 40 mm long, pale greenish-brown with black spots and 5 orange coloured longitudinal lines and a few scattered short hairs. Pupae are lustrous reddish-brown. Adults are stout moths having greyish-brown forewings bearing wavy white markings and hind wings membraneous white with brown margins: wing expanse is 32 to 40 mm.

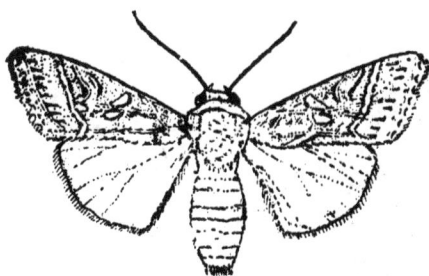

Spodoptera litura (Fabricius)

Eggs are laid in clusters of 50 to 300 on ventral surface of leaves covered with buff coloured hairs. A single female can lay 1500 to 2000 eggs during her life span of about 8 days. Egg, larval and pupal periods last for 3 to 4, 13 to 19 and 6 to 10 days respectively (Patel, 1977).

Hand-picking and mechanical destruction of caterpillars in early stage of attack is quite effective In case of severe infestation spray with 0.1% carbaryl or 0.5% endosulfan or dichlorvos or dust 10% BHC or 5% carbaryl. In case of leaf rollers, prompt removal and destruction of rolled parts of leaves with larvae or pupae within will check the pest population. Spraying with 0.2% BHC (W.P.) or carbaryl to run-off point over the leaf rolls also helps in killing the pest but the treatment should be given only in case of serious infestation.

LEAF SUCKING BUGS

Banana lace-wing bug, *Stephanitis typicus* Distant is widely distributed in India and has also been reported from Sri Lanka, Bangladesh, South China, Taiwan and Philippines. Though a minor pest of banana in South India, it is a serious pest of coconut, cardamom and turmeric. Nymphs and adults congregate on ventral leaf surface sucking the cell sap therefrom. The feeding spots become greyish yellow and appear as whitish blotches from dorsal side of leaves; ultimately the affected patches become brown. In case of severe infestation, plant growth is retarded, yielding fewer and poor sized fruits.

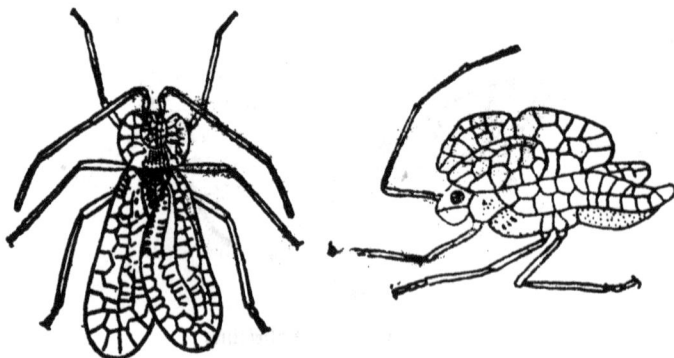

Stephanitis typicus Distant

The adults are small, dull coloured bugs with transparent lace-like wings. Eggs are laid singly on ventral surface of leaves. A single female lays from 25 to 50 eggs. These hatch in 10 to 12 days and nymphal duration lasts for 13 to 15 days.

Spittle bug, *Phymatostetha deschampsi* Lethierry has been reported from South India (Vevai, 1971a). Nymphs feed on various weeds and are found in froth or foamy mass. Adults appear in large number during rainy season and suck the juice from banana leaves.

Other bugs found feeding on banana leaves include, a leaf-hopper, *Eurybrachys tomentosus* Fabricius and a capsid bug, *Prodromus subviridis* Distant. None of these cause severe damage to banana.

To control these bugs, spray, if and when necessary, with 0.1% carbaryl or 0.05% fenitrothion or formothion.

COCCOIDS

Perissopneumon tamarindus (Green), *Ferrisia virgata* (Cockerell) *Abgrallaspis cyanophylli* (Signoret), *Aonidiella aurantii* (Maskell), *A.orientalis* (Newstead),*Aspidiotus destructor* (Signoret),*Chrysomphalus ficus* Ashmead (*aonidum*), *Hemiberlesia lataniae* (Signoret), *Coccus acutissimus* (Green), *C. discrepans* (Green), *C. hesperidum* Linnaeus and *Saissetia depressum* Cockerell have been recorded from India as minor pests of banana. These are polyphagous and only a few of them are of economic importance on other host plants. These insects suck cell sap from leaves and occasionally from fruits as well, resulting in general loss of vitality. Normally, attack by these insects seldom has any serious effect on fruit production, except when they attack the fruits; the affected fruits lose their market value. These pests are active during Summer and hibernate during Winter.

No sooner the attack of any scale insect is located, the affected leaves must be pruned and the affected fruits, if any, removed and destroyed immediately. This will prevent the pests from spreading.

BANANA APHID

Pentalonia nigronervosa Coquillett (*caladii*) has been recorded from India, Sri Lanka, Taiwan, Philippines, East Australia, Fiji, Tonga, Samoan, Hawaii, Honduras, West Indies, Brazil (Colombia,) Canary Island, Sierra Leone, Egypt and Mauritius as also from greenhouses in France and England (Simmonds, 1966). Its distribution is probably coexistent with banana cultivation (CIE map No. A–242). It has a limited host range and prefers warm and moist climate; thus in Australia it is found during Summer while in Egypt it is active during Winter. In India, it has been observed congregating under the outer leaf base on the pseudostems and around the crown of the plants. In case of severe infestation theer is progressive leaf dwarfing and leaf curling; fruit bunches become small and sometimes even the fruits are distorted. These aphids also act as vectors of some important virus diseases. Bunchy top – the most destructive disease of banana, is transmitted by this aphid. The disease is widespread in eastern and southern India. In Kerala alone this disease of banana causes an estimated loss of 60 million rupees annually (Reddy, 1968). Besides, in India, Australia and Brazil, this aphid also transmits chlorosis or heart rot – another virus disease of banana, though of minor importance.

The aphids reproduce by agamic viviparity. A new colony produces 7 to 10 apterous generations which are followed by a generation of alate form, the latter migrate to infest new plants while the apterous ones die away. Adults are 1 to 2 mm long, brown in colour with antennae as long as body and siphunculi slightly clavate and quite long. Alate forms have a prominent dark brown to black wing-venation.

Pentalonia nigronervosa Coquillett

Cotton aphid, *Aphis gossypii* Glover has also been recorded on banana, transmitting yet another virus disease – the banana mosaic. This is a minor disease and is quite common in Gujarat and Maharashtra. As a result of the disease, the growth of infested mats (plants) is stunted and there is seldom any fruit formation.

Spraying with nicotine sulphate (1:800) or 0.03% phosphamidon or oxydemeton methyl or 0.05% dichlorvos is effective. It is rather desirable to rouge out the virus affected plants before the spraying.

THRIPS

Thrips occurring on banana in India are, *Helionothrips kadaliphilus* (Ramakrishna and Margabandu), *Panchaetothrips indicus* Bagnall, *Scolothrips asura* Ramakrishna and Margabandu, *Thrips florum* Schmutz (*hawaiiensis* Morgan), *Astrothrips parvilimbus* Stannard and Mitri, *Chaetanaphothrips signipennis* (Bagnall) and *Hercinothrips bicinctus* Bagnall (Ananthakrishnan, 1971) – the first three lacerate the leaves whereas last four damage the fruits.

Helionothrips kadaliphilus – the banana thrip, was first recorded by Ayyar and Margabandu (1931) and is very common in South India. It has reddish-brown body and yellowish-grey wings. Nymphs and adults are found congregated on ventral surface of young leaves, lacerating the tissues and rasping the sap that oozes out from these wounds. The feeding spots turn yellowish-brown and in case of severe infestation the affected leaves dry up. Egg and nymphal stages occupy 4 to 6 and 9 to 12 days respectively (Murthy, 1958).

Panchaetothrips indicus, commonly known as turmeric thrip, is a serious pest of turmeric. It is yellowish-brown to brown in colour with forewings having dark setae. Females have saw-like ovipositor curved downwards. Infested trees invariably have a high population of this species with the result the leaves turn pale yellow and start curling, rolling and drying.

Thrips florum (*nigricornis* Schmutz) is a flower thrip, having a very wide range of host plants (Ananthakrishnan, 1971). In case of banana, nymphs and adults feed within the developing buds or on perciarp of young fruits covered by the bracts, thereby causing minute light brown rust-like spots and affecting adversely the quality of fruits. The affected fruits develop a corky scabbing. The pest is active throughout the year except during severe cold. Adults are uniformly brown with pale yellow antennae; wings are dark grey, paler at the base.

Astrothrips parvilimbus – adults of which are yellowish-brown to brownish, was first recorded and described by Stannard and Mitri (1962). Like *T. florum*, this species is also responsible for rust-like spots on banana fruits. It is often found in association with *T. florum*, feeding on pericarp of young fruits resulting in powdery blotchings on the pericarp of affected fruits.

Chaetanaphothrips signipennis, commonly known as banana rust thrip, was considered synonymous with *C. orchidii* (Moulton) till Stannard (1956) pointed out their distinct identities. This thrip is native of old world and has been reported from India, Sri Lanka, Philippines, New Guinea, N. E. Australia (Queensland), Fiji, Hawaii, Honduras, Costa Rica, Panama, Brazil, Trinidad, etc. It is specific to banana. Adults are golden-yellow in colour, rather minute in size (1.2 to 1.6 mm long) and have wings fringed with long fine hairs. These remain congregated in between the fruits or underneath the leaf sheaths. Eggs are laid either on fruits or on pseudostems just beneath the surface of plant tissues. On hatching, nymphs crawl and feed on leaves and pupate in the soil. As a result of their feeding rough brown spots are seen on leaves and fruits. When immature fruits are severely attacked, their skin becomes rough and pulp may dry inside, making the fruits inedible. In case of mature fruits the skin splits, exposing the pulp of the fruit (Roy and Sharma, 1952).

Chaetanaphothrips orchidii is pantropical in distribution and causes similar damage as *C. signipennis*. *C. clarus* (Moulton) has been reported as banana rust thrip in St. Lucia (West Indies) but has been recorded only on grasses in India (Feakin, 1977). It has also been reported from Guam and Marshall Islands as predaceous on red spider (Mitri and Stannard, 1962).

FRUIT FLIES

Oriental fruit fly, *Dacus dorsalis* (Hendel) has been reported from India, Sri Lanka, Indonesia, Taiwan, Bonin Island, Hawaii, etc. The pest is polyphagous and active throughout the year, migrating from one host to other. Its main host is mango. The flies occasionally attack the ripe banana fruits, lay eggs on or inside the pericarp; grubs emerging from these eggs bore inside and feed on the pulp. When full fed, the grubs pupate in the soil.

To control these flies, plough around the trees to expose and kill the pupae. Bait-sprays containing protein hydrolysate, molasses or sugar and diazinon or malathion have also proved effective.

Mediterranean fruit fly. *Ceratitis capitata* Wiedemann has been recorded from South Europe, North Africa, South-West Australia, Hawaii, Bermuda, Central and South America while *Dacus musae* (Tryon) – specific to banana – *D. tryoni* Froggot and *Rioxa pornia* (Walker) have been found to occur in Queensland, but none of these have been so far intercepted in India. Simmonds (1966) has struck a note of caution as regards to *C. capitata* and *D.musae* which might

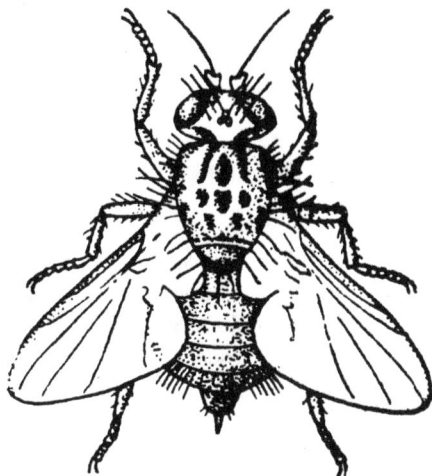

Ceratitis capitata Wiedemann

become serious pests in other parts of tropics and therefore, these merit special attention by quarantine organisation.

BANANA FRUIT SCARRING BEETLE

Colaspis hypochlora Lefevre is a major pest of banana in Central America (Mexico, Guatemala, British Honduras, Nicaragua, Costa Rica and Panama). It has also been recorded from India though as a minor pest.

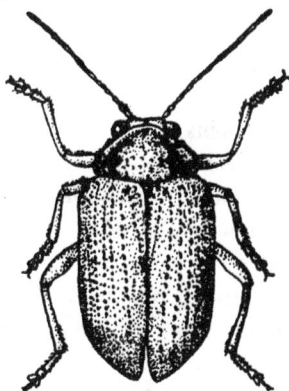

Colaspis hypochlora Lefevre

Eggs are laid in the soil, on hatching the grubs feed on roots of banana and other weeds around. Pupation takes place in the soil. Adult beetles are nocturnal and feed on unfurled leaves of banana as also on the skin of fruits. The affected fruits show oval scars specially on the lower proximal surface and as such these fruits do not fetch a good market price.

Egg, larval and pupal stages last for 6 to 9, 21 to 22 and 7 to 10 days respectively while the adults live for several months (Hill, 1975).

Clean cultivation coupled with soil application of 5% aldrin or chlordane dust keeps the pest under check. In case of severe infestation, spray with 0.05% dichlorvos. Frogs, lizards and spiders have been found preying upon these beetles but reduction in pest population has not been appreciable.

FRUIT BORING CATERPILLER

Heteromicta latro – a major pest of banana in Queensland (Jarris, 1974), has also been reported from India as a minor pest. The larvae occasionally bore in distal ends of banana fruits and the entry holes are covered with webbed excreta. Roy and Sharma (1952) reported that the attack by this pest is followed by a fungus (*Gleosporium* species) which causes ripe-rot of fruits. To check the damage, remove and destroy all the infested fruits and if necessary spray with a good contact insecticide like carbaryl or endosulfan.

FRUIT SUCKING MOTHS

Anua coronata (Fabricius) and *Othreis fullonia* (Clerck) (*fullonica* Linnaeus) have been reported damaging banana fruits (Nair,1975). Both these are polyphagous and among fruits prefer *Citrus* spp. Eggs are laid on leaves of various crops, the caterpillars feed on foliage of these crops but cause no economic loss. Only the adult moths are harmful; they puncture the fruits for feeding and the affected fruits subsequently rot, decay and become unfit for human consumption. As these are minor pests of banana, normally, no control measures are adopted against these pests.

GUAVA

GUAVA, *Psidium guajava* Linnaeus is a native of tropical America but is now pantropical in distribution. In India, it was introduced in early 17th century and at present it occupies an area of 60,000 hectares. Of this more than half is in Uttar Pradesh and about 10,000 hectares in Bihar, the remaining 20,000 hectares are spread all over India. The fruit is very rich in vitamin C and A and contains fair amount of calcium as well. It is eaten as such or cooked and is pre-eminently a fruit for making jam and jelly. It can also be canned in sugar syrup or made into fruit butter. Immature fruits are not edible being hightly astringent in taste. The trees are tolerant to varying soil conditions and resistant to drought.

About 80 species of insects have so far been recorded on guava trees affecting yield and quality of fruits. Of these less than twenty occupy the status of pests – major or minor. The major pests include, fruit flies, bark eating caterpillars, castor capsule borer and coccoids (scale insects and mealy bugs) while those of minor importance are aphids, whiteflies. thrips, tea mosquito bugs, cockchafer beetles, grey weevils, castor semilooper, stem borers, fruit borers, etc.

FRUIT FLIES

Guava fruits are attacked by a number of fruit flies, including *Dacus diversus* (Coquillett), *D. dorsalis* (Hendel), *D. cucurbitae* (Coquillett) and *D. zonatus* (Saunders). These are polyphagous pests and breed on a large number of fruits. The adult flies are strong fliers and can fly 1 to 2 km in search of suitable host.

Eggs are laid in soft skin of ripening fruits. On hatching, the maggots bore further into the fruits and feed on soft pulp. Unripe fruits are seldom attacked as the flies are unable to puncture the hard skin of these fruits for oviposition. The infested fruits show depressions with dark greenish punctures and when cut open the wriggling maggots are seen inside. Later, the affected fruits get malformed and in conjunction with bacterial activity, the fruits rot and ultimately fall down.   Full fed maggots come out of these fruits to pupate in the soil.

Dacus diversus – the guava fruit fly, is a pest of potential importance.   It is widely distributed all over India, Pakistan, Bangladesh, Burma and Sri Lanka.   Besides guava, it also attacks banana, citrus, *jamun*, mango, melons, papaya, etc.   The pest is active throughout the year except for the duration of severe cold. In Summer it breeds exclusively in lower buds of cucurbitaceous plants and migrates to orchards during Winter.   The adult flies subsist on honeydew secreted by melon aphid (*Aphis gossypii* Glover) but are often deterred by the presence of black ants, *Camponotus compressus* Fabricius which also feed on honeydew.   Overwintering is in adult stage.   These adults are seen congregating in the folds of loquat and guava leaves and become active again in Spring.   The activity of the pest in the orchards dwindles during Summer as the pest migrates back to cucurbit fields.

Eggs are smooth, shiny white, 1.0 to 1.5 mm long and slightly curved.   Full grown maggots are pale cream in colour, cylindrical in shape and 5 to 8 mm long.   Pupae are barrel-shaped, ochraceous and 5 to 6 mm long.   Adults are smoky brown with greenish - black thorax having yellow markings; wing expanse is 9 to 11 and 12 to 14 mm in case of male and female flies respectively.

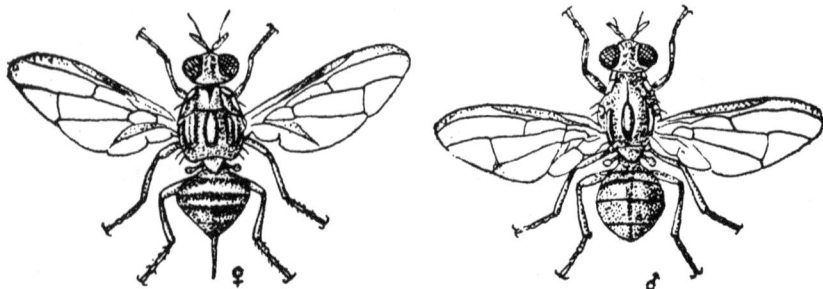

Dacus diversus (Coquillett)

Incubation period is 1 to 4 days during July, larval stage lasts for 4 to 5 days and pupal period extends from 7 days in August to 13 days in November (Batra, 1933).

Dacus dorsalis is the major pest of mango, but guava is its next preferred host.   Adults of *D. dorsalis* and *D. diversus* look alike, both being of same size, shape and having similar type of wing-venation; but thorax in case of *D. diversus* is greenish - black with yellow middle stripe whereas that of *D. dorsalis* is ferruginous and without any yellow middle stripe. *D. cucurbitae*, the melon fruit fly, is larger in size than other fruit flies (except *D. tau)* and has distinctly different wing-venation.

Dacus zonatus – the peach fruit fly, attacks guava usually during rainy season.

To check the infestation from increasing and to prevent the carry over of the pest, collect all the fallen and infested fruits and destroy these immediatly either by burning or deep burrying. Ploughing around the trees is also helpful in checking the pest population as it kills the pupae in the soil by exposing the same to sun's heat and other predators. Chemical control of these flies is rather difficult because maggots are invariably inside the fruits, pupae in the soil and adult flies are active on wings. Poison-baiting with protein hydrolysate, malathion and water (1:10) has been suggested (Mehta and Varma, 1968). This bait attracts and kills not only the fruit flies but also fruit sucking moths and adults of castor capsule borer. Spraying with 0.03% phosphamidon, 0.03% oxydemeton methyl or 0.06% dimethoate also brings down the population to some extent (Talgeri, 1967).

BARK EATING CATERPILLARS

Indarbela tetraonis (Moore) and *I. quadrinotata* (Walker) have been recorded boring into trunk, main stems and thick branches of guava trees in India. Both are polyphagous and widely distributed in India. *I. tetraonis* is more common and destructive on guava trees, specially in Panjab, Uttar Pradesh and South India as also in neglected orchards where crop sanitation is lacking The varieties *Seedless guava* and *Apple guava* are comparatively more susceptible while *Allahabad safeda, Strawberry* and *Cattley* are tolerant to the attack of these caterpillars. Besides guava, this species has been reported damaging *aonla, ber*, citrus, *falsa*, jack-fruit, *jamun*, litchi, mango, etc.

Eggs are laid in cuts and crevices in the bark. On hatching the larvae bore into the bark or main stems. The larvae remain within the bored holes during day and come out at night tofe ed on the bark. They feed sheltered under the silken galleries. The attacked trees show the presence of winding silken galleries full of frass and faecal matter on the bark surface extending from the bored holes downwards. Generally one larva will be there inside each hole and in case of severe infestation, 15 to 30 larvae may bore the same tree. Pupation takes place within these galleries.

Full grown larvae are dirty brown and 38 to 45 mm long. Adults are stout, pale brown moths with brown spots and streaks on forewings and whitish hind wings. These moths are slightly bigger than those of *I. quadrinotata* ; wing expanse being 35 to 38 and 46 to 50 mm in case of males and females respectively. Egg and pupal periods

Indarbella tetraonis (Moore)

occupy 8 to 10 and 15 to 25 days respectively while the larvae live for 10 to 11 months - thus there is only one generation in a year.

To prevent infestation of these borers avoid growing the varieties that are highly susceptible and keep the orchards clean. If and when any infestation is noticed, kill the caterpillars mechanically by inserting an iron spike into the holes made by these caterpillars. In case of severe infestation, clean the affected portion of the trunk or main stem and insert into the holes swab of cotton wool soaked in a good fumigant like carbon bisulphide, chloroform, petrol, etc. Patel *et al.* (1964) found even kerosene oil to be effective whereas Srivastava (1964) suggested using a mixture of ethyl glycol and kerosene (3 : 1). The holes, after treatment, should be sealed with mud and such of the holes as may reopen, should be retreated.

CASTOR CAPSULE BORER

Dichocrocis punctiferalis (Guenée) is a major pest of castor and is often found on guava throughout the plains of India. Ansari (1945) recorded it as a serious pest of guava in the Punjab. It has also been reported from Burma, Sri Lanka, China, Japan, Malaysia, Indonesia and Australia. Beside castor and guava,it is a major pest of ginger and turmeric and of minor importance on citrus, jack-fruit, mango, mulberry, peach, pear, plum, etc.

Eggs are laid on fruits but in the absence of fruits, these may be deposited on buds and tender shoots. On hatching the caterpillars bore into fruit, bud or shoot and feed within, on pulp and seeds or soft tissues. The affected shoots wilt and dry away, buds do not open and the fruits fall prematurely. Eggs are flat, oval (0.5mm in diametre) and pinkish in colour. Full grown caterpillars are 20 to 30 mm long, dark pinkish-brown with spiny wrats all over. Moths are medium sized, uniformily brownish-yellow with numerous black dots on the wings.

Dichocrocis punctiferalis (Guenee)

Incubation period is 5 to 7 days while larval and pupal durations vary from 14 to 20 and 7 to 10 days respectively. Total life-cycle occupies 28 to 35 days and as usual it is longer. during Winter than in Summer. Longevity of moths is 3 to 5 days.

To control this borer, destroy all the infested shoots, buds and fruits in the initial stage of attack. In case of severe infestation, dust 5 to 10% BHC or spray with 0.1% BHC + 0.1% DDT or 0.1% carbaryl or fenthion.

COCCOIDS

A number of scale insects and mealy bugs have been reported

damaging guava trees. These are mostly polyphagous pests that suck cell sap from leaves, twigs and even the fruits and devitalize the same. These bugs also exude sweet sticky substance, honeydew on which sooty mould fungus grows rapidly, covering the affected areas with a superficial black coating. In case of severe infestation, the thick black coating interferes with the photosynthetic activity of the tree, resulting in poor growth as also small and fewer friuts. The species recorded damaging guava trees include, *Drosicha dahlbergiae* (Green), *D. mangiferae* (Green), *Hemiaspidoproctus cinerae* (Green), *Icerya aegyptiaca* (Douglas), *I. purchasi* Maskell, *I. seychellarum* (Westwood), *Labioproctus poleii* (Green), *Ferrisia virgata* (Cockerell), *Nipaecoccus viridis* (Newstead), *Planococcus lilacinus* (Cockerell), *Aonidiella aurantii* (Maskell), *A. orientalis* (Newstead), *Aspidiotus destructor* Signoret, *Hemiberlesia lataniae* (Signoret), *Chrysomphalus ficus* Ashmead, *Insulaspis pallidula* Williams, *Lindingaspis greeni* (Brain and Kelly), *L. rossi* (Maskell), *Parlatoria pergandii* Comstock, *Pseudanoidia trilobitiformis* (Green), *Ceroplastes ceriferus* (Anderson), *C. floridensis* Comstock, *Ceroplastodes cajani* Maskell, *Chloropulvinaria psidii* (Maskell), *Coccus hespridum* Linnaeus, *C. viridis* (Green), *Parasaissetia nigra* (Nietner), *Saissetia coffeae* (Walker), *S. oleae* (Bernard), etc. Of these hardly half a dozen are of regular occurrence and can be classified as pests of major or minor importance; the rest are of sporadic occurrence in different parts of India, hardly causing any economic loss.

Chloropulvinaria psidii is a dominent species of guava (Ayyar, 1963) and has also been found on citrus, *gular*, jack-fruit, *jamun*, litchi, loquat, mango, sapota, etc. It is a major pest of guava, more pronounced in South India, Maharashtra, Punjab and Uttar Pradesh.

Chloropulvinaria psidii (Maskell)

Besdes India, it has also been reported from Bangladesh and Sri Lanka. Small, flat, motionless, broadly oval, yellowish-green, scale-like insects are found in large number, sticking to leaves on ventral side, tender twigs and shoots specially during Summer. The females (6 to 8 mm long) feed voraciously and also exude copious quantity of honeydew.

Saissetia oleae (Bernard) – the olive soft scale or black scale is almost cosmopolitan in distribution, widely distributed in warm climatic regions (CIE map No. A – 24) as also in green-houses in temperate regions. Its main hosts are *olive* and *Cirtus* spp. besides a wide range of alternate hosts, including, fig, guava, sapota and tamarind, A female lays on an average about 2000 eggs in 2 to 4 weeks. On hatching, the crawlers start feeding within a few hours on shoot tips and the underside of the leaves whereas the adults prefer shoots and twigs.

Eggs are oval and pinkish. Freshly hatched nymphs are also pinkish, similar to those of *C. hesperidum*. The second stage nymphs or immature females are flat yet distinctly marked with dorsal H-ridges. These are light brown in colour and become dark brown with age. Adult females are apterous, 4 to 6 mm long elongate convex with a distinct H - shaped crest on the back and black in colour.

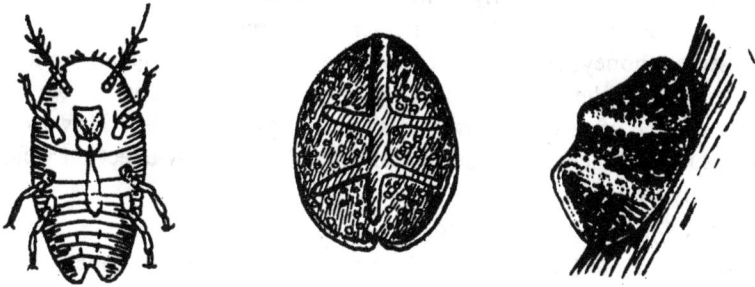

Saissetia oleae (Bernard)

Males are winged but very rare. Reproduction usually occurs without fertilization. Egg period is 2 to 3 weeks and nymphal 2 to 3 months (Hill, 1975).

Saissetia coffeae (Walker) (*hemisphaerica* Targioni-Tozzetti) also known as red coffee bug is almost cosmopolitan in distribution widely found throughout the tropics and subtropics (CIE map No. A–318). Besides guava, it also attacks citrus, loquat, mango, etc.

Eggs are ovate and white, later turning brown. Immature females are flat, hemispherical soft scales, yellowish-green when young and dark brown when older. They also possess three carniae, forming H. This carination disappears completely in adult females and is therefore a diagnostic character. Adult females are strongly convex, dome-shaped, 2 to 3 mm long, smooth, shiny and reddish-brown to dark brown in colour. Reproduction is mostly parthenogenetic as males are rare.

To control these scale insects, prune the affected parts and burn the same in the initial stage of infestation. In case of severe infestation, pruning and destruction of the affected parts may be followed by spraying with 0.04% diazinon or dichlorvos or 0.1% BHC+0.1% DDT. To keep the pest under check, 2 to 3 sprayings at an interval of 10 to 12 days may be necessary.

APHID

Aphis gossypii Glover – the melon aphid, is a polyphagous pest. It is a minor pest of guava. Colonies of nymphs and adults may sometimes appear specially during August to October, sucking cell sap from ventral leaf surfaces, tender twigs and shoots. The aphids also secrete honeydew. In case of severe infestation, drops of honeydew and/or patches of black coating (sooty mould) may be visible on upper surface of leaves below the aphid infested leaves. However, there is no evidence that these sporadic attacks have any effect on yield or quality of guava fruits.

To control this aphid, if and when necessary, spray with 40% nicotine sulphate (1:600) or 0.03% dimethoate, oxydemeton methyl or phosphamidon.

WHITEFLIES

Aleurocanthus rugosa Singh, *A. woglumi* Ashby, *Aleyrotuberculatus psidii* (Singh) and *Pealius misrae* Singh have been recorded on guava leaves but none of these are pests of any importance. If and when these appear, they suck cell sap from ventral leaf surfaces and thereby devitalize the trees to some extent. Spraying with 0.02% phoshamidon or 0.03% dimethoate or 0.04% dichlorvos is quite effective in controlling the whiteflies.

THRIP

Selenothrips rubrocinctus (Giard) – a major pest of cocoa specially in Africa and South America, has not yet been reported on cocoa in India. The pest is pantropical in distribution (CIE map No. A – 130). In India, it is found in large numbers on leaves of cashewnut, guava, mango, pear, etc. A sporadically serious pest in nurseries, it very rarely damages the grown up or mature trees. Nymphs are comparatively more destructive than adults due to their enormous number. Both nymphs and adults may be found feeding on ventral leaf surfaces, depressions or grooves adjacent to the main veins are favoured sites. The nymphs often carry a small shiny reddish globule of excreta on the tip of their upturned abdomen. These globules are dropped every now and then on leaf lamina where these dry and turn into shiny black spots.

Eggs are kidney shaped and about 0.25 mm long. Nymphs are yellow with bright red band around the base of abdomen; when full grown these are about one mm long. Adults are dark brown, 1.0 to 1.5 mm long, fragile, slender insects with dark grey heavily fringed wings.

Selenothrips rubrocinctus (Giard)

Males are rare. Egg, nymphal and pupal periods are, 12 to 18, 6 to 10 and 3 to 6 days respectively (Hill, 1975).

TEA MOSQUITO BUGS

Helopeltis antonii Signoret is not a mosquito but a bug. It is widely distributed in South India, Sri Lanka and S.E. Asia (CIE map No. A–296) and is a polyphagous pest with a wide range of host plants. In India, besides tea, it is a major pest of cashewnut and minor pest of apple, avocado, grapevine, guava, tamarind, etc. Nymphs and adults feed on tender leaves, shoots and fruits; as a result leaves and shoots turn brown, dark brown, and then black while blisters and scales appear on the fruits. Pest population is generally low during the dry weather. Heavy and continuous rains with no sunshine is favourable for its rapid multiplication. Eggs are white, 1.5 to 2.0 mm long and sausage-shaped. Young nymphs are ant-like, hairy and amber coloured, later becoming orange red. The diagnostic feature in both nymphs and adults is the pin-like knobbed process (1 to 2 mm long), arising vertically from the thorax. Adults are small, active, elongated (7 to 9 mm long) bugs having a mixture of red, black and white colouration, with long legs and very long antennae. Eggs are thrust into tissues of green succulent shoots, buds as also midribs and petioles of leaves. A female can lay upto 500 eggs. Preoviposition period is 2 days. Egg period is 5 to 7 days during Summer and 20 to 27 days in Winter. Nymphal period is about 3 weeks during Summer and extends upto 6 weeks in Winter. Adult females live for 6 to 10 weeks.

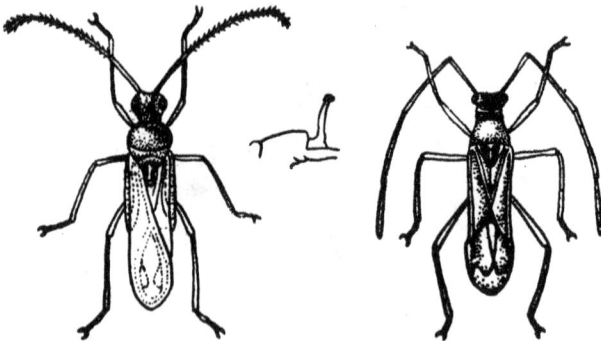

Helopeltis antonii Signoret   H. febriculosa Bergroth

Another allied species, *H febriculosa* Bergroth (*theivora* Waterhouse) commonly found in eastern India. is also an important pest of tea and a minor pest of guava. Its nymphs are dirty yellow.

Adults are of same shape and size as those of *H. antonii* but these have olive green head, pale yellowish-black thorax and greenish-black abdomen.

Spraying with 0.1% BHC or carbaryl twice at an interval of 12 to 14 days during flowering season gives satisfactory control of this pest.

COCKCHAFER BEETLES

The beetles found occasionally damaging guava in India, include melolonthids, namely, *Holotrichia consanguinea* (Blanchard), *H. insularis* Brenske, *H. serrata* Fabricius and *Schizonycha rufficollis* Blanchard as also rutelids, *viz., Adoretus duvauceli* Blanchard, *A. horticola* Arrow, *A. lasiopygus* Burmeister, *A. versutus* Harold and *Anomala bengalensis* Blanchard. All these are minor pests of guava, though some of these are of major importance on other host plants. The overwintering adults are stimulated to activity by a good shower of rain whether pre-monsoon (David and Kalra, 1966) or monsoon (Srivastava and Khan, 1963). The activity declines with heavy showers of rain. Thus the adult beetles are found in guava orchards from March-April to June-July. Being nocturnal in habit, these appear soon after dusk, feeding on leaves and making nuptial flights. Just before day break the beetles go down in the soil just below the trees. Grubs usually feed on roots and other organic matter available in the soil.

Eggs are laid in the soil, 50 to 150 mm deep, when the soil is sufficiently moist (June-July). In Rajasthan incubation period is 7 to 12 days, larval 8 to 21 weeks, pupal 2 to 3 weeks and total life-cycle 11 to 16 weeks (*H. insularis* – Srivastava and Khan, 1963; *H. consanguinea* – Rai *et al.*, 1969). Adults emerge by November but remain inactive in the soil for the entire Winter and Spring seasons.

These pests are difficult to combat as they live deep in the soil. Raking of soil around the trees coupled with mixing thoroughly in the soil heptachlor dust @ 8 kg a.i. per hectare (Joshi *et al.*, 1969) during the active period of the pest help to minimize the pest population. For adult beetles, foliar spraying with 0.2% BHC or carbaryl, twice at 7 to 10 days' interval is suggested.

GREY WEEVILS

Myllocerus blandus Faust, *M. undecimpustulatus maculosus* Desbrocher, *M. sabulosus* Marshall and *M. viridanus* Fabricius have been reported damaging guava leaves. These are polyphagous pests having a wide range of host plants. The grubs feed on roots of all sort.while the adult weevils nibble the leaves starting from margins and eat away large patches. The loss caused is usually of minor importance. Ovoid light yellow eggs are laid in the soil, 60 to 80 mm deep. The grubs are stout, fleshy, apodous, curved, 6 to 8 mm long and light yellow in colour. Pupation takes place in the soil. The adults are small weevils, 5 to 7 mm long with greyish-white elytra having dark lines.

To control these weevils, if and when necessary, spray with 0.2% BHC or carbaryl or 0.05% dichlorvos. Dusting with 5% BHC or carbaryl is also effective.

CASTOR SEMILOOPER

Achaea janata (Linnaeus) is widely distributed all over India. Damage is caused by the caterpillars as well as adults. The larvae feed voraciously on leaves of various grasses as also some economic crops like, *ber*, castor, pomegranate, etc., while the adult moths suck juice from various fruits including, citrus, grapevine, guava and mango. The moths are nocturnal and appear only after dusk. They puncture only the fruits. These punctures also serve as entry for various bacteria and fungi. As a result the attacked fruits start rotting and drop down prematurely. It is not a serious pest of guava in India and generally calls for no control measures. Moths of *Achaea* spp. have been reported from Nigeria attacking ripe and ripening fruits of fig, guava, mango and tomato (Golding, 1945).

STEM BORERS

Guava stem borer, *Microcolona leucosticta* Meyrick, commonly found in Assam (Chowdhury and Majid, 1954) is a minor pest of guava. The caterpillars bore into tender twigs and damage the pith generally near the tips. The attacked stems become hollow and filled with black frass.

Another species, *Microcolona technographa* Meyrick has been reported from North-West India, causing similar damage to guava trees (Pruthi and Batra, 1960).

Microcolona technographa Meyrick

Cherry stem borer, *Aeolesthes holoserica* Fabricius – a major pest of cherry has also been recorded boring guava stems (Fletcher, 1920a). This is a polyphagous pest, found all over India. Eggs are laid in cuts and crevices in bark of the trees. On hatching, grubs feed on inner layers of bark and make zigzag galleries; later the grubs bore into trunk or main branches and thick stems, feed within the same, making irregular cavities. The grubs are active for 2 to $2\frac{1}{2}$ years and pupation takes place in these cavities. A single life-cycle occupies $2\frac{1}{2}$ to 3 years.

Coloborhombus hemipterus Olivier is another cerambycid found boring guava stems in North India. Grubs of a buprestid beetle, *Belionota prasina* Thunberg have also been found boring into the stems and branches of guava trees in Maharashtra (Nair, 1975). Both are minor pests.

Belionota prasina Thunberg

To control these lepidopterous and coleopterous borers, collect
and kill the larvae and adults in the initial stage of attack. Fumigate
the holes with paradichlorobenzene (Alam, 1972) or insert in the
holes cotton wick (about 150 mm long) soaked in a fumigant like
carbon bisulphide or chloroform or 0.1% dichlorvos and seal the
treated holes with mud.

FRUIT BORER

Pomegranate butterfly, *Virachola isocrates* (Fabricius) – a
polyphagous pest is widely distributed all over India. It is a major pest
of pomegranate and has also been reported attacking a number of
other fruits, including guava. It is a minor pest of guava. Eggs are
laid on calyx of flowers or young fruits. On hatching the larvae
puncture the fruits, bore inside the same and feed within on pulp and
immature seeds. Pupation usually takes place inside the fruits. The
affected fruits rot, emit offensive smell and ultimately fall down
prematurely. The pest is active throughout the year.

Virachola isocrates (Fabricius)

To check the population of this pest, collect and destroy
promptly, all the infested and fallen fruits with caterpillars/pupae
within. A prophylactic spraying with 0.15 to 0.2% BHC or 0.03%
phosphamidon, when fruits are just beginning to form, also helps in
preventing the attack of this pest.

CUSTARD APPLE

CUSTARD APPLE, *Annona squamosa* Linnaeus or sugar apple commonly known as *Sitaphal* or *Sharifa* is native of tropical America and is said to have been introduced in India by Portuguese, several hundred years ago. The sculptural designs at Ajanta and Ellora and its mention in *Ain-i-Akbari* (1590) prove that the fruit has been in India for quite a long time. At present, it occupies an area of about 46,000 hectares; of which over 40,000 hectares are in the forest region of Andhra Pradesh (around Hyderabad), 2500 hectares each in Assam and Bihar and a few hundred hectares in Uttar Pradesh, Madhya Pradesh, Maharashtra and Gujarat (Hayes, 1966 ; Vevai, 1971b). The crop is most suited for dry and hot climate, being sensitive to cold and frost. It tolerates drought to a considerable extent but cannot withstand water-logging. Traces of hydrocyanic acid have been found in roots, bark, leaves and seeds – hence it is not eaten by goats or cattle.

About 20 insect pests have been found feeding on various parts of these trees, some of them causing substantial loss. Mealy bugs, fruit borer and fruit fly are the pests of major importance (Ayyar, 1938) while some sucking pests like, scale insects, thrips, whitefly and cow-bug cause minor damage.

MEALY BUGS

These are small, soft bodied insects covered with mealy wax, secreted by the insect themselves. They infest leaves, shoots, buds and even the fruits and suck the cell sap therefrom. As a result, the affected plant parts get deformed in shape and reduced in size. These insects also secrete honeydew on which a non-parasitic fungus, *Capnodium* species grows very rapidly, covering the plant parts with sooty mould. This black coating in turn interferes with the photosynthetic activity of the tree and as a result, its growth and fruiting capacity are adversely affected. The honeydew also attracts the ants such as, *Camponotus compressus* Linnaeus, *Oecophylla smaragdina* (Fabricius), etc. These ants help in dissemination of the pest from tree to tree.

The mealy bugs commonly found attacking custard apple in India are, *Ferrisia virgata* (Cockerell) and *Planococcus lilacinus* (Cockerell) (*crotonis* Green); the former being comparatively more common and destructive than the latter.

Ferrisia virgata commonly known as white-tailed or stripped mealy bug is widely distributed in tropical and subtropical countries (CIE map No. A–219) causing serious damage to a number of economic crops including, *aonla*, banana, citrus, custard apple, guava, grapevine and jack-fruit. Nymphs and adult females remain clustered on ventral leaf surfaces, terminal shoots and sometimes even on fruits sucking the sap. The affected spots turn yellow, wither and ultimately dry away. In case of severe infestation, there is premature shedding of fruits. The pest prefers dry weather and heavy infestation often occurs following periods of prolonged drought.

Adult females are apterous, oval in shape (4×2 mm) covered with a number of waxy filaments all over the body. They also have a pair of conspicuous longitudinal submedian dark stripes and two pronounced long waxy processes at posterior end.

Ferrisia virgata (Cockerell)

They are active and mobile throughout their life. Reproduction is sexual as well as parthenogenetic, the latter being more common. Mating takes place only once and a fertilized female lays 100 to 300 eggs in 3 to 4 weeks. The egg-masses remain under the females till the young ones hatch out. Incubation period is about 3 to 4 hours. The development period of male and female nymphs varies from 31 to 57 and 26 to 47 days respectively. Longevity of males is only 1 to 3 days while that of females extends from 36 to 53 days (Nayar *et al.*, 1976).

Planococcus lilacinus (Cockerell) is widely distributed in South and South-East Asia (CIE map No. A-101). In India, this is a major pest of coffee specially in Karnataka; the other host plants include, citrus, custard apple, fig, guava, pomegranate, sapota and tamarind. These mealy bugs feed on roots as well as leaves. More individuals are found in the root zone, 400 to 600 mm deep in the soil, than on the leaves. The infested roots develop spongy tissues and affected leaves become chlorotic and ultimately fall down.

To check the infestation of these mealy bugs, remove and destroy the affected leaves and twigs. If necessary, spray with 0.1% diazinon or monocrotophos. To check the subterranean population of *P. lilacinus*, plough around the trees atleast 600 to 800 mm deep and mix thoroughly with soil 5% aldrin or chlordane dust.

FRUIT BORER

Anonaepestis bengalella (Ragonot) is another sporadic but major pest of custard apple specially in South India. Besides India, the pest has been reported from Bangladesh, Vietnam, Philippines, etc. Eggs are laid singly in the sutures or on the peduncles of immature fruits. On hatching, the caterpillars bore inside these fruits, making irregular tunnels and causing damage to mesocarp. As a result, the development of infested fruits is arrested and ultimately these fruits fall down. Another clear symptom of the attack, is the excreta of caterpillars seen lying near the entry holes on the affected fruits.

The moths have olive green head and collar, purplish thorax and ochreaceous abdomen; costal half of forewings is olive green and white while the inner half is vinous red, hind wings are semihyaline; wing expanse is 24 mm (Hampson, 1896). The biology and bionomics of the pest have been studied in Philippines where incubation, larval and pupal periods have been recorded as 4 to 5, 12 to 19 and 12 days respectively (Estalilla, 1921). Pupation takes place in the tunnels within the fruits near the epidermis of fruits, so as to make it easier for the moths to emerge.

To check the pest population, collect and burn the attacked fruits. Alam (1962) recommended spraying with 0.2% BHC or DDT or 0.03% phosphamidon. Spraying 0.1% BHC+0.1% DDT at early fruit formation is also effective (Butani, 1976 a).

FRUIT FLY

Dacus zonatus (Saunders) – the peach fruit fly, is widely distributed in Indian sub-continent and some African countries. It is a polyphagous pest, peach being its main host. The pest is active throughout the year except during very cold months of January and February when they hibernate as pupae.

Eggs are laid inside the fruits and on hatching the maggots feed on the pulp. The characteristic symptom of attack is the oozing out of a fluid from the small fine punctures made by the flies on the fruits during oviposition. The infested fruits rot, show brown patches and utimately fall down (Butani, 1976a). The full fed maggots come out from these fruits and pupate in the soil; sometimes pupation may also take place within the infested fruits.

All the infested and fallen fruits should be collected and destroyed, this will kill the maggots inside. Ploughing around the trees specially in Jaunary-February kills the hibernating pupae and thereby reduces the pest population. Poison-baiting consisting of protein hydrolysate, malathion and water has been found very useful in trapping and killing the adult flies.

SCALE INSECTS

Aonidiella citrina (Coquillett) and *Ceroplastes floridensis* Comstock have been occasionally recorded damaging leaves of custard apple trees. Both are polyphgaous. *A. citrina* – the yellow scale of citrus, prefers *Citrus* spp., while *C. floridensis*, commonly called Florida wax scale is a minor pest of apple, avocado, cashewnut, custard apple, fig, guava, *gular*, jack-fruit, mango, mulberry, peach, pear, plum, pomegranate, etc.

Ceroplastes floridensis has been reported from India, Sri Lanka, Japan, Taiwan, Australia, Brazil, British Guiana, West Indies, Bermudas, Honduras, Mexico and USA (Hawaii). In India, it is commonly found in Assam, Bengal, Karnataka and Kerala. A female lays about 75 eggs under its own covering, usually on ventral surface of leaves. The nymphal period is about 3 months and there are three generations in a year (Pruthi and Mani, 1945).

Kerria communis (Mahdihassan) – the lac insect, has also been found on custard apple trees. The resinous secretion of the females provides stick-lac, which is of some commercial value. These insects are often preyed upon by the caterpillars of *Anatrachyntis falcatella* (Stainton).

Anatrachyntis falcatella (Stainton) – Predator of *Kerria communis*

THRIPS

Grapevine thrip, *Rhipiphorothrips cruentatus* Hood and castor thrip, *Retithrips syriacus* (Mayet) have been recorded on custard apple. Both are polyphagous having a wide range of host plants and are often found in association of one another. Besides custard apple, these thrips have also been recorded on grapevine and pomegranate. Nymphs and adults lacerate the leaf tissues and suck the exuding sap. Severe infestation often results in leaves getting almost completely bleached, filled with excreta and gradually drying up.

The two species are more or less alike except that forewings of *Rhipiphorothrips cruentatus* are yellowish and those of *Retithrips syriacus* are dark grey. Their life-cycle is of short duration lasting for 11 to 14 days.

Generally these are minor pests of custard apple and require no control measures. If there is a severe attack, spray with 0.03% dimethoate, oxydemeton methyl or phosphamidon or 0.05% dichlorvos or endosulfan. This will also control the whitefly and cow-bug infestation, if any.

WHITEFLY

Dialeuropora decempuncta (Quaintance and Baker) has been occasionally found attacking custard apple trees in South India. Nymphs suck the sap from ventral surface of leaves. They also excrete honeydew on which sooty mould grows, covering the affected areas with superficial black coating. As this is a minor pest, generally no control measures are needed.

COW-BUG

Otinotus oneratus Walker is a minor pest of custard apple. It is polyphagous pest with a very wide range of host plants. Ananthasubramaniam and Ananthakrishnan (1975) have listed as many as 38 host plants. Nymphs and adults are found in clusters near the midrib of leaves on adaxial sides. They suck sap from the leaves but loss caused is seldom of any serious nature. Often the common black ant, *Camponotus compressus* Linnaeus is seen attending these nymphs and adults, for want of honeydew excreted by the bugs.

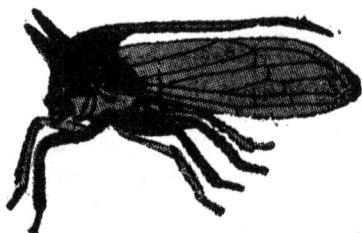

Otinotus oneratus Walker

Adults are small jumping bugs, 10 to 12 mm long, castaneous or piceous-brown in colour; characterised by a pair of ocelli, pronotum extending backward over the entire abdomen and having two horn-like processes. Being a minor pest, no control measures are normally adopted.

PAPAYA

PAPAYA or *papita*, *Carica papaya* Linnaeus is native of Mexico. It was introduced into India in 16th century and is by now grown all over the country in frost free areas. It occupies about 10,000 hectares (Singh, 1969) of which 4000 hectares are in Bihar. It is essentially a tropical crop and is highly susceptible to frost and needs well-drained soil, sun and high temperatures. The fruit is an important source of vitamin A and B and is rich in sugar and digestive enzymes. It also contains papain – a proteolytic enzyme, used for tenderizing meat. The seeds and pulp of unripe fruits have anti-fertility property while use of ripe fruits is a cure for various stomach disorders.

Hitherto, no plant protection measures were being carried out in case of papaya as these were considered to be uneconomical. The trees usually do not suffer any substantial loss from insect pests but mites and virus diseases (mostly transmitted by insects) do take a heavy toll. With consumers becoming quality conscious and the high price that the papaya fruits now fetch, it has become imperative to protect this crop as well, against the ravages of insect pests and diseases. The major pests are, *ak* grasshopper and mites. However, aphids and whitefly though sporadic in occurrence may cause serious loss by transmiting virus diseases. Other pests recorded on papaya trees include, fruit flies, coccoids, stem borer, grey weevil and namatodes. In addition to these, monkeys, bats, crows and other birds also cause damage to ripe papaya fruits.

AK GRASSHOPPER

Poekilocerus pictus Fabricius has been reported from India, Bangladesh, Pakistan and Africa. In India it is widely distributed throughout the plains. It is primarly a defoliator of *ak* plants (*Calotropis* spp.), but has also been reported defoliating *ber*, banana, citrus, fig, grapevine, guava, melons, papaya, peach, various ornamental plants and vegetables, etc. In case of severe infestation even the bark of papaya trees is not spared. The attack may start as early as April and continue till the onset of Winter, maximum damage is done during July-August.

Pruthi and Nigam (1939) have described its immature stages. Eggs are elongate, curved and yellowish-orange in colour. Nymphs are yellowish-white with orange and black stripes and dots all over their bodies. Adults are stout, yellowish with.broad bluish-green stripes on head and thorax; antennae bluish-black with yellow rings; abdomen yellowish with transverse blue-black bands. Forewings are bluish-green with yellow veins and reticulations, hindwings hyaline.

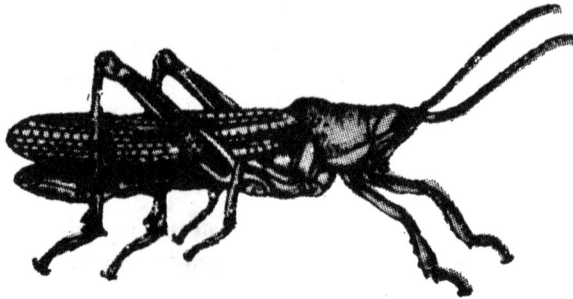

Poekilocerus pictus Fabricius

Mating lasts for 5 to 7 hours and pre-oviposition period is 3 to 4 weeks. The female thrusts its abdomen into the soil and lays about 150 to 180 eggs at a depth of 120 to 200 mm, depending upon the texture of the soil. There are atleast two broods in a year – a short one with incubation period one month (June-July) and nymphal period two months and a long generation when eggs laid during September to November overwinter and hatch around the end of March or early April and become adults in another $2\frac{1}{2}$ months.

To control these grasshoppers, Batra (1955) suggested dusting the trees with 5% BHC. Spraying with 0.03% lindane, 0.05% chlordane or 0.1% malathion is also effective (Verma *et al.*, 1970).

MITES

Red spider mite, *Tetranychus cinnabarinus* (Boisduval)[*]has world wide distribution and has been reported from Central and South

---

[*] *T. telarius* Linnaeus has been taxonomically sunk and replaced by two species, viz., *T. cinnabarinus*, the pantropical species and *T. urticae*, the temperate species.

America, Africa, Middle East, Pakistan, India, Sri Lanka, S.E. Asia, Japan, Indonesia, Philippines and Australasia. It is a polyphagous pest having a very wide range of host plants including, almond, apple, cherry, citrus, grapevine, melons, papaya, peach, pear and strawberry. On papaya, it has been occasionally reported as serious. Clusters of yellow spots may be seen on dorsal surface of leaves. On turning these leaves innumerable silken webs, which are spun by adult males, are visible sheltering minute reddish creatures – the mites. The larvae, nymphs and adults suck the cell sap from plant tissues directly.

Eggs are spherical, about 0.1 mm in diameter and whitish in colour. Larvae are pinkish, about 0.2 mm long and have 3 pairs of legs. Nymphs are greenish-red and have 4 pairs of legs like adults. Adults are ovate and reddish-green; males being 0.3 to 0.4 mm long and females 0.4 to 0.5 mm.

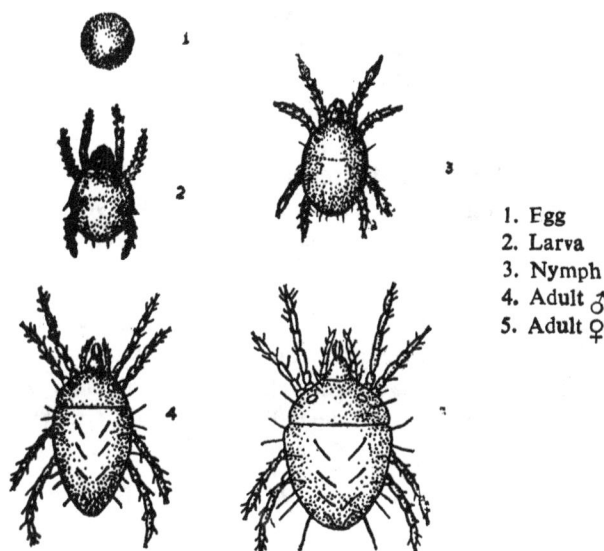

1. Egg
2. Larva
3. Nymph
4. Adult ♂
5. Adult ♀

Tetranychus cinnabarinus (Boisduval)

A female lays on an average 200 eggs singly on ventral leaf surface struck on the strands of silken webs. These hatch in 4 to 7 days. Larval stage is 3 to 5 days, nymphal 6 to 10 days and adult females live for about 3 weeks (Hill, 1975).

Another mite *Brevipalpus phoenicis* (Geijskes) has also been reported damaging papaya and other fruit trees though tea and *Citrus* spp. are its main hosts. This species is widely distributed and occurs throughout the tropical and subtropical regions (CIE map No. A–106) and is a minor pest of papaya.

To control these mites, remove the infested leaves and burn the same. Chemical control measures are not usually required but if and when there is a heavy infestation, the trees may be dusted with sulphur or sprayed with wettable sulphur or 0.05% chlorphenamidine or dicofol.

APHIDS

Aphis gossypii Glover and *Myzus persicae* (Sulzer) have been found feeding on papaya leaves (Singh, 1972); the former is a major pest of melons and latter of peach. Nymphs and adults inject their saliva in the plant tissues and suck the cell sap, but do not cause any serious damage. In additon, these aphids act as vectors by transmitting various virus diseases and the loss thus caused is often phenomenal. The affected trees seldom recover from the disease and gradually die away. Capoor and Verma (1948) reported the virus disease – mosaic, form Maharashtra, transmitted by this aphid. Reddy (1968) and Singh (1972) listed, *Aphis gossypii*, *A medicagenis*, *Myzus persicae* and *Macrosiphum sonchi* as vectors of papaya mosaic virus.

The early symptoms of the disease are appearance of necrotic dots on the leaves; later blistered patches of green tissue may be seen distributed indiscriminately over the yellowing leaf lamina. Affected fruits become smaller in size, elongated and develop circular water soaked lesions with solid spot in the centre.

It is quite easy to control the aphid as pest but far more difficult to control it as vector. By the time the insects are noticed and dusting or spraying done to check their population, the function as vector of transmitting the virus disease has been accomplished and then killing the vector is like, locking the stable after the horse has been stolen. Therefore, only prophylactic sprayings can save the trees from attack of aphids and subsequent appearance of the disease. It is suggested to spray 0.03 % dimethoate, fenitrothion, monocrotophos or phosphamidon, etc. If and when any virus disease is noticed, cut and destroy immediately the affected leaves, to prevent the disease from spreading.

WHITEFLY

Bemisia tabaci (Gennadius) (*gossypiperda* Misra and Lamba) is found in most parts of tropics and subtropics (CIE map No. A-284). Its main hosts are tobacco, cotton and some winter vegetables on which the infestation is sporadically severe. Besides, this whitefly species also attacks other trees growing around, both wild and cultivated, including papaya and *falsa*.

Tiny, white scale-like objects may be seen clustering inbetween the veins on ventral surface of leaves. Shake the leaf slightly and a herd of tiny whiteflies flutter out for a few seconds and resettle rapidly on the leaf. The pest activity is more common during dry season and declines rapidly with the onset of rains. As a result of their sucking the sap from leaves, the affected leaves become yellowish, curl downwards, wrinkle and there is early shedding of such leaves. In addition, this whitefly also acts as vector by transmitting the virus, causing leaf curl disease.

(Courtsey : S.P, Singh)

Leaf curl disease of papaya

Eggs are pear-shaped, about 2 mm long and stand upright on leaves being anchored by tail like appendage inserted into the stoma of leaves. On hatching the nymphs move a little and settle down on the same leaf, never to move again. Nymphs are oval, scale-like and greenish-white in colour. Adults are minute flies (about one mm long), covered completely with a white waxy bloom; wing expanse is 4 to 5 mm.

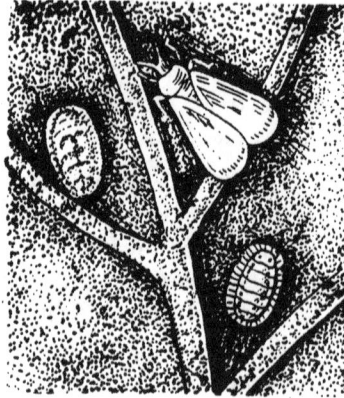

Adult

Nymph

Bemisia tabaci (Gennadius)

Incubation period is 3 to 5 days in Summer and 7 to 16 days during Winter. Nymphal and pupal periods are 17 to 73 and 2 to 8 days respectively; longer during Winter than in Summer. Longevity of adults is 2 to 5 days during Summer and extends up to 24 days in Winter. There are about 12 overlapping generations in a year.

The control measures suggested for aphids are good enough for whitefly also.

FRUIT FLIES

Male flies of *Dacus diversus* (Coquillett) have been observed in large number visiting female flowers of papaya (Pruthi and Batra, 1960) but these cause practically no damage. Most species of fruit flies do not have that long ovipositor with which they can thrust their eggs inside papaya fruits. However, *D. cucurbitae* (Coquillett) – a major pest of melons has been recorded on papaya (Narayanan and Batra, 1960) so also *Toxotrypana curvicauda* Gerst. (Dutta, 1966) and *Dacus pedestris* Bezzi (Hill, 1975). But all these are of minor importance. Unripe or ripening fruits are seldom attacked by any speciesof fruit fly whereas ripe and over-ripe fruits may be occasionally attacked but the infestation is never severe.

Outside India, Mediterranean fruit fly, *Ceratitis capitata* (Wiedemann) has been recorded as major pest of papaya.

COCCOIDS

Among mealy bugs and scale insects *Drosicha mangiferae* (Green), *Planococcus citri* (Risso), *Aspidiotus destructor* (Signoret) and *Pseudoparlatoria ostriate* Cockerell have been occasionally reported as minor pests sucking cell sap from leaves of papaya trees. But none of these cause any major damage. Clipping off the affected leaves is recommeded to check the spread of these coccoids.

STEM BORER

Cerostoma rugosella (Stainton) has been recorded as a minor pest of papaya from Bihar, Maharashtra and Tamil Nadu (Fletcher, 1921; Hayes, 1966). The pest has earlier been reported feeding on bark of mango and *gular* trees (Lefroy and Howlett, 1909). The caterpillars bore into the stems, feed within and pupate in silken cocoons either under the bark of the trees or in the soil.

To check the infestation of this pest, hand-picking of caterpillars and pupae and their destruction is suggested.

GREY WEEVIL

Myllocerus viridanus Fabricius has been recorded from South India by Jayaraj *et al.* (1960), skeletonizing the leaves and sometimes causing premature death of the tree. However, this is a minor pest and can be controlled by dusting with 5% BHC or carbaryl.

2.6
PINEAPPLE

PINEAPPLE, *Ananas comosus* Merr. is native of Brazil. It was brought to the old world by Columbus during the end of 15th century and was introduced in South India by Portuguese in 1550 (Collins, 1960). There is mention of pineapple even in *Ain-i-Akbari* published in 1590 during the regime of Akbar the Great. Today the important pineapple growing countries are Singapore, Malayasia, Indonesia, Australia, USA (Hawaii), Brazil, West Indies and South Africa. The plantations can be raised successfully in tropical lowlands, but require fertile soil.

Pineapple fruit is a multiple organ formed from the coalescence of a large number of individual flowers. It contains high sugar percent and is rich in vitamin A and C. Besides, it also possesses medicinal properties. It rouses appetite (Oviedo, 1520), relieves throat and nasal congestion, stomach inflammation and indigestion (Jaramillo, 1946) and the juice of green fruit can also be used as contraceptive (Laszio and Henshaw, 1954). The fruit is eaten as dessert as well as used for canning and making jam, jelly, juice, confectionary, etc. It is also exported in unripe conditon to foreign markets.

In India, pineapple occupies an area of about 10,000 hectares confined mostly to Assam, Tripura and Bengal in the East and Karnataka, Kerala and Tamil Nadu in South India. It does not suffer much from direct ravages of insect pests but insects as vectors of virus diseases take a heavy toll which can be avoided. Fortunately, the major pests of pineapple occurring in various parts of the world, like *Castnia licus* (Drury) in Brazil, *Thecla echion* (Linnaeus) in Hawaii, Central and South America and *Metamasius ritchiei* (Marshall) in West Indies have not yet been reported from any part of India. Mealy bugs are the only destructive pests of pineapple reported from India. Besides, thrip, slug caterpillar, black palm beetle and termites also cause some damage.

MEALY BUGS

Pineapple mealy bug, *Dysmicoccus brevipes* (Cockerell) occurs in almost all tropical and subtropical regions (CIE map No. A-50).

Besides pineapple it has also been found feeding on banana and other crops. Colonies of these insects live underground and feed on roots; only a small number may be found on basal portions of the leaves and fruits, sucking the cell sap and injecting toxins into the plants through their Saliva. The individuals inhabiting aerial parts cause only a minor loss whereas the subterranean ones' not only damage the roots but also act as vectors by transmitting the virus responsbile for Pineapple wilt disease, causing there by a substantial loss. The symptoms of this disease are discoloration of leaves which turn reddish-yellow and stunted growth which ultimately affect the fruit size and weight. The mealy bugs also secrete honeydew which attracts ants which in turn help in spreading the pest

Citrus mealy bug, *Planococcus citri* (Risso) – a polyphagous pest, has also been reported damaging pineapple in India. The main host of this mealy bug is citrus. The nature of damage is more or less same as that of *D. brevipes*. *Pseudococcus bromilliae* Bouché is another mealy bug reported from South India. This is a small reddish bug with white waxy covering. Colonies of this bug are found sucking sap from fruits and occasionally from leaves as well (Ayyar, 1932).

To control infestation of these mealy bugs, plant only healthy pineapple cuttings (Abraham, 1953) or dip the cuttings in aldrin solution just before planting. Fumigation of infested pineapple suckers has also been suggested (Rao, 1972). Against aerial individuals, cut and destroy the affected parts and spray with 0.05% diazinon, dimethoate or monocrotophos.

THRIP

Onion thrip, *Thrips tabaci* Lindemann is a polyphagous pest and vector of virus diseases of tobacco, tomato, pineapple, etc. (Hill, 1975). Its main host is onion; pineapple is one of the alternate hosts. The pest is almost cosmopolitan in distribution (CIE map No. A-20), the range extending from Canada and Scandinavia (60° N) to South Africa and Newzealand (50° S) and is abundant in areas having rather hot dry climate.

Nymphs and adults rasp the epidermis of leaves and imbibe the oozing sap, thereby devitalizing and distorting the leaves and affecting the growth adversely. In case of severe infestation the leaves show silvery sheen and bear small spots of faecal matter. This thrip has also been recorded as a vector transmitting yellow spot virus disease of both plant and fruit (Linford, 1932). Luckily this disease has not yet been reported from any part of India.

Adults are small, about one mm long, yellowish-grey to brown in colour with darker transverse bands across the thorax and abdomen; forewings are shaded and possess short setae.

Thrips tabaci Lindemann

Reproduction appears to be thelytokous parthenogenetic. Males are extremely rare, the ratio being 3000 females to one male (Ananthakrishnan, 1971). Eggs are inserted into tissues of green leaf through a cut made by the ovipositor. A single female lays 40 to 50 eggs. These hatch in 4 to 9 days and the entire life-cycle is completed in 11 to 21 days (Rahman and Batra, 1944). The life-cycle is completed in less than 14 days during April although it takes 24 days in December (Singh 1960).

The wide range of host plants together with the ease with which the insects are dispersed by wind and the rapidity with which they breed makes this thrip an unpredictable pest which can be difficult to control (Mound 1977). Spraying with 0.1% DDT suspension or 0.04% diazinon is quite effective. Repetition may be necessary.

SLUG CATERPILLAR

Latoia lepida (Cramer) is a polyphagous pest and along with many other hosts has also been reported damaging pineapple plantations. Newly hatched caterpillars feed gregariously on tender leaves for some time and then migrate to adjoining leaves and feed voraciously leaving behind the midribs. It is a minor pest and usually requires no control measures. However, hand-picking and mechanical destruction of caterpillars is suggested at early stage of infestation. Dusting with 5% BHC or carbaryl is also effective and should be done when the caterpillars are still feeding gregariously.

BLACK PALM BEETLE

Oryctes rhinoceros (Linnaeus) is one of the most destructive pest of various palmae all over Asia. Besides, in India, it has also been reported as a minor pest of pineapple and banana. Eggs are laid in decaying vegetation, compost or manure heaps or pits. On hatching, the grubs feed on the decaying organic matter and pupate there.

Oryctes rhinoceros (Linnaeus)

Adult beetles emerging from the breeding places start feeding on economic crops growing nearby. Maximum damage is caused during monsoon, but on the whole the damage caused is not serious so as to warrant any control measures.

TERMITES

These subterranean enemies are found in almost all orchards including pineapple plantations specially if the soil is sandy or sandy-loam. Pineapples are generally not grown in lighter soils as such they do not suffer much from the ravages of termites. Anyway, if and when their presence is detected, mix thoroughly with soil 5% aldrin, chlordane or heptachlor dust @ 22 to 25 kg per hectare or if convenient, aldrin emulsion concentrate may be added to irrigational water @ one kg a.i. per hectare.

Cet insecte est tres mechant
Quand on l'attaque il se defend

SAPOTA

SAPOTA, *Achras sapota* Linnaeus or *chickoo* as it is commonly called, is native of Mexico (tropical America) and is grown in India for the past 7 to 8 decades. In India its cultivation occupies about 800 hectares confined mostly to the moist coastal tracts of peninsular India and Bengal. Of late its cultivation has also spread to the drier tracts of Deccan plateau and submontane areas of Uttar Pradesh. The crop requires warm and moist weather, hot dry Summers being unfavourable.

Sapota trees are attacked by about 25 insect pests. Of these, *chickoo* moth is the most destructive one; scale insects, mealy bugs and fruit flies being other pests of major importance. The minor pests include, sucking insects like mango hoppers, green plant hopper, whitefly and thrip; bark eating caterpillar, leaf miner, leaf eating caterpillar, bud borers, fruit borer as also chafer beetle and grey weevil.

CHICKOO MOTH

Nephopteryx eugraphella Ragonot is the worst enemy of sapota. It was first reported by Hampson (1896) on *Mimusops elengi* Linnaeus and later by Misra (1920) on sapota. The caterpillars generally feed on leaves but are often found on buds and flowers and sometimes on tender fruits as well. The caterpillars web together a bunch of leaves and feed within on chlorophyll, leaving behind a fine net work of veins. They also bore inside the flower-buds and tender fruits which wither away and the caterpillars move on to the next bud or fruit. The infestation of this pest can be easily detected by the presence of clusters of dried leaves hanging on webbed shoots. The pest is active throughout the year but the activity increases with the appearance of new shoots (April-May) and flower-buds (May-June); maximum activity is observed during June-July after which it decreases gradually but the low level of activity continues even during winter months.

Eggs are oval in shape and pale yellowish-white in colour. Caterpillars are pinkish in colour with three dorso-lateral pinkish-brown stripes on each side; when full grown these are about 25 mm long. Adult moths have grey body; forewings slightly suffused with brown and irrorated with black, hind wings are semi-hyaline.

Eggs are laid on tender shoots or ventral surface of leaves glued firmly on the side of midribs. These hatch in 2 to 4 days in Summer and 4 to 11 days during Winter. The larval, pupal and total life-cycle durations are 13 to 26, 8 to 13 and 26 to 38 days in Summer and 31 to 60, 15 to 29 and 49 to 92 days during Winter respectively. There are 7 to 9 overlapping generations in a year (Gupta and Gangrade, 1955).

To control this pest, remove and destroy all the infested clumps and spray with 0.05% endosulfan, chlorpyriphos, quinalphos or phosphamidon.

COCCOIDS

The leaves of sapota trees are invaded by quite a few speices of scale insects and mealy bugs. Those of common occurrence are, *Drosicha dalbergiae* (Green), *Hemiaspidoproctus cinerea* (Green), *Planococcus citri* (Risso), *P. lilacinus* (Cockerell), *Rastrococcus iceryoides* (Green), *Aspidiotus destructor* Signoret (*A.transparens* Green), *Chloropulvinaria psidii* (Maskell), *Coccus elongatus* (Signoret), *C. longulum* (Douglas) and *Saissetia oleae* (Bernard). These are all polyphagous and some of these are major pests of sapota as well as other preferred hosts. *P. citri* and *S.oleae* prefer *Citrus* spp., *A. destructor* is more common on palmae and *C. psidii* is a major pest of guava.

Individualy the various species of coccoids recorded on sapota may not cause serious loss but as a group, these insects are the dominant pests, collectively causing economic loss to sapota trees. Nymphs and adult females are seen congregated on tender leaves, shoots and inflorescences and sucking the sap there-from. As a result the entire tree is devitalized and in case of severe infestation fruiting capacity of the tree is adversely affected. These insects are more harmful during Summer when they also excrete large quantity of honeydew, which attracts the ants and favours development of sooty mould, covering rapidly the leaves and twigs with a superficial black coating which interferes with the photosynthetic activity resulting in stunted growth.

Drosicha dalbergiae has been reported from Punjab and Uttar Pradesh. Though polyphagous in nature, it is only an occasional pest of minor importance. There is no record of this species from outside India.

Hemiaspidoproctus cinerea has been reported from South India and Sri Lanka. It feeds on various host plants including citrus, guava, pomegranate and sapota (Ayyar, 1930). Since it does not cause any appreciable damage and occurs only occasionally, very little information is available on this species.

Rastrococcus iceryoides is found all over India but is more common in North India on mangoes. Besides mangoes, it has also been found feeding on citrus, sapota, etc. Eggs hatch in 6. to 9 days. Nymphal period of males and females is 18 to 26 and 20 to 31 days respectively. Adult males live for one to two days, unmated females for 13 to 23 days and mated females upto 80 days (Nayar *et al.*, 1976).

To check the infestation of these pests, remove and destroy all the infested parts in the initial stage of infestation. If the attack is severe, spray with 0.04% diazinon or monocrotophos, which is quite effective in controlling the young nymyhs.

FRUIT FLIES

Dacus correctus (Bezzi), *D. dorsalis* (Hendel), *D. tau* (Walker) (*caudatus* Fabricius, *hageni* de Meijere) and *D. zonatus* (Saunders) have been reported attacking sapota fruits in India (Butani, 1975 f). The attack is generally more severe in orchards or on trees grown in the vicinity of mango orchards. These are polyphagous pests. The preferred host of *D. dorsalis* and *D. zonatus* is mango and peach respectively.

D. tau has been reported damaging citrus, mango, melons, mulberry, sapota and a large number of vegetables specially cucurbitaceous. The species is widely distributed in India and has also been reported from Nepal, Burma, Sri Lanka, Malaysia, Taiwan, Philippines and Indonesia (Kapoor, 1970).

Female flies puncture the ripening or over ripe fruits by means of their ovipositor which is thrust inside the fruit and 2 to 10 eggs are laid below the epidermis of fruit. These eggs hatch in a couple of days and the white legless grubs start feeding on soft pulp of the fruits. In case of partially developed fruits, latex is seen oozing out of the small punctures made for oviposition. Affected fruits get wrinkled, rot and fall down.

Eggs are elongated, about 3 mm long and pointed at micropyle end. Full grown maggots are pale white and 8 to 11 mm long. Pupae are reddish-brown, cylindrical and 5 to 6 mm long. Adult flies are dark ferruginous in colour and are slightly bigger than *D. cucurbitae*. Eggs hatch in 24 to 40 hours; maggots complete development in 3 to 7 days and pupal period lasts for 6 to 12 days – thus a life-cycle is completed in 12 to 18 days and there are several generations in a year (Batra, 1968). Generally there is no diapause and under otherwise favourable conditions the pest breeds during Winter also. However, if there is a severe Winter the pest may hibernate in adult stage.

To prevent the attack of fruit flies, harvest the fruits before ripening. Collect all the affected and fallen fruits and destroy the same. In case of severe attack, apply bait-spray consisting of 150 mg yeast hydrolysate, 150 gm molasses and 100 mg malathion (50% WP) or 250 ml diazinon (20% EC) in 100 litres of water. Spraying with 0.03% phosphamidon or 0.03% oxydemeton methyl or 0.06% dimethoate has also been suggested (Talgeri, 1967). This brings down the population of fruit flies to some extent.

LEAF HOPPERS

Mango hoppers, *Amritodus atkinsoni* (Lethierry), *Idioscopus clypealis* (Lethierry) and *I. niveosparsus* (Lethierry) as also green plant hopper, *Falka ocellala* Fabricius have been occasionally found on leaves of sapota trees, sucking the cell sap. Generally none of these cause any serious loss to warrant control measures. However, if at any time they appear in epidemic form, spray with 0.03% dimethoate, phosphamidon or oxydemeton methyl. This will also control the incidence of whitefly and thrip, if any.

WHITEFLY

Castor whitefly, *Trialeurodes ricini* Misra is primarily a pest of castor and a sporadic pest of sapota leaves. It breeds throughout the year, though the activity slackens during November to January and is maximum during March to June. A large number of nymphs may be seen sticking on ventral surface of leaves and sucking the sap. The affected leaves show yellow patches. The nymphs also excrete honeydew which favours development of sooty mould, covering dorsal surface of leaves with black coating; such leaves gradually dry away and drop off from the trees.

Eggs are laid in large clusters or scattered on ventral surface of leaves, each egg anchored to the leaf by means of a short slender petiole. Eggs are pale yellow ' in colour. Nymphs are transluscent light yellow with entire margin encircled by a broad fringe of closely set thick waxen filaments. Adults have pale yellow body and white wings covered with waxy powder.

THRIP

Frankliniella dampfi Priesner (*sulphurea* Schmutz)*, the common blossom thrip, has been recorded infesting flowers of citrus, melons and sapota. Innumerable nymyhs and adults may be seen rasping and sucking the sap from flowers, causing drying and premature shedding of infested flowers, which in turn affects the fruiting capacity of the trees. Adult thrips are slender, fragile insects, body and legs sulphur yellow in colour, antennae greyish-brown and wings clear.

BARK EATING CATERPILLAR

Indarbela tetraonis (Moore) has been reported attacking sapota trees (Bhat and Patil, 1933) in Maharashtra. This is a polyphagous pest. guava and *falsa* being its preferred hosts. The incidence of attack is higher in neglected orchards. The typical symptom of

---

* Mound (1968) considers *sulphurea* Schmutz and *dampfi* Priesner as synonyms of *Frankliniella schultzei* (Trybom).

attack is the presence of fine, ribbon-like, silky webs containing chewed wood particles and excreta of the larvae, seen plastered on tree trunk and main branches specially near the forkings.

Crop sanitation and avoidance of over-crowding of trees prevents the occurrence of this pest. Keep a regular watch from July onwards and no sooner the attack is observed, clean the affected parts and insert in each hole, cotton-wool soaked in carbon bisulphide, petrol or 0.05% endosulfan and seal the holes with mud.

LEAF MINER

Acrocercops gemoniella (Stainton) has been reported as a pest of sapota (Fletcher, 1920 b) from Bihar. The tiny caterpillars mine young and tender leaves and feed within. The affected leaves show glistening galleries, get distorted, dry and ultimately fall down.

To prevent the infestation from spreading, remove and destroy the affected leaves in the initial stage of attack. Spraying with 0.03% oxydemeton methyl or phosphamidon or 0.1% malathion is effective (Sandhu, 1973).

LEAF EATING CATERPILLAR

Metanastria hyrtaca (Cramer) is commonly found in India and Sri Lanka. It is a polyphagous pest recorded feeding on leaves of cashewnut, country almond, *jamun*, sapota, etc. The caterpillars feed gergariously and voraciously at night and remain crowded on the bark of the trees during day.

Eggs are spherical, dull white with two brownish-black markings. Full fed caterpillars are stout, cylindrical, 65 to 75 mm long greyish-dark-brown with hairs of different colours and sizes. Moths are big, stout with greyish-brown wings. Males possess pectinate antennae and forewings having a chocolate brown patch with a white spot in the centre. Females have longer and broader wings with wavy transverse bands of light and dark brown colour and no dark patch. Wing expanse is 50 to 60 mm and 70 to 85 mm in case of males and females respectively.

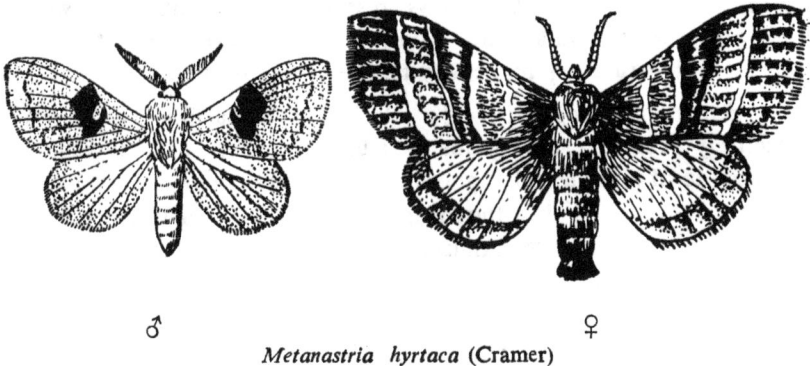

♂   ♀

Metanastria hyrtaca (Cramer)

In South India egg, larval and pupal periods occupy, 9 to 12, 45 to 100 and 9 to 18 days respectively and the life-cycle is completed in 75 to 110 days (Beeson, 1941).

To control this pest, dust the trees including the tree trunks with 10% BHC or spray with 0.1% BHC+0.1% DDT.

BUD AND FRUIT BORERS

Acrobasis remonella, Anarsia species and *Chelaria* species are not reckoned as pests of consequence, though their larvae may occasionally bore into flower buds of sapota trees and affect the fruit setting capacity of the trees. Caterpillars of pomegranate butterfly, *Virachola isocrates* (Fabricius) have also been reported boring sapota fruits and feeding on the pulp. This too is a minor pest and generally no control measures are adopted for any of these borers.

CHAFER BEETLE

Holotrichia consanguinea (Blanchard) – a polyphagous pest, has wide range of host plants including, citrus, *falsa*, fig, guava, *jamun*, mango, sapota and tamarind. The grubs are subterranean and feed on organic matter, rootlets, etc., causing negligible damage. The adults feed on leaves during night, hiding in the soil during day. In case of severe infestation the entire tree may be defoliated. The activity of the pest continues for about two months. The damage is more pronounced in sandy and sandy-loam soils while the orchards having clay or heavy soils are generally free from this pest.

In South India, incubation, larval and pupal periods last for **8** to 10, 114 to 120 and 16 to 18 days respectively (Husain, 1974) with only one generation in a year. Adults emerge in early November but remain in diapause in pupal cavity till the advent of monsoon.

Soil application with 5% aldrin or heptachlor dust can effectively control the grubs ; Husain (1974) has suggested application of phorate 10% granules or endosulfan 2% dust. For the control of adults, dust the foliage with 10% BHC or spray 0.05% endosulfan, as soon as the adults appear.

GREY WEEVIL

Myllocerus undecimpustulatus Faust has been reported from Bihar (Fletcher, 1917) nibbling the leaves of sapota trees. The grubs live underground and feed on roots of all sort. The loss caused is not much and as such no control measures are usually required. However, dusting with 5% BHC or spraying with 0.2% BHC is effective in checking the population of this weevil.

Myllocerus undecimpustulatus Faust

JACK-FRUIT

JACK-FRUIT, *Artocarpus heterophyllus* Lamarck is native of India. It is also grown in Sri Lanka, Bangladesh, Burma, Malaysia, Indonesia, Philippines, Brazil, etc, In India, the estimated area under this fruit is 26,000 hectares. Of this a large chunk is in Karnataka and Kerala where it is grown in mixed plantation with arecanut and coconut along the western coast. There are about 8500 hectares under its cultivation in Assam, 6200 hectares in Tripura, 4500 hectares in Bihar and smaller areas in Uttar Pradesh, Maharashtra, Andhra Pradesh and Tamil Nadu. Inspite of so much area under jack-fruit, it is not classed as a commercial fruit and is seldom grown in regular plantations. The trees flourish in humid climate specially on hill-slopes and are sensitive to cold, frost or drought. As such these trees are grown from sea level to about 1500 metres elevation in the South and at the foot-hills of the Himalayas in the North (Hayes, 1966). The fruits contain some minerals as also vitamin A and B. Beside being used occasionally as table fruit, it is widely used as a vegetable, as also for making pickles and other forms of culinary preparations.

Like any other crop, jack-fruit trees also sustain some loss due to ravages of insect pests. In nature, it is attacked by over 35 species of insects. Among these, those of major importance are, shoot borer, bark borers, bud weevil, spittle bugs and mealy bugs while the minor pests include, scale insects, aphids, whiteflies, thrips, leaf webbers, and longicorn beetles.

SHOOT BORER

Diaphania caesalis (Walker) was originally described by Walker (1859) as *Glyphodes caesalis* sp. nova; subsequently, it was transferred to genus *Margaronia*, then to genus *Palpita* and finally Wang (1963) placed it under genus *Diaphania*. Fletcher (1914) reported *G. caesalis* as a pest of jack-fruit in Karnataka and Maharashtra; Chowdhury and Majid (1954) recorded it as a major pest in Assam. It is also recorded from Sikkim, Bihar, Uttar Pradesh, Andhra Pradesh and Tamil Nadu. Outside India, it has been reported from Borneo,

Burma and Sri Lanka (Hampson, 1896) as also Bangladesh (Alam *et al.,* 1964).

The pest is active from May to October. Eggs are laid on tender shoots and flower-buds. On hatching the caterpillars bore into tender shoots, flowering buds and developing fruits and tunnel through the same. As a result, the shoots wilt and droop, buds dry and drop down and the fruits start rotting (Butani 1978 b). Pupation takes place inside the tunnels.

The caterpillars have yellowish head and prothorax with reddish-brown body having numerous black flattened horny wrats each bearing one short bristle like hair. Pupae are reddish-brown. Adults are whitish-brown moths having wings with greyish elliptical patterns and a marginal series of black specks; wing expanse is 26 to 34 mm, females being slightly bigger than males.

Diaphania caesalis (Walker)

To protect the fruits from ravages of this pest cover the fruits with alkathene bags (Madhava Rao, 1965). Remove and destroy the affected shoots, flower buds and fruits in the initial stage of attack. Spraying with 0.03% phosphamidon or 0.2% BHC or DDT is also effective (Alam, 1962).

BARK BORERS

Indarbela tetraonis (Moore) is the only lepidopterous borer found boring into trunk and main stems of jack-fruit trees in India. Besides, a number of coleopterous beetles, namely, *Apriona rugicollis* Chevrolat, *Batocera rufomaculata* (de Geer), *B. rubus* (Linnaeus), *Epepeotes luscus* Fabricius, *Glenea belli* Gahan, *Sthenias grisator* Fabricius and *Platypus indicus* Strohmeyer have been recorded boring the jack-fruit trees. These are mostly polyphagous pests and in case of jack-fruit, *I. tetraonis* and *Batocera* spp. are comparatively more destructive. These are widely distributed all over India and their presence is conspicuously felt in areas where the surroundings are not kept clean or where the trees are not well looked after.

Indarbela tetraonis has been recorded damaging various fruit trees. The young caterpillars nibble the bark of the trees for couple of days and then bore inside the trunk or main stems, making a short tunnel downwards. Often more than one borer attacks the same tree, each making its independent hole. A severe infestation arrests the growth and affects the fruiting capacity of the tree.

Batocera spp. are widely distributed in Oriental region and besides jack-fruit, they also attack a number of other tropical and temperate fruit trees. The female lays about 200 eggs, placed one by one in incisions, cut in the bark with her mandibles. On hatching the grubs tunnel into the main stem or trunk and feed on meristem and then penetrate deeper; the excavations made in the early stage of infestation are extensive, irregular and deep in sapwood. In case of severe infestation, translocation of cell sap is interrupted, affecting adversely the growth and fruiting capacity of the trees; often the bark may also split open exposing the inner tunnels. Adult beetles are nocturnal in habit and feed by gnawing the bark of living twigs.

Apriona rugicollis is a main pest of forest trees (Beeson, 1941). Adult beetles are about 50 mm long, yellowish-grey with elytral

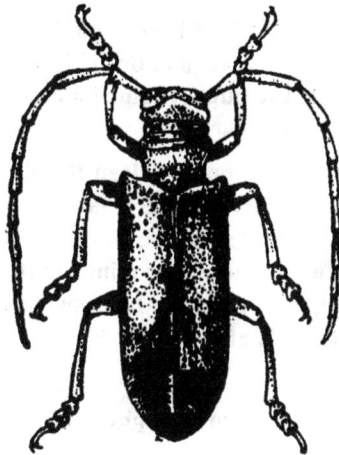

Apriona rugicollis Chevrolat

sutures and margins bluish-grey. These are active during March to October and feed on bark of living shoots; the affected shoots are girdled and gradually die away.

Epepeotes luscus has been reported as minor pest of jack-fruit and mango (Nair, 1975). It has also been reported on various forest trees including *Ficus* spp. The adult beetles are 17 to 27 mm long, reddish-brown with vague grey markings and two black shoulder spots on elytra. It has three generations in a year ; shortest life-cycle being 2½ to 3 months and prolonged one 8 to 10 months. Females live for over 4 months and each lays as many as 1400 eggs (Beeson, 1941).

Glenea belli is a longicorn beetle reported attacking jack-fruits in South India. Adult beetles are shining black and 12 to 15 mm long. Life-cycle is annual, with emergence starting early in May and continuing up to July.

Sthenias grisator is commonly known as girdler beetle as it girdles the stems of the host trees and the affected stems are often killed. The main host of this beetle is grapevine but occasionally, citrus, jack-fruit and mulberry trees are also attacked. The beetles have only one generation in a year and adults are commonly found during August to October in South India.

Platypus indicus – the pinhole borer, is a main pest of forest trees (Beeson, 1941). It has also been reported as a minor pest of jack-fruit (Nair, 1975). These are tiny beetles, 3.5 to 3.8 mm long and have a very peculiar habit of boring into the bark or wood. The main and branch tunnels curve sinuously through the soft wood in a horizontal plane and the pupal chambers are offset vertically on both sides, spaced slightly apart.

To control the main borers, clean the affected portion of the trunk or main stem and insert into the holes a wick of cotton-wool soaked in carbon bisulphide, chloroform, petrol or even kerosene oil and seal the holes with mud. For minor pests, specially *Apriona rugicollis* and *Sthenias grisator* hand-picking and mechanical destruction of the beetles is also helpful .

BUD WEEVIL

Ochyromera artocarpi Marshall is an specific pest of jack-fruit.

It is found all over India and has been recorded as a major pest from Assam and western coast of Karnataka. The small whitish grubs bore into tender flower buds and fruits. With the result the affected buds and fruits fall prematurely. Adults are greyish-brown weevils found nibbling the leaves.

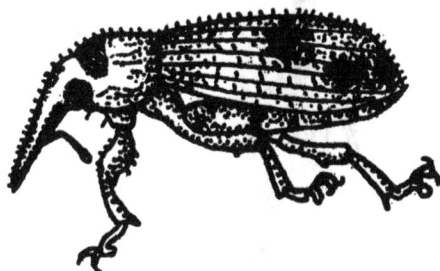

Ochyromera artocarpi Marshall

Two more curculionid weevils, namely, *Onychocnemis careyae* Marshall and *Teluropus ballardi* Marshall have been reported feeding on leaves of jack-fruit trees in South India (Nair, 1975), but not causing any appreciable loss.

To control these weevils, collect and kill mechanically the grubs and adults in the initial stage of attack. Later, remove and destroy the affected shoots as also infested and fallen flower-buds and fruits. Spraying with 0.2% BHC, DDT or carbaryl may be done in severe cases of attack.

SPITTLE BUGS

Cosmoscarta relata Distant is a serious pest of jack-fruit trees, specially in South India (Beeson, 1941). These bugs appear in swarms and feed on young shoots and leaves. The affected leaves often get curled. Nymphs are wingless, dark purple with yellowish head, pronotum and legs. These are found living inside common mass of froth generally on stalks and young shoots and ripening fruits. The frothy mass consists of the fluid voided from the anus and mucilaginous secretion from the epidermal glands. After the last moult the nymphs leave this frothy mass and become adults. Adults are large sized bugs, about 15 mm long with reddish head and pronotum and forewings with reddish markings on the tagmina. The bugs are very active and difficult to catch.

Two more spittle bugs, namely, *Clovia lineaticollis* (Motschulsky)
and *Ptyleus* species have been reported from South India (Nair, 1975).

Clovia lineaticollis (Motschulsky)

These bugs are smaller than *C. relata.*   Adults of *C. lineaticollis* are
about 9 mm long, chestnut brown with yellowish stripes while those
of *Ptyleus* species are 10 to 12 mm long and greyish in colour.

To control these bugs, collect the masses of froth and destroy
mechanically the young nymphs therein.

MEALY BUGS

Nipaecoccus   viridis   (Newstead)   [*corymbatus*   (Green ,
flamentosus (Cockerell),   *vastator*   (Maskell)], is a polyphagous pest
recorded on **ber**, citrus, fig, grapevine, guava, mango, mulberry,
tamarind, etc.   It is found all over India.   Peak period of activity is
around November when clusters of these mealy bugs may be found
on leaves and tender shoots, sucking the cell sap. Eggs are roundish
cylindrical, flattened at both ends, chestnut brown in colour and
fully covered with fine cretaceous material forming the ovisac.
Nymphs are deep chocolate in colour, having their dorsum covered
thinly with   a whitish mealy material.   Adult females are dark
castaneous covered completely with sticky cretaceous white ovisac.
If removed from this ovisac, the females are seen covered with
whitish mealy substance.   Eggs are laid in ovisacs and the females
die soon after egg-laying. A single female lays 400 to 700 eggs. The
eggs hatch in 7 to 10 days and the nymphal development takes
about 15 and 20 days in case of males and females respectively.

Ferrisia virgata (Cockerell), the white tailed mealy bug, is also a polyphagous pest and is widely distributed all over India (Ali, 1962, 1968). It feeds on leaves and tender shoots and during dry weather it moves down and inhabit the roots. A prolonged period of drought may result in a severe outbreak of this pest. The peak period of activity is August to November, when the affected parts turn yellow, wither and ultimately die away.

To control these mealy bugs, spraying may be done with 0.05% diazinon, dichlorvos or monocrotophos. This sparying will also check to some extent the infestation of mango mealy bugs and scale insects.

Mango mealy bugs, *Drosicha mangiferae* (Green) and *D.stebbingi* (Green) are the two allied species often confused with one another. Both have been reported as minor pests of jack-fruit. Apparently, these species look alike and show same symptoms of damage, habits, The gravid females descend down the trees and enter into the soil for egg-laying. On hatching, the nymphs come out from the soil and climb up the trees, to suck sap from succulent and tender leaves and twigs.

Raking the soil around the trees during Summer exposes and destroys the eggs; soil application of 5% aldrin, chlordane or heptachlor dust @ 300 to 500 gm per tree, just before Winter sets in, checks the nymphal population and tying the alkathene band around the tree trunks is effective in preventing the nymphs from climbing up the tree (Srivastava and Butani, 1972).

LEAF WEBBERS

Leaf webbers or leaf eating caterpillars commonly found damaging jack-fruit trees in India, include, *Perina nuda* Fabricius and *Diaphania bivitralis* (Guenée). Both are sporadic pests usually of minor importance. The caterpillars roll, fold or web together the leaves and feed within.

Perina nuda has been reported from China, India and Sri Lanka. Besides jack-fruit, it has also been recorded as a minor pest of fig and mango trees in India. Eggs are about 0.7 mm long, cylindrically round tapering towards the end attached to the leaf. These are light pink in colour when freshly laid, later becoming red and finally brick red. Full grown caterpillars are 22 to 25 mm long, having short erect tufts of dusky grey to brownish hairs. Pupae are hairy and brightly coloured brownish-green dorsally and pale yellowish ventrally. Male pupae are on an average 16 mm long and female pupae are 18 mm long. In the moths, there is extreme sexual dimorphism. Male moths are smaller than female moths. A single female lays 60 to 400 eggs. These hatch in 4 to 6 days. Larval and pupal periods are 16 to 20 and 5 to 9 days respectively. The adult longevity is 3 to 11 days and the total life-cycle is completed in 27 to 39 days (Cherian and Israel, 1939).

Diaphania bivitralis cause considerable damage to jack-fruit foliage specially in Tamil Nadu (Muthukrishnan et al., 1958). It has also been reported from Sikkim and Assam and outside India from Taiwan, Borneo, Burma and Sri Lanka. Full grown caterpillars are about 30 mm long, olive-brown in colour with conspicuous white markings. These pupate within the leaf folds in thin white silken cocoons. Pupae are red and 20 to 24 mm long. Moths are chestnut brown in colour dorsally and white ventrally. Forewings are chestnut brown having two sem-hyaline white blotches with a small black discocellar spot between the two. Hind wings are iridescent hyaline white having a broad chestnut marginal band with a black line on its inner edge. The eggs hatch in 5 to 6 days; caterpillars pupate after 14 to 16 days and pupae become adults in 8 to 10 days (Nagaraja Rao and Subramaniam, 1958).

To control these pests, hand-picking and mechanical destruction of caterpillars in the initial stage of attack is suggested. In case of severe infestation, dust 5 to 10% BHC. Spray with 0.1% BHC+0.1% DDT or 0.1% carbaryl is also effective in checking the pest population.

CASTOR CAPSULE BORER

Dichocrocis punctiferalis (Guenée) is a major pest of guava and a minor pest of various fruit crops including jack-fruit. Eggs are usually laid on young fruits and occasionally on flower-buds. On hatching the pinkish-brown caterpillars bore inside the flower buds of young fruits and feed within. The bored holes are plugged with excreta and in case of severe infestation, the excreta fall on the ground below and can be seen there in small heaps. The affected buds do not open, wither and dry away while the fruits fall prematurely.

To check the population of this pest, remove and destroy all the infested flower-buds and fruits.

SCALE INSECTS

The scale insects found damaging jack-fruit trees include, fluted scale, *Icerya (Crossotosoma) aegyptica* (Douglas); hard scales, *Hemiberlesia* (*Aspidiotus*) *lataniae* (Signoret), *Parlaspis papillosa* (Green), *Pinnapis* (*Chionaspis*) *aspidistrae* (Signoret) and *Semelaspidus artocarpi* (Green) *(Aspidiotus cistuloides* Green, *triglandulosus* (Green) as also soft scales, *Ceroplastes rubens* Maskell, *Chloropulvinaria psidii* (Maskell) and *Coccus (Lecanium) acutissimus* (Green). Except *S. artocarpi* and *P. papillosa*, the others are all polyphagous pests. Loss caused to jack-fruit trees by these insects is usually negligible. It is only *C. rubens* that occur regularly, the rest of the species occur sporadically and that too only in certain pockets.

Ceroplastes rubens commonly known as waxy scale is a main pest of *Citrus* spp. and is often found on fig, jack-fruit, mango, pear, etc. Hill (1975) also mentions coffee and tea as its alternate host plants. The pest has limited distribution (CIE map No. A – 118) and has been reported from East Africa, India, Sri Lanka, parts of China, Japan, Malaysia, Philippines, East Australia, Pacific Islands, Newzealand, USA (Hawaii), etc. It is widely distributed in India. Colonies of these scales may be found on jack-fruit trees, covering the shoots and fruit stalks and sucking the cell sap therefrom. These insects also excrete copious amount of honeydew which keeps dripping on lower leaves and fruits. As generally happens, the honeydew attracts the ants and also favours the development of sooty mould which covers the affected parts with a black superficial coating. Adult females are convex in shape, 3 to 4 mm long and covered with a pink waxy shell often with white vertical stripes. Males are rare.

Ceroplastes rubens Maskell

Life-cycle is not known in detail, however, there is only one generation in a year.

Generally, chemical control measures are not required against these insects on jack-fruit. If and when necessary, spray with 0.05% dia ziron ichlorvos or dimethoate.

APHIDS

Jack-fruit aphid, *Greenidia artocarpi* (Westwood) and black citrus aphid, *Toxoptera aurantii* (Fonscolmbe) have been recorded feeding on jack-fruit trees; the former is an specific pest of jack-fruit whereas the latter is a polyphagous pest, *Citrus* spp. being the main hosts. Clusters of nymphs and adults may be seen on tender shoots and leaves sucking the cell sap and giving out honeydew. As a result, the entire tree is devitalised and the affected parts are distorted and covered with sooty mould. A dry spell of weather followed by rains favours the rapid multiplication of the aphids resulting in severe infestation.

Greenidia artocarpi is a pale green aphid having long and hairy cornicles. It has been reported from Tamil Nadu where the pest is active from January-February to September-October (David, 1956). Reproduction is usually parthenogenetical, as a result, the population build-up is very rapid.

To control these aphids, spray with 0.03% dimethoate, monocrotophos or phosphamidon. Normally couple of sprayings given at an interval of 10 to 12 days, are quite sufficient to keep the pest under check. These sprayings will also control whiteflies and thrips, if present.

WHITEFLIES

Aleurotrachelus caerulescens Singh, *A. rachipora* Singh and *Pealius schimae* Takahashi have been recorded on jack-fruit but not as pests; the first two are polyphagous while the last one is specific to jack-fruit. Occasionally the nymphs of these whiteflies may be seen sucking the sap from leaves and giving out honeydew but the loss caused is seldom serious.

THRIP

Pseudodendrothrips dwivarna (Ramakrishna and Margabandhu)

was originally described by the authors (1931) as *Dendothrips dwivarna* but subsequently, Singh (1942) placed it under genus *Pseudodendrothrips*. This is typically monophagous species infesting tender leaves of jack-fruit (Ananthakrishan, 1971). It is confined to peninsular India ond has been recorded from Karnataka, Kerala and Tamil Nadu. Each affected leaf may have 15 to 20 adults and numerous eggs and nymphs on it (Ananthakrishnan, 1955). Concentrated feeding of nymphs and adults produces number of whitish spots and patches on the leaves and the affected leaves ultimately fade and die away.

Eggs are minute, elongated oval, slightly curved and 0.2 mm long. Freshly hatched nymphs are very small, less than 0.5 mm long and orange yellow with dark grey antennae and pale grey legs. Adults are also small, females being 0.9 to 1.1 mm long and males 0.6 to 0.7 mm, having head and thorax reddish-brown and abdomen yellow with heavily fringed wings. Females are always more in sumber; sex ratio being five females to one male.

RED ANT

Oecophylla smaragdina (Fabricius) has been reported from various orchards, plantations and forests all over India. Innumerable workers of these ants are seen going up and down the trees specially on sunny days. They cause no direct injury to the trees except webbing together a few apical leaves for constructing their nests. On the contrary being carnivorous, they prey upon a number of insect pests. But they also relish honeydew for which they carry the nymphs of aphids, mealy bugs and scale insects from tree to tree – thus helping the dissemination of these harmful insects. Besides they are a nuisance to persons who have to climb the trees.

To control these ants, their nests should be removed and destroyed mechanically or the same may be sprayed to run off point with 0.1% BHC (Butani and Tahiliani, 1974).

2.9
WATER NUT

WATER NUT, *Trapa bispinosa* Roxburgii – popularly known as *singhara*, is grown all over India though more extensively in northern States. Besides India, it is grown in Pakistan, Sri Lanka, Bangladesh, Burma, South-East Asia as also in tropical Africa. It is grown like lotus on stagnant water of lakes, ponds and tanks. Till recently its fruit was generally eaten as such but of late it is gaining more importance economically as staple diet of the poor. The fruit is dried, powdered and used as flour. The crop is severely damaged by singhara beetles and to lesser extent by blue beetle, aphid and aquatic weevil (Batra, 1961). Among the non-insect pests, tortoise has been reported causing 12.5 to 25 per cent damage in Uttar Pradesh (Srivastava, 1956 a); crabs have been reported damaging the crop in Assam (Srivastava, 1976) and snails in Delhi (Batra, 1961).

S NGHARA BEETLES

Pyrrhalta birmanica (Jacoby) and *P. rugosa* (Jacoby) have been recorded so far from India. Of these, *P. birmanica* is more common and destructive. This is an specific pest of water nut (Khatib, 1934; Bindra, 1959) and has been recorded from Assam, Bihar, Uttar Pradesh, Madhya Predesh, Rajasthan, Haryana, Delhi and Kashmir (Srivastava, 1956 b; Reddy, 1968). The insect was first reported from Burma and described by Jacoby (1889) as *Lachmoea birmanica* and later the Indian specics was described by Lefroy (1910) as *Galerucella singhara*. Maulik (1936) gave the detailed description of this insect as *Galerucella birmanica* (Jacoby) (=*L.birmanica*). It is obvious that the original work of Jacoby was not consulted by Lefroy and again the work of Lefroy apparently escaped the notice of Maulik. Anyway, the species has now been placed under the genus *Pyrrhalta* and thus the correct name of *singhara* beetle found in India is *Pyrrhalta birmanica* (Jacoby).

Grubs and adults feed usually on leaves and sometimes also on petioles and integument of the fruits. Full grown grubs are oracious feeders and more destructive than adult beetles. The beetles are sluggish and not active fliers. According to Pruthi and Batra (1960) the pest causes a loss of 40 to 60 percent. Eggs are round, pale yellowish with reddish-brown tinge. Freshly hatched grubs are yellowish-brown but soon turn into dark greyish-brown colour. Pupae are 4 to 5 mm long, bright yellow and later become yellowish-brown. Adult beetles on emergence are bright yellow but soon become reddish-brown ; antennae being brownish-black. The beetles are 6 to 8 mm long. Hibernating beetles are darker in colour than those of summer brood. Functional and developmental anatomy of reproductive organs of male has been studied in detail by Verma (1969).

Pyrrhalta birmanica (Jacoby)

Life history and bionomics have been studied by Khatib (1934), Srivastava (1956 b) and Srivastava *et al.* (1969). A single female lays on an average 101 eggs in clusters of 5 to 8 eggs, glued firmly to the upper surface of leaves. Pre-mating period is about 2 days during July and extends upto 7 days by November. Copulation lasts for 5 to 15 minutes and pre-oviposition period is 15 hours to 4 days. Oviposition period lasts for 6 to 19 days. Incubation, grub and pupal periods are 4 to 9, 9 to 20 and 5 to 10 days respectively. Total life-cycle from egg to adult stage occupies 18 to 32 days and under favourable conditions there may be as many as nine generations in a year. Longevity of males and females is 11 to 28 and 13 to 33 days respectively. The entire life of the beetle and its immature stages is passed on the leaves of water nut plants. With the advent of cold and dry weather the activity of the pest declines and by December the beetles enter into hibernation in cracks and crevices on banks of lakes and ponds. The hibernation continues for 4 to 5 months and beetles become active again around April-May. Warm and humid weather increases the activity of these beetles. The maximum activity has been observed around August-September.

To control this pest, Mehta and Varma (1968) suggested mechanical collection of adults and egg-mases from the leaves and their destruction in early stages of crop. Dusting with 5% BHC was found effective by Bhatia and Sikka (1952), Purewal and Kang (1956) and Srivastava (1956 b). According to Pradhan *et al.* (1963) dusting with 5% BHC or DDT was effective in controlling this pest but after extensive use for 10 years these insecticides proved to be ineffective as the pest appears to have developed resistance to these insecticides. Yawalkar and Chitriv (1969) also observed that BHC dusting had no effect. Jotwani and Sarup (1966) found carbaryl, mevinphos and phosphamidon to be effective while Pradhan *et al.* (1964) supported by Verma *et al.* (1967) and Srivastava *et al.* (1969) have recommended dusting with 10% carbaryl @ 25 kg per hectare.

BLUE   BEETLE

Lefroy and Howlett (1909) reported *Altica cyanea* (Weber) as one of the perfectly harmless insect. Fletcher (1920 a) reported these beetles feeding on a variety of weeds but breeding only on *Ammania* species. Batra (1961) recorded gregarious presence of adult beetles on wheat and barley without causing any damage to these crops. Similarly, Ayyar (1963) found hundreds of these beetles on paddy but causing no damage to the crop. The only economic crop damaged by these beetles is water nut.

The grubs feed on upper epidermis of the leaves while the adult beetles nibble the leaf margins eating away the peripheral border. Soon after the epidermal damage, the tissues below die away and the leaves present a scorched appearance. The pest is more active during rainy season.

Grubs are dark brown in colour, 8 to 10 mm long and have thick short black hairs arising from tubercles which give the grubs a bristly appearance. When full grown the grubs leave the host and pupate in the soil in earthen cocoons. Overwintering is in pupal stage. Pupae are 42 to 48 mm long, orange-yellow when freshly formed and gradually becoming darker. Adults are steel-blue beetles, about 5 mm long with black antennae (Maulik, 1926).

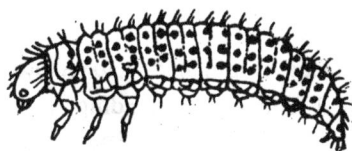

Grub   beetle

Altica cyancea (Weber)

To control these beetles, dust with 5% carbaryl @ 20 kg per hectare.

SINGHARA APHID

Rhopalosiphum nymphae is a minor pest reported from India (Prewal and Kang, 1956) and Bangladesh (Alam *et al.*, 1964). Nymphs and adults suck the cell sap and devitalize the plants. In case of severe infestation, the leaves get distorted and finally the crop yield is lowered considerably.

Dusting the crop against *singhara* beetles or blue beetles controls this aphid as well.

AQUATIC WEEVIL

Prashad (1960) recorded a small black weevil, *Bagous trapae* species nova and described its adult.

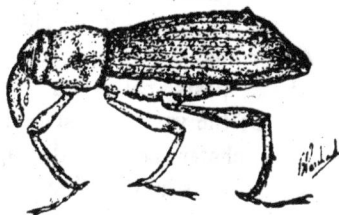

Bagous trapae Prashad

This weevil feeds on submerged soft stem of water nut crop, near the internodes, affecting adversely the branching of plant and making the crop patchy. In case of severe attack, the maturity of the crop is delayed resulting in low yield. The pest however, is of minor importance.

2.10
TAMARIND

TAMARIND, *Tamarindus indica* Linnaeus is indigenous to Africa (Watt, 1908). It is grown in tropical countries extending from Africa to South-East Asia, including UAR, Pakistan, India, Sri Lanka, Bangladesh and Burma. In India, its cultivation can be traced to first century (629 to 645) and at present it is grown all over the country. It is seldom consumed as fresh fruit, only dried tamarind is used as condiment and is favourite ingredient in *curries* and *chutnies*. It also has medicinal properties (Varma, 1955). Besides, leaves, flowers and fruits are used as auxilliaries in dyeing and dried tamarind is used at home for polishing the brass and copper wares. Tamarind trees are also nectar yielding plants and are favoured very much by the bees. Above all, dry tamarind is highly relished by pregnant women, all over the world.

Inspite of the fact that there is regular demand of dried tamarind in every house, the quantity consumed per capita is very low. That's why there are no tamarind orchards any where. The trees are mainly grown in the backyards. Tamarind trees being evergreen are planted on road-sides for shade but the villagers have an aversion to sleeping under its shade because of the supposed acid axhalation from the leaves.

Like any other crop, tamarind trees too are attacked by a number of insect pests. About 40 insect species have been recorded so far from India. According to Joseph and Oommen (1960) in Kerala State alone, these pests cause a loss, amounting to Rs. 10,000,000/- per year. Of the various pests recorded, coccoids are the most destructive ones; chafer beetles damage the roots and leaves, good many sucking insects and leaf eating caterpillars feed on foliage and flower-buds while a few damage the fruits as well. The worst enemies are those insects that attack the developing fruits and continue their ravages in stored and dried tamarind fruits.

COCCOIDS

A number of mealy bugs and scale insects have been recorded

devitalizing the tamarind trees, those commonly found, include, *Perissopneumon tamarindus* (Green), *Drosicha* spp., *Nipaecoccus viridis* (Newstead), *Planococcus lilacinus* (Cockerell), *Aonidiella orientalis* (Newstead), *Aspidiotus destructor* Signoret, *A. tamarindi* Green, *Cardiococcus castilleae* (Green), *Hemiberlesia lataniae* (Signoret), *Howardia biclavis* (Comstock), *Pinnaspis strachani* (Cooley), *P. temporaria* Ferris, *Unaspis articolor* (Green), *Saissetia oleae* (Bernard) and *Kerria lacca* (Kerr). All these are polyphagous pests, having a wide range of host plants and most of these are major pests of one or the other crop. Among scale insects, those that prefer tamarind, are, *A. orientalis*, *A. tamarindi* and *S. oleae* – the last one is also a main pest of guava. Both young and adults of these insects suck the cell sap – the affected parts are devitalized and there is premature shedding of flower-buds, affecting ultimately the fruit setting capacity as also the size of fruits. (Butani, 1978 d)

Ber mealy bug, *Perissopneumon tamarindus* and mango mealy bugs, *Drosicha* spp. have been occasionally reported as serious pests. The gravid females descend the trees and enter into the soil for depositing their eggs. On hatching, the nymphs again crawl up the trees to feed on succulent and tender foliage. They have only one generation in a year. Plough around the trees to expose and kill the nymphs that are there. Soil treatment with 5% aldrin, chlordane or heptachlor dust @ 500 gm per tree further checks the nymphal population. Tie alkathene bands round the tree trunks to prevent nymphs from ascending the trees (Srivastava and Butani, 1972).

The mealy bugs, *Nipaecoccus viridis* and *Planococcus lilacinus* are sporadic pests of tamarind; the former prefers citrus and jack-fruit while the main hosts of latter are custard apple, pomegranate and tamarind. These insects are characterised by a mealy or waxy coating over their bodies. Their nymphs and mature females may be found in large number on ventral surface of leaflets, base of leaf petioles, tender shoots and even fruits, sucking the cell sap. As a result of their attack, the leaflets become chlorotic and fall down. In case of severe infestation, there may also be immature fruit fall. *P. lilacinus* also attacks, the roots and thereby devitalizes the trees. In addition, these mealy bugs exude large quantity of honeydew which attracts the ants and favours the development of sooty mould.

To control these mealy bugs, in the initial stage of attack, remove the infested twigs along with the bugs thereon and destroy the same. In case of severe infestation spray with 0.1% diazinon or monocrotophos. This will also control the scale insects, if present on these trees. For subterranean population of *P. lilacinus*, plough around the trees and mix thoroughly with the soil 5% aldrin, chlordane or heptachlor dust @ 500 gm per tree.

Aonidiella orientalis (Newstead) (*Aspidiotus coccotiphagus* Marlatt) commonly known as Oriental yellow scale has been reported attacking arecanut, avocado, *bael*, banana, *ber*, citrus, coconut, guava, *jamun*, mulberry, peach, pomegranate, tamarind, etc. Eggs are laid under a protective armour. On hatching, the crawlers move freely and after findings suitable succulent spot, insert their needle like mouth parts into the plant tissues and start feeding. Simultaneously, they also secrete a waxy covering over their dorsal surface and gradually they lose their legs and become immobile. The crawlers grow into adult females while in case of males, there are prepupal and pupal stages. Males are winged and have legs whereas females are wingless, legless, circular, about 2 mm in diameter and have yellowish-brown body.

Kerria lacca – the important lac insect is widely distributed in India. The resinuous secretion of females covers the twigs of host tree and yields stick-lac, which has great commercial value. The insects are polyphagous reported on a large variety of wild and cultivated host plants, including, *ber*, citrus, grapevine, mango and tamarind. Females are apterous having irregular globose body with vestigial antennae. They live enclosed in a resinous mass secreted by the lac-resin glands that are found all over their dermis. Males are either winged or wingless. A single female produces 200 to 500 eggs which are extruded into the cell in a space formed by the contraction of female body. On hatching, the tiny red crawlers escape from the cell and are found actively wandering about in large number on host plants in search of succulent tender shoot. Once suitable spot is located they insert their proboscis and suck the sap, remaining stationary thereon. Once settled, they start secreting resinuous fluid which at first appears as a shining uniform coating over their bodies but gradually it incraeses in thickness and extent untill the secretion of adjoining larvae eventually meet and coalesce in a continuous encrustation. There are two generation in a year. This is seldom a major pest of tamarind.

To check the infestation of scale insects and lac insects, remove and destroy promptly the affected parts along with the insects in the initial stage of attack. This will prevent the infestation from spreading. In case of severe infestation, which may be rare, spray with 0.1% diazinon or monocrotophos.

CHAFER BEETLES

Holotrichia insularis Brenske is one of the most abundant and injurious species found feeding on a large number of host plants, including, *ber, falsa,* guava, *jamun,* mango, pomegranate and tamarind. In India, the pest is more common in Maharashtra, Gujarat, Rajasthan, Haryana and Punjab.

Adult beetles emerge in June after break of monsoon. The beetles are active at night and strongly phototropic during early hours of night. They feed on foliage and in case of severe attack the seedlings and young trees may be completely defoliated. The beetles can fly as high as 14-15 metres for about a kilometre. Eggs are laid in moist sandy soils, 30 to 150 mm deep. A single female lays 40 to 65 eggs. On hatching, the grubs feed on roots. If the attack is heavy the saplings are killed outright whereas the young plants wither and gradually dry up. Pupation takes place in earthen cells deep in the soil.

Eggs are shiny white in colour and oval in shape. Full grown grubs are dingy white, fleshy, curved, 38 to 44 mm long and 6 to 9 mm wide. Adults are brownish-black, convex beetles, 18 to 20 mm long and 8 to 10 mm wide. Mating usually takes place after the females have fed for about a fortnight. Pre-oviposition and incubation periods last for 4 to 6 and 8 to 12 days respectively while grub and pupal stages occupy 11 to 16 and 2 to 3 weeks respectively. There is only one generation in a year. Pupae and adults hibernate in the soil from November to June.

Another species, *Schizonycha ruficollis* Blanchard has been reported damaging tamarind in Karnataka (Veeresh, 1974) and citrus in Gujarat (Nair, 1975). Symptoms of damage and habits etc., are same as those of *H. insularis*.

Plough around the trees during Winter to expose and kill the hibernating pupae and adults.   Soil treatment with heptachlor or thiodemeton is effective againts grubs (Srivastava and Khan, 1963). To kill the adults, spray with 0.1% diazinon or fenitrothion specially during breeding season.

APHID

Black citrus aphid, *Toxoptera aurantii* (Fonscolombe) is a polyphagous pest. Its main hosts are *Cirtus* species though it has been recorded on over 50 host plants (Essig, 1949).   Clusters of nymphs and adults my be seen clinging on tender shoots and leaflets and sucking the sap therefrom.   Normally fresh infestation occurs at flush of new leaves   As a result of aphid attack, the leaflets get distorted in shape and twigs get covered with sooty mould which develops on honeydew excreted by the aphids.   Severe outbreak occurs when a spell of dry weather is followed by rains.

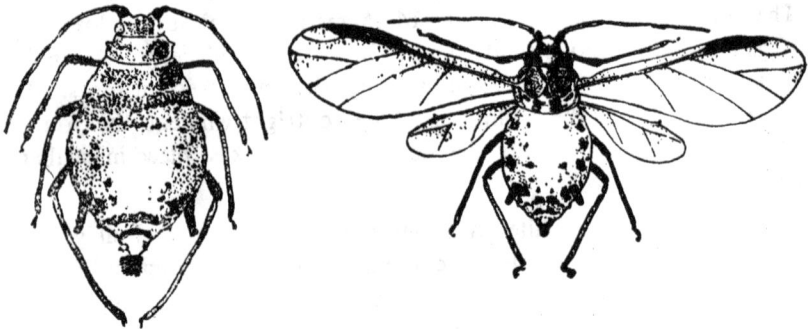

Toxoptera aurantii (Fonscolombe)

Spraying   with   0.05%   dimethoate,   monocrotophos   or phosphamidon is effective in controlling the aphid. It may be necessary to repeat the spraying after an interval of 8 to 12 days. These sprayings will also control the whitefly infestation, if any.

WHITEFLY

Acaudaleyrodes rachispora (Singh) has been reported as a pest of tamarind (Nayar *et al.*, 1976).   It sucks the cell sap and excretes honeydew on which sooty mould grows – but the magnitude of loss is negligible.

THRIPS

Scirtothrips dorsalis Hood, *Ramaswamiahiella subnudula* Karny and *Haplothrips ceylonicus* Schmutz have been reported infesting flowers of tamarind. Nymphs and adults infest flower-buds; acerate the tissues and suck the oozing cell sap therefrom. The affected parts gradually fade and shrivel; a severe infestation may even result in malformation of flower-buds – thus affecting adversely the fruit setting capacity of the tree.

Scirtothrips dorsalis – the chilli thrip, is a polyphagous pest with a wide range of host plants. Its adults are minute, less than one mm long, slender, yellowish-straw coloured insects having heavily fringed wings (Hood, 1919).

Scirtothrips dorsalis Hood

Reproduction is sexual as well as parthenogenetic. Females are more abundant than males; their average ratio being 6:1. A single female lays 40 to 50 eggs @ 4-5 eggs per day. In Assam, this thrip is a major pest of tea and there its incubation, nymphal and pupal periods during Summer last for 6.2 to 7.5, 4.3 to 6.0 and 2.9 to 4.1 days respectively and life-cycle 13.4 to 17.6 days (Dev, 1964). Adult longevity is 10 to 15 days and there are 25 overlapping generations in a year (Raizada, 1965).

Ramaswamiahiella subnudula is a small yellowish species. It is also polyphagous, having a wide range of host plants, including the flowers of citrus, mango, pomegranate and tamarind.

Haplothrips ceylonicus is blackish-brown species with rich red pigments. It was earlier reported infesting flowers of lantana, jasmine and mango (Ayyar, 1928); Sivagami *et al.* (1964) from South India, recorded it on tamarind flowers. The infestation is usually noticed during May to July (maximum during June).

To control the thrips dusting with 5% BHC (Nagaraja Rao, 1955) or spraying with 0.03% dimethoate or monocrotophos or 0.2% carbaryl (Rathore *et al.*, 1970) are effective. If necessary, the insecticidal application may be repeated after a fortnight.

COW–BUGS

Oxyrhachis tarandus (Fabricius) is found all over India. Though red gram is its main host, it is also found on apple, citrus, tamarind, etc. Eggs are found superficially glued to bark of main stems in a particular pattern in two rows. On hatching, the nymphs and adults suck the cell sap from tender shoots and leaflets. They also secrete honeydew on which sooty mould developes.

Other species of cow-bugs, include, *Otinotus oneratus* Walker and *Leptocentrus obliquis* Walker; the former is a sporadic pest of *ber*, custard apple, mango, tamarind, etc., while the latter has been reported as a minor pest of *ber* and tamarind.

To control these cow-bugs spray with 0.05% monocrotophos or dichlorvos

LEAF EATING CATERPILLARS

Thosea aperiens (Walker) is found in India and Sri Lanka as a minor pest of tamarind and various pulses. The caterpillars feed on leaves during October to December. They are sluggish and often cause annoyance by their very presence. The caterpillars are green in colour and have stinging hairs. When full fed, they enter into the soil, construct cocoons during December-January and stay there to diapause or aestivate for 8 to 9 months. Pupation takes place during August-September and moths emerge in October-November. Thus there is only one generation in a year.

Thalassodes quadraria Guenée has been recorded feeding on leaves of litchi, mango and tamarind causing some minor damage. The caterpillars are slender, brownish-green loopers and often assume a characteristic pose on the twigs that these are mistaken for part of the twig or leaf petioles. Full grown caterpillars are about 35 mm long. Pupae are slender, green, about 15 mm long and remain attached to the leaves. Adult moths are also slender, delicate and green in colour; wings are semi-hyaline with numerous indistinct, pale, straight and oblique lines; wing expanse of males is 24 to 28 mm and that of females, 30 to 34 mm.

Stauropus alternus Walker, commonly called lobster or crab caterpillars, are curious looking caterpillars. These have been recorded feeding on leaves of mango, tamarind and number of forest trees (Beeson, 1941), but the loss caused in case of mango and tamarind is of minor importance. Eggs are small and flattened. Caterpillars have long and slender thoracic legs – first pair being very much shorter than the other two pairs; the terminal abdomen segments are swollen and bear a pair of slender curved appendages instead of anal claspers. Full grown caterpillars are 40 to 45 mm long, body colour varying from brownish-black to greyish-black, specked with white dots and more or less pubescent.

Stauropus alternus Walker – caterpillar

Pupae are 18 to 20 mm long and shiny dark brown in colour. Moths are stout, pale greyish-brown, suffused with darker shades and a marginal row of reddish-brown and pale spots on both wings; wing expanse is 35 to 45 and 50 to 60 mm in case of males and females respectively. Male moths have pectinate antennae. Egg, larval and pupal periods occupy 4 to 20, 21 to 42 and 10 to 23 days respectively

(Nair, 1975). Pupation takes place in loosely woven cocoons o yellow fibrous silk covered over by the leaflets.

Caterpillars of *Chaliodes vitrea* Hampson and *Pteroma plagiophleps* Hampson commonly called bagworms, have also been noted feeding on tamarind leaves (Nair, 1975). The larval case of *C.vitrea* is cylindrical, made of pale brown silk and 25 to 30 mm long tapering distally whereas that of *P. plagiophleps* consists of fragments of leaves and is only 9 to 12 mm long. The damage caused by these pests is negligible.

To control these leaf eating caterpillars, spray with 0.2% carbaryl or 0.1% BHC+0.1% DDT.

BUD AND FRUIT BORERS

Eublemma angulifera Moore has been reported damaging tamarind flowers in South India (Sivagami *et al.*, 1964). It has also been reported as a minor pest infesting mango inflorescences (Aiyer, 1943 a). The stalks of infested flowers become weak and the flowers subsequently dry up. Eggs are laid singly on pedicels and sepals of flower-buds. Freshly hatched larvae feed on sepals and later they spin webs beneath the flower buds and flowers and feed within. Pupation takes place in cocoons near the base of flowers. Full grown caterpillars are 16 to 20 mm long, pale green in colour and devoid of any markings. Moths are greyish-buff coloured with an oblique wavy line on the wings; wing expanse is 18 to 22 mm. Egg, larval and pupal periods last for 4, 16 to 18 and 13 days respectively.

Caterpillars of *Cydia* (*Laspeyresia*) *palamedes* (Meyrick) have been observed boring into tender buds of tamarind and feeding on the inner contents. These larvae web together the buds and flowers – causing thereby much more loss than what they actually eat. Full grown caterpillars are about 4 mm long and cream coloured. Pupae are about 3 mm long and reddish-brown. Moths are small in size, dark in colour with shiny forewings having white lines at upper edge and lower corner; wing expanse is 6 to 8 mm.

Virachola isocrates (Fabricius) – pomegranate butterfly is a polyphagous pest. Eggs are laid singly on calyx of flowers or on tender fruits. On hatching the caterpillars bore into the fruits and feed on the seeds, plugging the entry holes with their anal segment or excreta. The affected fruits, give offensive smell and lose their market value.

Castor capsule borer, *Dichocrosis punctiferalis* (Guenée) – a primary pest of castor, has been recorded damaging various fruits, including tamarind, but only as a minor pest. Eggs are laid on developing fruits. On hatching the larvae bore inside the fruits and feed on developing seeds. Pupation takes place in silken cocoons inside the bored seeds.

Phycita orthoclina Meyrick is another minor pest of tamarind reported from Kerala (Oommen, 1962). Eggs are laid singly on the fruit pulp through the cracks often found on fruit pods. On hatching the caterpillars feed on the pulp. They also construct silken webs and galleries within which they live and pupate. Besides feeding damage, the cast skins and excreta of these caterpillars also get mixed up with the pulp – thus making the entire pulp unfit for human consumption. Eggs are broadly oval (0.6 × 0.4 mm) and flat. Full grown caterpillars are cylindrical in shape, 8 to 10 mm long and pink in colour. Pupae are 8 to 10 mm long and yellowish-brown. During Summer the egg, larval and pupal stages last for 4 to 5, 17 to 40 and 6 to 8 days respectively.

Cryptophlebia (*Argyroploce*) *illepida* (Butler) – a pest of *bael*, *ber*, citrus, litchi and wood apple has also been reported attacking tamarind (Usman and Puttarudriah, 1955). The caterpillars bore into the fruits and feed on inner contents, filling the hollowed seeds with their excreta on which some fungi often grow. The entry holes on the fruits are plugged with the excreta. The pest is active from August to November. Adults are small blackish-brown moths with brownish wings and a dark brown spot at lower margin of forewings; wing expanse is 15 to 18 mm.

Aphomia gularis Zeller – a major pest in South India attacks the ripening pods on trees during January to April and the infestation continues in the stored tamarind. Full grown caterpillars are stout, 16 to 18 mm long and greyish in colour with small brownish spots. Pupae are 8 to 10 mm long and yellowish-brown. Moths have dark grey forewings with minute blue black specks and a distinct blue black spot in the centre; hind wings are light grey. Wing expanse is 17 to 24 mm, females being bigger than males. Incubation, larval and pupal periods last for 4 to 5 days, 7 to 11 weeks and 2 to 3 weeks respectively.

Assaria albicostalis Walker is a minor pest of tamarind. The caterpillars bore inside the fruits and feed on pulp within. They also construct loose silken galleries on the pods and occasionally bore into the seeds as well. Besides tamarind, it has been recorded boring dry mango seeds. Full grown caterpillars are about 15 mm long, apple green in colour having fine short setae arising from minute pale brown wrats. Moths are medium sized, purplish-brown in colour with forewings greyish-brown and hind wings semi-hyaline; wing expanse is 22 to 25 mm and 25 to 30 mm in case of males and females respectively.

Oecadarchis species is another minor pest of tamarind reported from Kerala. Its caterpillars bore into ripening fruits during January-February and make loose silken galleries that are often found full of excreta. The caterpillars are slender, elongate, 7 to 8 mm long and greenish-white in colour. Moths are small with yellowish-brown forewings having light brown spots and pale yellowish hind wings with fringed margins (Joseph and Oommen, 1960).

Araecerus suturalis Boheman has been reported breeding on fruits of *gular* and various forest trees, dead inflorescences of mango and dry stems of papaya (Beeson, 1941). It has also been recorded from South India on dry plantain and tamarind fruits (Joseph and Oommen, 1960). Grubs and beetles bore into the fruits and feed within. Full grown grubs are thick, fleshy, convex in shape dorsally, about 4.0 mm long and pearly white in colour. Pupae are naked about 4.5 mm long, slightly curved and whitish in colour. Adult beetles are stout, 5 to 7 mm long and dark brown.

Besides, grubs of *Caryoborus gonagra* (Fabricius), *Alphitobius laevigatus* (Fabricius). *Echocerus maxillosus* Fabricius and *Uloma* species have also been reported burrowing the ripening pods of tamarind; their feeding often continues in the stored products as well. Except *C. gonagra*, which is a major pest, of dried tamarind, in South India and Bangladesh, others are of minor importance.

3
PESTS OF SUBTROPICAL FRUITS

CITRUS

CITRUS spp. are the most cherished and highly prized fruits throughout the tropical and subtropical regions of the world (40° latitude North and South of equator). Most of the wild species as also citron (*Citrus medica* Linnaeus), lemon (*C. limon* Burmann), rough lemon or *khatti* (*C. jambhiri* Lushington), seville or sour orange (*C. aurantium* Linnaeus), sweat lime (*C. limettioides* Tanaka) and wild orange (*C. indica* Tanaka) are indigenous to India; kagzi lime (*C. aurantifolia* Swingle) and pomelo (*C. grandis* Osbeck) came from South-East Asia; mandarin (*C. reticulata* Blanco) and sweet orange or malta (*C.sinensis* Osbeck) from South China and only grapefruit (*C. paradisi* Macfadyen) is from West Indies (New World). These fruits are a rich source of vitamin C and also contain vitamin P which keeps the small blood vessels in our body in a healthy condition and helps in assimilation of vitamin C (Singh, 1969).

In India, citrus is commercially grown in over 140,000 hectares which is about 15% of the total area under citrus in the world. The main citrus growing states in India are, Andhra Pradesh (29,000 hectares), Maharashtra (26,000 hectares) and Punjab (21,000 hectares). Ebling (1959) has listed as many as 823 species of insects and mites reported on citrus trees around the world. In India only 250 species of insects have been reported on various citrus species (Fletcher 1921; Pruthi and Mani 1945; Wadhi and Batra, 1964; Nayar *et al.*, 1976). Of these only a few pests are of major importance that cause regular heavy loss, namely, lemon butterflies, citrus leaf miner, citrus psylla, whiteflies, scale insects and mealy bugs. Besides, bark eating caterpillars, bark boring beetles, fruit sucking moths and fruit flies also cause serious damage but occur occasionally. Among minor pests mention may be made of aphids, thrips, hairy caterpillars, lea eating beetles and weevils, blossom midges, flower moths, fruit sucking bugs, rind borers, pomegranate butterfly, wasps, termites, etc.

LEMON BUTTERFLIES

Papilio spp. or swallow-tail butterflies are amongst some of the most beautiful butterflies found all the year round in gardens and orchards visiting various flowers but causing no damage. The caterpillars of course feed on foliage and cause some economic loss. Eight species of these papilionids have been reported feeding citrus leaveson in India, namely, *Papilio demoleus* Linnaeus* *P. helenus* Linnaeus, *P. h. daksha* Moore, *P. polytes* Linnaeus, *P. polyctor* Boisduval, *P. polymnestor* Cramer, *P. machaon asiatica* Ménétriés, *P. memnon* Linnaeus and *P. protenor* Cramer. Of these only the first one is the major pest of citrus, others are of minor or no importance as these are either sporadic in occurrence or confined to certain pockets.

Papilio demoleus described originally by Linnaeus (1758) is widely distributed from North Australia to Arabia, including, Iran, Pakistan, India, Sri Lanka, Bangladesh, Burma, China, Taiwan, South-East Asia, Indonesia and Philippines. It attacks almost all *Citrus* species but prefers *C. sinensis* and *C. grandis*. Besides citrus, it has been recorded on, *bael* (*Aegle marmelo* Correa*)*, *ber* (*Zizyphus* spp.), wood apple (*Feronia* species) and some ornamental and medicinal plants. These are found throughout the year, though rare during Winter. Their attack is more pronounced in nurseries and young plantations where the seedlings and trees may be completely defoliated.

The nuptial flights and matings are rather complicated; after which the female butterflies hover swiftly over their host plants laying one or two eggs at a time on dorsal surface of leaves, sometimes even on ventral surface or leaf stalks. A female lays 75 to 120 eggs in 2 to 5 days during Summer and as many as 180 eggs in 3 to 6 days in Winter. On hatching, the young larvae first eat the empty egg shells and then start feeding voraciously on leaves. Usually they are found on dorsal surface of leaves feeding by biting and gnawing the leaf lamina from edges inwards. The fully fed caterpillars, remain motionless in an exposed condition.

Eggs are pale yellow, tough, smooth, round and big (1.0 to 1.25 mm in diameter). Young larvae are tuberculate, blackish-brown with milky white conspicuous markings and are often mistaken for birds' excreta. Full grown caterpillars are 28 to 35 mm long, brilliant rich green in colour, smooth and velvety (without tubercles) and have dusky brown oblique bands on lateral abdominal segments that do

---

Papilio demodecus Esper – a closet relative occurs on *Citrus* spp. in Africa.

not meet on dorsum. The caterpillars on being disturbed throw out a bifid, red, Y-shaped osmaterium from behind the head which also emits strong odour. Pupae are 25 to 32 mm long and of three different colours – parrot green, dull yellowish-brown (dry straw) or dark grey. The colour variation is not associated with the sex of the adults (Ghosh 1914). Adult butterflies are beautifully patterned, body ventrally yellow and wings ornamented with yellow and black markings; hind wings without tail-like projection and having a brick red oval spot near upper margin and a blue spot near lower margin. Wing expanse is 80 to 100 mm.

Egg, larval. prepupal and pupal durations last for 3 to 8, 11 to 40, 2 to 3 and 8 to 36 days respectively. Life-cycle is completed in 3 to 6 weeks in Summer and takes 13 to 15 weeks during Winter. Overwintering is in pupal stage. There are four to five overlapping generations in a year in North India and five to six in the plains of South India, where there is no hibernation.

Papilio polytes is common in plains of India and has also been reported from West China and Indonesia. Caterpillars are rich glaucous green dorsally and slightly yellowish laterally. There is sexual dimorphism in the butterflies. Male butterflies are black having forewings with white marginal spots and hind wings with a white post dorsal band. Female butterflies have three distinct variable forms – one resembling male butterflies; the second having

Papilio polytes Linnaeus

white discal patches on hind wings and the third with red discal patches on hind wings. Wing expanse is 70 to 90 mm.

Papilio helenus is common in South India and Assam. It has also been reported from Burma, Thailand and Malaysia. These caterpillars are yellowish-green and have oblique bands on lateral sides which unite dorsally. The butterflies have rich velvety brownish-black wings. The hind wings have three white spots forming a conspicuous upper discal patch. The patch is more prolonged anally in females than in males. Wing expanse is 110 to 120 mm.

Papilio polyctor is less common than *P. helenus* and *P. polytes*. It is found at the foot of Himalayan range; from Pakistan to Burma through Kashmir hills. The pest is more common during rainy season. Caterpillars are green with four yellowish-grey oblique stripes on abdomen. Butterflies have wings which are dull black dorsally and chocolate brown ventrally; forewings have broadsubmarginal golden green band and hind wings a prominent bluish-green patch. Wing expanse is 90 to 120 mm. Sexual dimorphism exists in the adults.

Papilio polymnestor has been recorded from Bengal, Maharashtra and Karnataka. Caterpillars are rich velvety dark grass green dorsally with a yellowish shade above segments 2 to 5 and are covered with minute erect hairs. Butterflies have forewings with pale blue discal band that is obsolescent anteriorly; hind wings have distal area pale blue and 5 irregular small red patches at the base. Wing expanse is 120 to 140 mm.

Papilio machaon asiatica is common in Kashmir and Pakistan. These butterflies are smaller in size with dull black wings. Hind wings have creamy yellow area and the wing expanse is 75 to 95 mm.

Papilio memnon – the Oriental species, is widely distributed from South Japan to North India and Sri Lanka, including Borneo and Burma. Caterpillars are dark velvety green with bluish tinge having two lateral oblique bands which meet dorsally. Male butterflies have forewings which are deep indigo dorsally and pale black ventrally; hind wings are dull opaque black. Female butterflies are polymorphic and exhibit great variety of pattern; forewings are sepia coloured dorsally and hind wings black with 5 to 7 white or yellowish discal patches. Wing expanse is 120 to 150 mm.

Papilio protenor is not a common species. It has been reported from Kashmir to Kumaon hills, Sikkim and Assam only during

Winter. It is also found in Burma, West China, Thialand, etc. The caterpillars are green with a conspicuous band on thorax and two oblique bands on lateral sides of abdomen which do not meet on the dorsum. Butterflies have velvety indigo wings; forewings paler than hind wings and latter with a broad pale yellowish-white sub-costal stripe. Like *P. demoleus*, the hind wings are without swallow-tail projection. Wing expanse is 100 to 130 mm.

To control lemon butterflies, hand-picking of various stages of the pest and their destruction has been suggested. This is very useful in mitigating the pest problem specially in nurseries and new orchards. In case of severe infestation, spray with 0.04% monocrotophos or quinalphos or 0.05% chlorpyriphos or phosalone. Spraying the entomogenous fungus, *Bacillus thuringiensis* Berliner or nematode DD-136 strain also gives very high mortality of the caterpillars.

Besides *Papilio* species, there are other lepidopterous larvae often found feeding on leaves of various *Citrus* species. These include, *Psorostica zizyphi* (Stainton), *Latoia lepida* (Cramer), *Tortrix epicyrta* (Meyrick), *T. pensilis* (Meyrick), *Ulodemis trigrapha* Meyrick, *Chilades lajus* (Cramer), *C. putli* Frey, *Clania crameri* Westwood, *Jamides bochus* Cramer, *Tarucus theophrastus* (Fabricius), *Euproctis fraterna* Moore, *Macroglossum belis* (Linnaeus) and *Notolophus postica* (Walker). Most of these are sporadic pests of citrus.

Microglossum belis (Linnaeus)

Dusting with 5 to 10% BHC has been found effective in controlling these caterpillars.

CITRUS LEAF MINER

Phyllocnistis citrella Stainton is found all over the Orient from Africa to Australia (CIE map No. A–274). Almost all *Citrus* spp. are attacked though the varieities with succulent leaves and thin cuticle are preferred (Latif and Yunas, 1951). *Citrus medica* and *C. sinensis* which have succulent leaves are heavily damaged while *C. aurantium* having thick and coarse leaves are least damaged (Pandey and Pandey, 1964). Besides *Citrus* spp., it has various other hosts also, including *bael*.

The caterpillars attack only young and tender leaves; old leaves are avoided. The larvae make serpentine mines in the leaves feeding on epidermal cells of the leaf leaving behind the remaining leaf tissues quite intact. The mining larvae feed actually more on sap than on solid tissues of the leaf (Pruthi and Mani, 1945). The mined leaves turn pale, get distorted and may dry up. Occasionally, when the pest population increases rapidly, the succulent tender shoots specially those of *C. sinensis* and *C. medica* are also mined. As a result the growth of young trees is retarded considerably and in the nurseries, saplings may even die away. Attack by this pest also encourages the development or citrus canker (Huston and Pinto, 1934). The pest is active throughout the year except during severe cold (December to February); though much of the injury is caused early in Spring and the population of the pest decreases during hot summer months.

Eggs are minute, about 0.3 mm long, broadly oval, flattened and greenish-yellow in colour. Full grown caterpillars are cylindrical, about 5 mm long, apodous and dull greenish-yellow in colour. Adults are tiny, silvery white moths with heavily fringed wings; fore wings having brown stripes and a prominent black spot near apical margin and hind wings pure white. Wing expanse is 4 to 5 mm. Pairing taken place mostly at night or at dawn, 14 to 24 hours after the emergence of female moths. Eggs are laid singly and are attached to leaves and twigs. A female lays 36 to 76 eggs in 2 to 6 days. The eggs hatch in 2 to 10 days. Larval duration is 5 to 10 days.

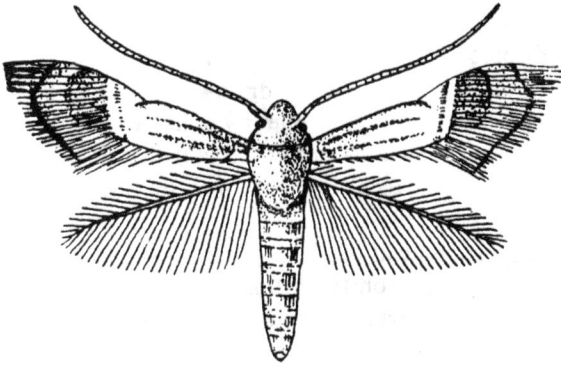

Phyllocnistis citrella Stainton

The pupae are found in white cocoons lying near the margin of mined leaf, the edge of which is turned over and the inside lined with silken webbing. Pupal period is 5 to 25 days and the entire life-cycle occupies 20 to 60 days depending upon the climatic conditions. There are 9 to 13 overlapping generations in a year

As the larvae are inside the mines, these cannot be easily killed by insecticidal applications. For effective control, prune heavily the affected parts during Winter and burn the same. Fumigation with hydrocyanic acid gas is quite effective but requires great care and technical aid. Spraying with 0.2% BHC or DDT (Atwal and Josan, 1962) or 0.25% neem cake extract (Nayar *et al.*, 1976) has also been suggested.

CITRUS PSYLLA

Diaphorina citri Kuwayama is found throughout tropical and subtropical Asia and the Far East. It has been recorded from Pakistan, India, Sri Lanka, Bangladesh, Burma, Malaysia, Indonesia, Philippines, Taiwan, southern China, etc. In India, it is a serious pest in Punjab, Delhi and Haryana and attacks almost all *Citrus* species.

Eggs are laid singly or in clusters in the folds of half opened leaves or pushed inbetween flower-buds and stems or petioles of leaves and axillary buds. On hatching, the nymphs crowd on terminal shoots and buds and suck the cell sap. After feeding for 2 to 3 days, the nymphs disperse and attack tender leaves as well as the older foliage. The affected parts gradually dry away. Though both nymphs and adults suck the plant sap, the nymphs are more destructive than adults. The insects also inject certain toxins along with the saliva, as a result the adjacent branches that are not actually attacked also dry and die away. Besides, these bugs also excrete honeydew resulting in the superficial black coating on the affected parts and stunted growth. In addition to all this damage, the psylla also acts as a vector of citrus virus causing citrus decline (Bindra, 1966; Capoor *et al.*, 1967).

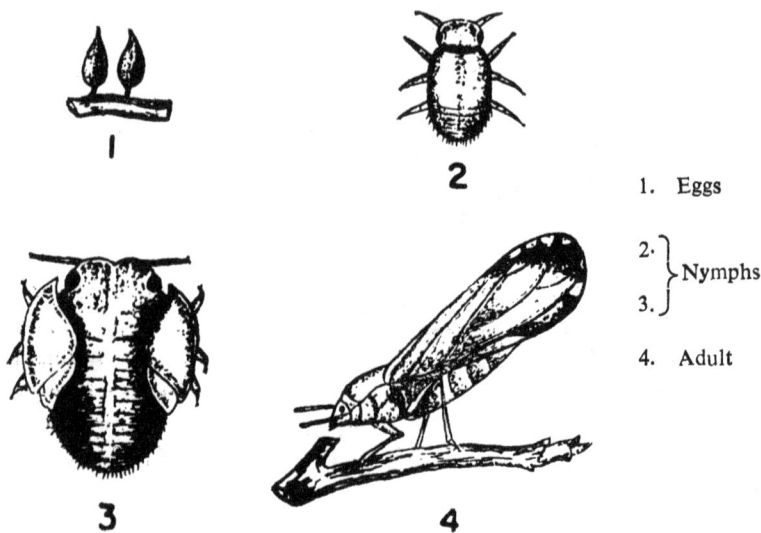

1. Eggs

2. ⎫
3. ⎭ Nymphs

4. Adult

Diaphorina citri Kuwayama

The pest is most active during Spring; population decreasing with the rise in temperature, but increases rapidly soon after the heavy monsoon showers (August) and the activity continues till the temperature again falls down substantially (December-January).

Eggs are orange coloured and almond-shaped. Full grown nymphs are flattened, circular in shape and yellowish-orange in colour. Adults are brownish bugs, about 2 to 3 mm long and may be seen sitting on ventral side of leaves with their heads almost touching the leaf surface and rest of the body raised up. Mating takes place soone aftr emergence of the adults and the oviposition follows almos immediately. A single female may lay as many as 800 eggs in about two months. The eggs are anchored in the tissues by means of short stalks. Incubation period lasts for 4 to 6 days in Summer extending upto 22 days during Winter, whereas nymphal period is 9 to 12 days during Summer and 34 to 37 days in Winter (Mangat, 1966). Developmental period from egg to adult takes 15 days in Summer and 47 days in Winter and there are 9 to 10 overlapping generations in a year. Overwintering is in adult stage and the overwintering adults may live as long as 189 days (Husain and Nath, 1927).

To control this pest spray with 0.05% malathion or 0.1% carbaryl (Sethi, 1967) or 0.025% phosphamidon (Atwal and Varma, 1968). Spraying with nicotine sulphate (1:600) or 0.05% diazinon or endosulfan is also effective (Butani, 1973 d). Bindra *et al.* (1970) reported that soil application with dimethoate @ 16 ml a.i. per tree was effective in checking the infestation of this pest.

WHITEFLIES

Of the thirty species of Aleyrodidae recorded on *Citrus* spp. throughout the world, atleast 15 have been reported from India. These include, *Aleurocanthus citriperdus* Quaintance and Baker, *A.husaini* Corbett, *A.spiniferus* (Quaintance), *A.punjabensis* Corbett, *A.woglumi* (Ashby), *Aleurocybotus setiferus* Quaintance and Baker, *Aleurolobus citrifolii* Corbett, *A.marlatti* (Quaintance), *Aleurothrixus flossosus* Maskell, *A.howardi* (Quaintance), *Aleurotuberculatus murrayee* (Singh), *Bemisia giffardi* (Kotinsky), *Dialeurodes citri* (Ashmead), *D.citrifolii* (Morgan) and *Dialeurolonga elongata* (Dozier). Of these only two species are of major importance, namely, *A. woglumi* and *D. citri* – the rest are sporadic pests of minor importance.

The nymphs and adults of these insects reduce the plant vigour by sucking large quantities of cell sap. Severely infested leaves turn pale green and gradually become pale brown, get badly curled and are even shed. The nymphs also produce copious quantity of honeydew on which sooty mould develops covering the foliage with superficial black coating and interfering with photosynthetic activity of the plant. As a result, the growth of the tree is stunted and the affected trees produce few blossoms, most of which are shed and even the fruits, that are formed, are inspid.

Aleurocanthus woglumi, commonly known as citrus blackfly, has been reported from tropical Asia, East Africa, Central America, West Indies, etc. (CIE map No. A-91). Its original home is India. It is primarily a pest of citrus, but it has been recorded on more than 70 cultivated and wild plants including, avocado, grapevine, guava, mango, pear, plum, pomegranate, quince and sapota.

Eggs are laid in spiral pattern on ventral side of leaves. There may be 35 to 50 eggs in a batch and a single female may lay little over hundred eggs in her life time. The eggs are attached to the leaf by a short pedicel situated near the posterior end. Eggs are oblong

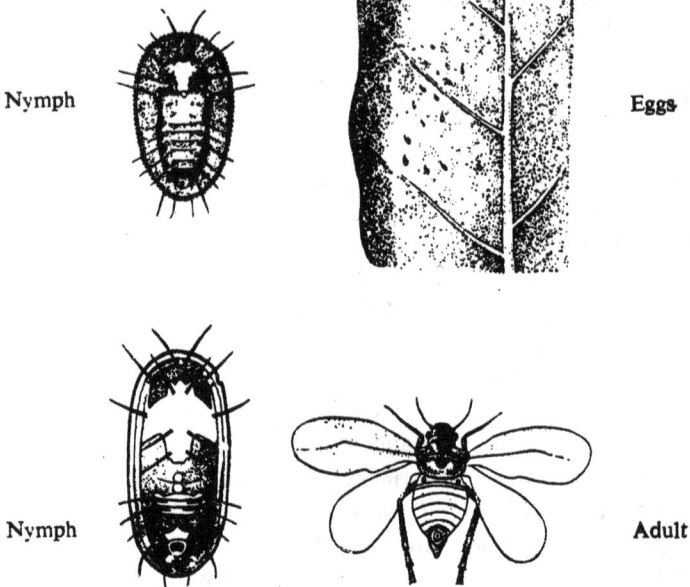

Nymph

Eggs

Nymph

Adult

Aleurocanthus woglumi (Ashby)

with rounded ends; creamy-white when freshly laid becoming brown and finally black. Crawlers are flattened, oval, scale-like in appearance, dark brown to shiny black in colour and conspicuously spiny, borderd by a white fringe of wax. When fully grown, these are 1 to 2 mm long. Pupae are black. Adults are about one mm long ($\male$ 0.8 mm and $\female$ 1.2 mm). When freshly emerged, head and thorax are bright red, eyes reddish-brown while antennae and legs are whitish (Dietz and Zetek, 1920). Within 24 hours of emergence, the adults become covered with a heavy pulverulence that gives the insects a slaty-bluish look. Wings have black patches on whitish background. Incubation period is 9 to 10 days whereas nymphal development is completed in 45 to 115 days depending upon climatic conditions (Pruthi and Mani, 1945). There are usually four to five generations in a year.

Aleurocanthus husaini is a minor pest of citrus reported from Punjab. The species is closely related to *A. woglumi* and was originally described by Corbett (1939) from specimen collected by Husain from Kulu valley. Adults are about one mm long having pale yellow body maculated with dark spots; wings are also maculated and slate-grey in colour. Egg, nymphal and pupal periods occupy 7 to 12, 26 to 52 and 64 to 108 days respectively (Nair, 1975). There are only two generations in a year.

Aleurocanthus spiniferus – the spiny whitefly, is common in India, Burma, Japan, Malaysia, Indonesia, Philippines, West Indies and West Africa (CIE map No. A-112). Besides *Citrus* spp., it has also been recorded in India on grapevine, pear, etc. Newly hatched nymphs are yellowish-green which gradually become brown and ultimately black. Adults are yellowish-orange to dark orange with darkened head and thorax and slate-grey wings. The wings have whitish spots and are reddish at the margins. In Japan, egg, nymphal and pupal periods last for 11 to 27, 21 to 84 and 12 to 60 days respectively with four generations in a year (Kodama, 1931); hibernation takes place in last nymphal stage. In India, the pest is active throughout the year, without any hibernation period and has several overlapping generations in a year.

Aleurolobus marlatti is yet another species found damaging citrus in India, Sri Lanka, Indonesia and Japan. It was earlier confused with *Aleurocanthus spiniferus*. This is a minor pest of *Citrus* species. In addition to *Citrus* spp., it is also known to attack several other plants, like, mulberry in India and *Ficus* spp. in Sri Lanka. All the stages of this pest are found in nature from March to December and only pupae are seen during January-February. Egg, larval and pupal stages last for 5 to 14, 12 to 35 and 36 to 77 days respectively (Nair, 1975).

Dialeurodes citri – the citrus whitefly, is another species of Indian origin widely distributed in South and East Asia as also in tropical

America (CIE map No. A-111). It is polyphagous, having a wide range of host plants including, banana, cherry, *jamun*, persimmon and pomegranate.

Eggs are oval and pale yellow; nymphs are pale yellow with purple eyes whereas pupae are pale yellow with orange tinge in the centre. Adults are 1.0 ($\male$) to 1.5 ($\female$) mm long, having pale yellow body and dark red medially constricted eyes. A female lays 150 to 200 eggs on underside of tender leaves. Reinking (1921) counted 20,000 eggs on one leaf. Eggs hatch in 10 to 20 days while nymphal and pupal stages last for 25 to 71 and 114 to 159 days respectively (Husain and Khan, 1945). Longevity of males and females is 2 to 3 and 7 to 10 days respectively and there are two generations in a year (under Punjab conditions). However, in Japan, where climate is much milder, three generations a year have been reported. In India, adults are found in abundance during Spring and early Autumn while all the stages of the pest are found during March to August but only pupae are present during October to February.

D.citrifolii is another but similar species commonly found in North India. It has cloudy wings and lays black reticulated eggs. Its development is slower than that of *D. citri*.

Avoid close planting and water-logging to prevent the attack of whiteflies. In case of localised attack, clip off and destroy the twigs bearing affected leaves; if infestation is severe, spray with 0.03% dimethoate, monocrotophos or phosphamidon.

COCCOIDS

More than 50 species of mealy bugs and scale insects have been recorded (appendix 7.1) feeding on various parts of citrus trees including roots, branches, leaves, flowers and fruits. They suck the cell sap and devitalize the plants. Most of these occur sporadically and cause only minor loss but atleast half a dozen species are of regular occurrence and if not checked in time cause collosal loss. These are, *Icerya purchasi* Maskell, *Planococcus citri* (Risso), *Aonidiella aurantii* (Maskell), *Chrysomphalus ficus* Ashmead, *Cornuaspis beckii* (Newman), *Orthezia insignis* (Browne nec. Douglas), etc.

Icerya purchasi, popularly known as citrus fluted scale or cottony cushion scale, is ubiquitous in tropical and subtropical regions (CIE map No. A – 51). It is native of Australia and was introduced in California in 1868; it now occurs in all the citrus growing areas of the world. Though a polyphagous pest; *Citrus* spp. are the main hosts. It has been recorded damaging almond, apple, apricot, fig, grapevine, guava, mango, peach, pomegranate, walnut, etc. Newly hatched larvae are very active and move about till they settle on the midrib or larger veins from which they draw their nourishment. The infested leaves become yellowish and fall prematurely. Young and tender shoots when heavily infested turn blackish-brown and die away. In nurseries the entire seedlings may be killed by the attack of this pest. The insects also excrete copious quantity of honeydew on which sooty mould grows.

Eggs are elongated oval (0.6 × 0.2 mm), smooth and pinkish. Newly hatched nymphs are also elongated oval (0.75 × 0.3 mm) having prominent eyes and long, slender antennae and legs. Soon after fixation on host, the body becomes covered with white waxy plates. Full grown larvae are broadly oval (3.0 × 1.5 mm), reddish-brown to brown to brick red in colour, having stout antennae and legs, and a few conspicuous patches of wax filaments on dorsal surface. Microscopic characters of various larval instars have been described by Bodenheimer (1951). Pupae are elliptical in shape, orange red in colour with dark purple eyes and yellowish-brown wing-pads and legs. Adult females (hermaphrodite) are broadly oval (4.5 × 3.3 mm), brown to reddish-brown, dorsal surface convex and ventral flat, with as many as 46 lateral wax bristles and body segmentation indistinct. Ovisac is white, as broad as the adult body; dorsally convex and distinctly fluted. It is the convex fluted mass of white wax underneath the scale that attracts the attention and characterises this insect (Talhouk, 1969). Males are slender, about 3 mm long, reddish-purple with long dark brown antennae and two short fleshy appendages with long bristles at caudal end; forewings are shiny metallic blue and halters foliated with 2-3 hooks at the tip; wing expanse is 5 to 7 mm. Life-cycle and bionomics have not been studied so far in India. In Japan, egg and larval stages last for 21 to 27 and 59 to 92 days respectively while in Palestine, preoviposition period lasts for 12 to 17 days, egg stage 16 to 35 days, larval 46 to 58 days and total life-cycle 79 to 98 days (Bodenheimer, 1951).

An effective control of this insect has been obtained by releasing its predator, the lady bird beetle, *Rodolia cardinalis* (Mulsant), introduced from Australia. *I.purchasi* and *R.cardinalis* are often quoted as an example of successful biological control all over the world.

Citrus mealy bug, *Planococcus citri* is almost completely pantropical in distribution extending well into subtropical regions (CIE map No. A–43) and is also found in greenhouses in temperate countries. Its main hosts are *Citrus* spp,, specially limes, lemons and oranges. It has also been recorded feeding on fig, pine apple, sapota, etc. Ayyar (1930) has listed about 50 host plants of this mealy bug whereas Bodenheimer (1951) has reported more than 26 hosts from Palestine, including grapes and melons. The pest attacks leaves, tender branches, fruits (at the base near the fruit stalk) and even the roots. They not only suck the sap but also excrete honeydew on which there is rapid growth of sooty mould. As a result, ultimately, the growth of the tree is arrested and the fruits fall off prematurely.

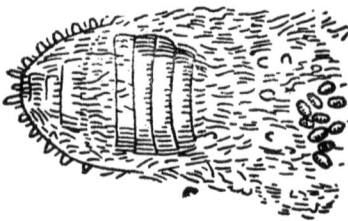

Ovipositing female   Neonate larvae

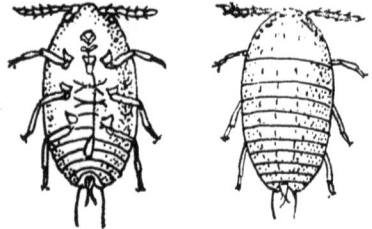

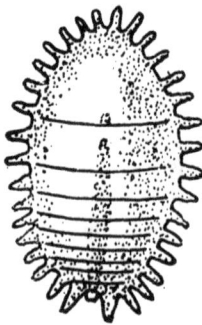

Adult ♀ covered with wax secrction and bare   Adult ♂

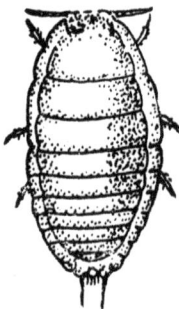

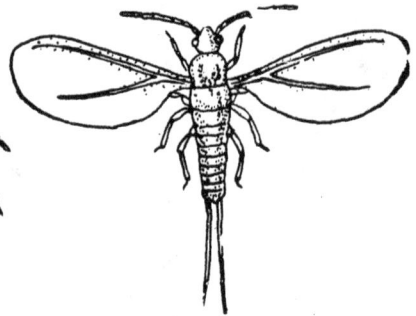

Planococcus citri (Risso)

Eggs are ellipsoidal, 0.3 to 0.4 mm long, and light creamy-yellow.Newly hatched nymphs are pale yellow having no waxy coating yet but slowly and gradually the waxy coating appears, covering the nymphs entirely. Adult females are slightly elongate ovate, 5 to 7 mm long, covered with white mealy wax secretion and having 34 wax covered appendages round the entire periphery. Eggs are deposited in loose cottony masses. A female lays 600 to 800 eggs in about a fortnight. Incubation period is one to two weeks in Summer. Female larvae undergo three moults and males four moults; there are usually two to three generations in a year (Pruthi and Batra, 1960).

Efficient control of this pest is rather difficult. Even fumigation with hydrocyanic acid gas remains rather unsatisfactory, though applied repeatedly. The killing concentration of mealy bugs is higher than the tolerence of the trees (Bindra, 1967). Spraying 1% diesel-oil is effective but expensive. A parasitic wasp, *Prospaltella perniciosi* (Tower) has been successfully established in Kashmir and gives fairly effective control of this mealy bug. Soil treatment with 5% aldrin heptachlor or chlordane has been effective in checking the pest, population to some extent.

Citrus red scale, *Aonidiella aurantii* is native of India and has now spread over all the citrus growing countries of the world, except West Africa (CIE map No. A–2). The pest thrives best in semi-arid climate. Quayle (1938) has listed 86 host plants of this scale from different parts of the world. In India, it is found all over the country infesting, *ber*, banana, citrus, fig, guava, *jamun*, mulberry, peach, pear, ornamentals, etc. Among *Citrus* species, the pest prefers malta or sweet orange *(C. sinensis)*, seville orange (*C.aurantium*), grapefruit (*C.paradisi*) and pomelo (*C.grandis*). The pest usually attacks leaves and tender shoots but in case of severe infestation even the fruits and sometimes the trunks of trees are not spared. Dispersion is usually by wind – the crawlers are blown off from tree to tree and from one orchard to another. The peculiar method of feeding i.e. injecting the toxic saliva in the cell sap before ingestion, often severely affects the host trees and occasionally the young trees are killed due to injected toxins. Branyovitis (1953) observed that these insects spend more time in moulting than in feeding. Besides, they feed only on cell sap in the parenchyma tissues and hardly ingest any sap from the phloem. As such even systemic insecticides do not have any appreciable effect in controlling these insects.

The various immature stages as well as adults of this pest are found all the year round though in different proportion in different seasons. There is no true diapause but the rate of growth slows down considerably during Winter (December to February) and the pest is most active from August to end of October. Morphology and binomics of the pest have been studied more in other countries than in India. Berlese and Leonardi (1896) were the first to describe the immature stages, The scale of females is more or less circular and flattened with raised centre while that of males is elongated. The scales appear reddish, which is actually the colour of insect seen through the scale covering (Ebling, 1959). Adult females are flat, legless, circular in shape, about 2 mm in diameter with conspicuous antennae and rostrum about twice the body length. Normally the females are reddish in colour but when gravid, these become kidney-shaped and turn yellowish-orange in colour. Males are somewhat elongated in shape and winged. They emerge from beneath the scales, fly about and copulate with virgin females (Talhouk, 1969). Females are ovo-viviparous and produce 2 to 3 nymphs per day for over two months. Sometime eggs are also laid. There is no parthenogenesis. Young nymphs remain under protection of their mother for one to three days, then crawl about a bit in search of suitable spot and within a few hours settle on leaves, twigs or fruits. Once a crawler settles, it seldom leaves that place. It starts secreting the wax that soon covers its entire body.

Fumigation with HCN gas has been found to be very effective and is extensively used in America. Since this method requires expert handling and sufficient technical knowledge, it is not being practised in India.

Chrysomphalus ficus Ashmead is native of Oriental region and is by now found in almost all tropical and subtropical countries (CIE map No. A-4). In temperate region it is restricted to greenhouses. Though widely distributed, it is a pest of economic importance only in hot and dry places. It is polyphagous and according to Bodenheimer (1951), 105 hosts are known from Cuba (West Indies) alone. In India, besides *Citrus* spp. it has also been recorded on almond, apple, avocado, banana, *ber*, coconut, date palm, grapevine, guava, mango, mulberry, pomegranate, etc. A female lays 50 to 150 eggs in her life time, 5 to 6 eggs per day during Summer and 1 to 2 in Winter. Egg production is higher on fruits than on leaves. Newly hatched nymphs are very active for about a day, search a suitable feeding spot and settle there. Soon after that they get covered with scaly substance secreted by them for protection. The pest prefers shade during Summer and sunny side of trees during Winter. Low temperature during Winter has been reported to cause 98% moratlity (Pruthi and Mani, 1945).

Eggs are long, oval (2.0 to 2.5×1.0 to 1.5) and lemon yellow in colour. Neoate nymphs are short and oval (0.25×1.2 mm) in shape and yellow in colour. Adult females are flat, and pear-shaped (1.2×1.0 mm), broadly rounded anteriorly; pale yellow when young, becoming darker and getting reddish tinge with advance in age. Their scaly covering is white, circular (1 to 2 mm in diameter) and nipple-shaped. It is generally worn off in older individuals. Males are 0.8 mm long, orange in colour, antennae and legs being brown and wings whitish; wing expanse being 1.8 to 2.0 mm. The life-history has not been studied in detail in India. Outside India, incubation period is recorded to be one to four days and egg to adult period takes about 6 and 10 weeks in case of males and females respectively (Quayle, 1938). There are atleast five to six generations in a year.

Cornuaspis beckii – a native of tropical America is now found all over the citrus growing regions of the world (CIE map No. A–49). It attacks almost all *Citrus* species, though it prefers sweet orange (*C. sinensis*).

Eggs are elongate (0.25 × 0.1 mm) and pearly white in colour. Neoate nymphs are elongate ovate (0.35 × 0.15 mm), white to pale yellow in colour with two small lobes and 4 spines protruding from posterior end. Pupae (males) are conspicuous, purple in colour with antennae, legs and wing-sheaths free from the body. Adult males are very delicate, less than one mm long, white to pale yellowish in colour, antennae and legs purple; wings are hyaline and wing expanse 2.0 mm. Adult females are elongate (1.3 to 1.5 mm), pale yellowish-white with terminal end reddish fulvous. Total life-cycle of females occupy 50 and 110 days during Summer and Winter respectively with four overlapping generations in a year.

Orthezia insignis, commonly known as lantana bug, has been reported from South India, Sri Lanka, Malaysia, some African countries, USA, Central and South America (CIE map No. A–73). The main host of this insect is lantana, a highly obnoxious weed found in abundance on the hills. The bug keeps this weed under check and attempts are being made to use this bug in developing the biological control of the weed. However, besides lantana, it has also been recorded feeding on a number of economic crops, including *Citrus* species. Both nymphs and adults infest leaves, shoots and fruits sucking the cell sap; but the loss caused is usually minor and goes unnoticed. According to Hill (1975), usefulness of this species, heavily outweighs its nuisance value of damaging flowering trees.

Adult females are olive green in colour with well developed legs and antennae; segmentation of the body is distinct with white waxy plates or laminae at the extremity. The egg-sac is situated between these waxen plates. There are also short waxy processes at the side and double row of similar processes along the back.

Orthezia insingis (Browne)

In India, no control measures are generally adopted against this scale.

APHIDS

Aphids as pests of citrus trees do not cause any major damage but their role as vectors for transmitting tristeza virus disease is most harmful and may result in tremendous loss. The species of aphids reported from India include, green apple aphid, *Aphis pomi* (de Geer), melon aphid, *A.gossypii* Glover, green peach aphid, *Myzus persicae* (Sulzer),black citrus aphid, *Toxoptera aurantii* (Fonscolombe) and brown citrus aphid, *T.citricidus* (Kirkaldy) (*traveresi* del Guerico). Of these only first two are of regular occurrence as pests whereas all except *A. pomi* have proved to be vectors of virus diseases.

Toxoptera aurantii is found in all the warmer parts of the world (CIE map No. A–131) as also in greenhouses in temperate regions whereas *T.citricidus* is predominantly a pest of humid tropical

regions (CIE map No. A –132) and does not occur in Europe,
North Africa and USA. *T. aurantii* is polyphagous though *Citrus*
species are its main hosts while *T.citricidus* is oligophagous and
feeds and breeds exclusively on Rutaceae. Normally, these aphids
attack at the time of flowering but occasionally severe outbreaks
also occur when rainy season is followed by dry weather. Distortion
of young leaves and tender twigs coupled with sooty mould
coating on older leaves are the usual symptoms of attack.

Both the species are similar in appearance. Young ones of
both the species are brown in colour, darker in case of *T. aurantii*
whereas adults, alate or apterous, are all shiny black in colour. The
only obvious diagnostic character is wing-venation – median vein of
forewing has one branch in *T.aurantii* and two in *T.citricidus*.

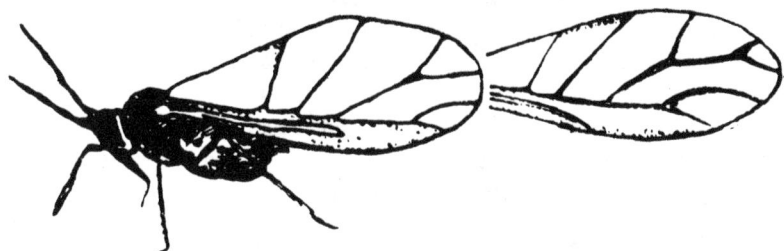

Toxoptera aurantii (Fonscolcmbe)   *T. citricidus* (Kirkaldy)

Reproduction is parthenogenetic. A generation normally takes
6 to 8 days but at 15°C it takes as long as 3 weeks and at 25°C
only 6 days. Development is retarded as temperature goes beyond
30°C. One virginpara can produce upto seven offsprings per day and
over 50 young ones in her life time.

Aphis pomi is a major pest of apple; citrus being one of its main
alternate hosts. Eggs are laid in large number on growing shoots
during Winter. The eggs are hardy and capable of withstanding
extreme low temperatures. The eggs hatch early in Spring and the
aphids attack leaves and tender branches of citrus. The affected
leaves get badly curled but not etiolated. As a result, ripening of
fruits is delayed and quality affected. In case of severe infestation
even the tender fruits are attacked causing premature fall of
unripe fruits.

To control these aphids, spray with 0.02% oxydemeton methyl
or monocrotophos (Singh and Rao, 1976).

THRIPS

Heliothrips haemorrhoidalis (Bouché) is the most common species specially in South India. Other thrips reported damaging citrus leaves, are *Rhopalandrothrips nilgiriensis* (Ramakrishna) and *Trips pandu* Ramakrishna. Besides, *Frankliniella dampfi* Priesner *Ramaswamiahiella subnudula* Karny, *Scirtothrips dorsalis* Hood, *Thrips flavus* Schrank and *T. florum* Schmutz (*hawaiiensis* Morgan) have been reported infesting the flowers, specially those of *Citrus medica*. *Scirtothrips aurantii* Fauré and *S.citri* (Moulton) the recent introductions from Africa and USA respectively, damage the developing fruits.

Heliothrips haemorrhoidalis – the common greenhouse thrip is cosmopolitan in distribution (CIE map No A–732). In temperate regions it is found only in glasshouses. Though a major pest of garden crotons and tea, it has also been reported attacking leaves of citrus, date palm, mango, passion fruit, etc. Lacerating of leaves and puncturing of tissues by several individuals in bleaching of green parts. Besides, nymphs also leave behind faecal globules that appear like black specks all over the leaves. Due to heavy infestation, the leaves get distorted, crinkled and mottled.

Eggs are bean-shaped and about 0.3 mm long. Freshly emerged nymphs are white gradually turning greenish-white and have red eyes. Pupae are yellow. Adults (females) are slender, about 1.5 mm long, body brown with head and throax dark brown, legs

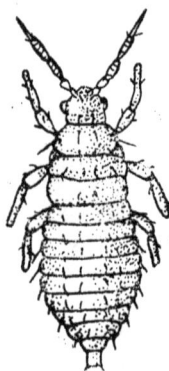 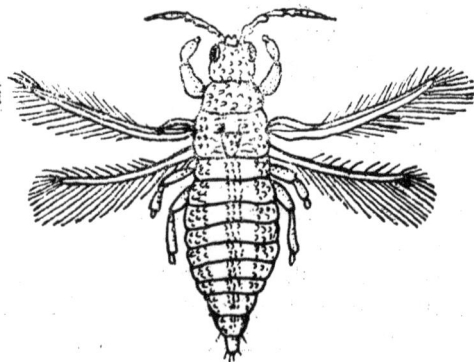

Nymph   Adult

Heliothrips haemorrhoidalis (Bouche)

uniformly yellow and wings yellowish with a median longitudinal pale grey band, paler at the base. Reproduction is parthenogenetic. Eggs are inserted singly in leaf tissues and covered with a drop of excreta. The eggs hatch in 2 to 7 days. The life-cycle is completed in 30 to 32 days at 26-28°C, 40 days at 23-25°C, two months at 20°C (Ananthakrishnan, 1971) and 12 weeks at 15°C (Hill, 1975). In South India, the life-cycle is completed within 20 to 30 days on coffee leaves (Sekhar and Sekhar, 1964).

Ramaswamiahiella subnudula is a polyphagous pest found feeding and breeding in flowers of several host plants, including *Cit us medica*. Adults are small sized, body orange yellow, antennae pale white, legs pale yellow and wings having greyish-yellow infumation.

Thrips flavus is a cosmopolitan species. In India, it is confined to high altitudes and is a major pest of ornamental plants particularly at hill-stations, besides it attacks apple and citrus. Body colour is uniformily yellow, antennae are dark grey distally and wings pale yellow.

Thrips florum is in abundance in flower buds of *Citrus* spp. specially *C. aurantia* and *C.medica*. It has also been reported damaging apple and banana. As a result, the affected flower-buds become smaller, the petals shrink and show feeding scars, ultimately fruit setting is adversely affected.

Scirtothrips aurantii, South African citrus thrip, is widyel distributed in Egypt, Sudan and South Africa (CIE map No. A-137). It is mainly the pest of *Citrus* species, though it also attacks various *Acacia* spp. Eggs are inserted into soft tissues of fruits. On hatching, nymphs usually feed on small fruits near the peduncle failing which they can feed on tender leaves as well. As a result a ring of scaly brownish tissues is formed on the fruits round the peduncle or irregular areas of scarred tissues may be seen on fruits. Eggs are bean-shaped and about 0.2 mm long. Nymphs are cigar-shaped, 0.6 to 0.8 mm long and orange yellow in colour. Adults are 0.8 to 1.0 mm long and reddish-orange in colour. Males are rare, thus reproduction is parthenogenetic. Egg period is about a week; nymphal and pupal (including pre-pupal) periods are one to two weeks each while adults may live for several weeks (Hill, 1975).

Scirtothips citri is a major pest of *Citrus* species in California (CIE map No. A–138). It has been reported from India causing blemishes on fruit surface (Dutta, 1966). The affected fruits lose their market value.

To control the thrips, spray with 0.2% BHC or DDT or 0.03% dimethoate, endosulfan, phosphamidon or thiometon.

BARK EATING CATERPILLARS

Indarbela quadrinotata (Walker) and *I. tetraonis* (Moore) occasionally occur in large number causing severe damage specially in neglected and unclean orchards. Older trees are more prone to their attack than the young ones. Eggs are laid on bark of trees during May-June. Freshly hatched caterpillars nibble the bark of trees for couple of days making irregular galleries, then bore inside the trunk or main branches, feed within, till end of December then pass through a quiescent stage and pupate in April. The feeding of the caterpillars causes interruption in translocation of cell sap, which adversely affects the growth and fruit setting capacity of the tree. Large webby masses comprising of chewed wooden particles and faecal matter are conspicuously seen plastered on tree trunks or main branches specially near the forking place.

STEM BORING BEETLES

Three bupretid beetles, *Agrilus grisator* Kerremans, *A. mediocris* Kerremans and *Belionota prasina* Thunberg and eight longicorn beetles, namely, *Chelidonium cinctum* (Guérin-Méneville), *Chloridolum alcmene* Thomson (*argentatum* Dalm.), *Demonax balvi* (Pascoe), *Gnatholea eburifera* (Thomson),*Oberea mangalorensis* Forster, *Peltotrachelus pubes* Faust, *Anoplophora* (*Monohammus*) *versteegi* (Ritsema) and *Stromatium barbatum* (Fabricius) have been reported as pests of citrus. Most of these are of minor importance but some do occur regularly, in certain pockets, causing appreciable loss.

Lime tree borer, *Chelidonium cinctum* is a major pest of citrus in South India. Initially the grubs bore and feed in the superficial layer and make corkscrew-shaped (spiral) tunnels round the twigs. This results on breaking of the bark from woody portions. Later, the grubs enter the pith region and cut off the sap flo·· which

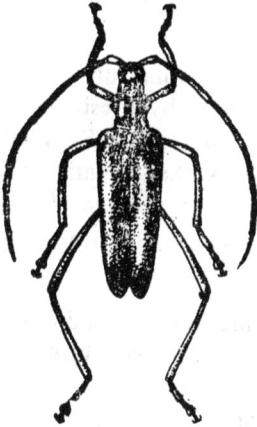

Chloridolum alcmene Thomson

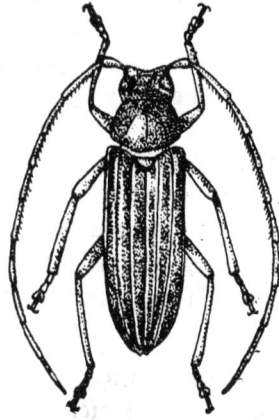

Stromatium barbatum (Fabricius)

causes wilting of the twigs. When infested twigs are dead, the grubs reverse the direction and move downwards, enter the thicker branches below and ultimately ramify in the main stem and feed on woody portion. Enroute small holes are also made for ventilation purpose and it is the appearance of these minute holes that indicate presence of the pest. Besides, gummy exudates from the entrance of the tunnel, withering of twigs and accumulation of chewed fibrous material at the base of the tree are other typical symptoms of attack.

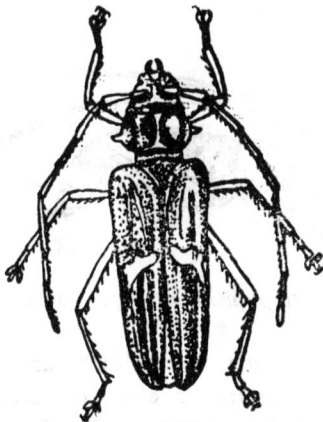

Chelidonium cinctum (Guerin-Meneville)

Eggs are flattened and yellowish. Full grown grubs are about 38 mm long. Adults are 25 to 30 mm long, dull metallic green to dark violet with yellow patch across the middle of elytra. Mating and egg-laying has been observed during June-July. Eggs are laid singly at angles of twigs or thorns and covered by a resinous fluid secretion. A female lays 30 to 50 eggs in her life time. Eggs hatch in 11 to 12 days. Grub stage lasts for 28 to 36 weeks. Pupation takes place in the tunnels and pupal period is on an average 3 weeks. Adults emerge from pupae in April-May but remain within pupal chambers till the onset of heavy showers.

The control measures recommended are: cutting and burning of the affected twigs, injecting fumigants like carbon bisulphide, chloroform, petrol, etc. into the tunnels and painting the stems with strong suspension of BHC or DDT (Kunhikannan, 1923, 1928).

Anoplophora versteegi (Ritsema) is another bark boring beetle reported as major pest from Assam and Sikkim (Hays, 1966; Dutta, 1966). Eggs are laid on trunks just above the soil level. Freshly hatched grubs feed beneath the bark and then bore inside the wood and tunnel through the pith taking a circuitous course. The entire grub and pupal periods are passed inside the trunk from where the adults emerge in Summer. The infested branches gradually dry up and the leaves wither away. Appearance of resinous exudation and saw-dust like powder on tree trunks indicate infestation of this pest.

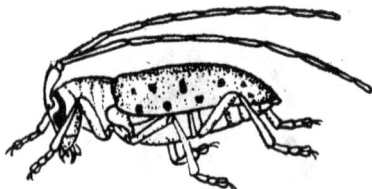

Anoplophora versteegi (Ritsema)

To control this beetle, Chowdhury and Majid (1954) have suggested smearing the tree trunks, about one metre from ground level with a suspension of 50% BHC or DDT (wettable powder) in water (1 : 100).

CITRUS FLOWER MOTH

Prays citri Milliere is found all over southern Europe, Syria, Israel, India, Sri Lanka, Malaysia, New South Wales and Philippines (Ebling, 1959). Citrus species are the only economic hosts recorded so far; flowers of citron (*C. medica*) and sour lemon (*C. jambhiri*) being preferred. Eggs are laid singly during night on buds and flowers, sometimes on fruits as well. On hatching, the caterpillars bore inside the fruits and feed below the rind, next to pulp, causing gall-like growth or swelling; more than 20 galls may be found on a single fruit. After feeding on the inside tissues of rind, the caterpillars come out to pupate. The openings in the galls or rind furnish an easy entry for bacteria and fungi. Such deformed and rotten fruits naturally have no market value. The young larvae also spin and web together the flowers, feed on calyces and other floral parts and pupate within these webbed-up flowers (Nair, 1975). The former type of damage is common in India and Philippines (Garcia, 1939) while in and around Mediterranean region only the latter type of damage has been observed. Colder climate accelerates the activity of this pest while dry heat is unfavourable; maximum activity has been observed during January-February.

Eggs are subelliptical, measuring on an average, 0.2 × 0.1 mm, colourless when freshly laid turning light yellow and gradually becoming deep yellow. Full grown caterpillars are 40 to 45 mm long, subcylindrical, semitransparent covered sparsely with fine hairs. Pupae are 5.0 to 5.5 mm long and chocolate brown in colour; their anal extremity is fastened to host by silken thread. Moths are greyish-brown with broadly fringed long narrow wings; forewings bear numerous irregular markings and a conspicuous marginal fringe of hair while hind wings are uniformily grey, membranous and without any spots; wing expanse is 9 to 11 mm (San Juan, 1924). The life-cycle and bionomics have not been studied in India. In Italy the total life-cycle was found to be 92 days at 20.5° C whereas in Philippines the life-cycle was completed in 69 days at 27°C with three and five generations in a year at the two places respectively.

Generally no control measures are required under Indian conditions. Hand-picking of caterpillars and infested blossoms and their mechanical destruction would be sufficient. According to Garcia (1939) red ants, *Oecophylla smaragdina* (Fabricius) prevent the infestation of the fruits on those trees where these ants have made their nests.

FRUIT SUCKING BUGS

Cappaea taprobanensis (Dallas), *Chrysocoris grandis* (Thunberg),
Dasynus antennatus (Kirby), *Nezara viridula* (Linnaeus), *Rhynchocoris
humeralis* (Thunberg),*Vitellus orientalis* Distant, *Leptoglossus australis*
(Fabricius) (*membranaceus* Fabricius), *Pendulinus antennatus* (Kirby)
and *Spilostethus* (*Lygaeus*) *pandurus* (Scopoli) have been recorded
sucking the juice from citrus fruits, but none of these cause any severe
damage.

Chrysocoris grandis has been reported from India, Burma,China
and Japan. In India, it is recorded as a minor pest of citrus, from
Assam. Eggs are laid in July and the nymphs of first generation suck
the sap from leaves and flower-buds. It is only the nymphs of second
generation that attack the fruits and are comparatively more
destructive than those of the first generation. Infested fruits fall
prematurely. There are only two generations in a year and the pest
overwinters in adult stage.

Nezara viridula, the green stink bug is almost completely
cosmopolitan in   distribution (CIE map No. A–27). It is found all
over India. It is a polyphagous pest causing severe damage to number
of fruits and vegetables. A female lays upto 300 eggs in batches of
40 to 60 eggs. These eggs are stuck together in rafts on ventral
surface of leaves.   On hatching, the nymphs stay in the egg shells
without feeding. It is only after the first moulting that the nymphs
leave the egg shells and attack unripe fruits. The feeding punctures
cause local necrosis with resulting brown spotting and deformation.
The affected fruits taste bitter. Subsequently bacterial and fungal

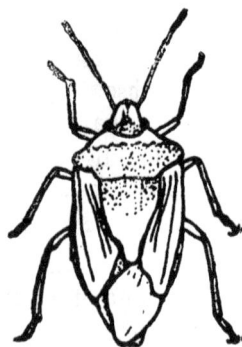

Nymph   Adult
Nezara viridula (Linnaeus)

invasions also follow and ultimately there is rotting of fruits and premature shedding. Eggs are barrel-shaped, 1.0 to 1.5 mm long, white when freshly laid turning pink before hatching. Adults are green shield bugs, 13 to 18 mm long with colour variation from apple green to reddish-brown. Incubation period is 4 to 7 days and pupal stage lasts for about a month. There are four to five overlapping generations in a year.

Rhynchocoris humeralis is a major pest of oranges in Assam. Nymphs and adults feed on developing fruits by sucking the juice. Freshly hatched nymphs feed gregariously on tender fruits but later they disperse and attack more and more fruits. The affected fruits also catch secondary infection of bacteria and/or fungi, start rotting and ultimately drop prematurely. Eggs are laid during May to September in 2 to 3 batches of 14 to 15 eggs on dorsal surface of leaves. Egg and nymphal periods are about a week and a month respectively. Overwintering is in adult stage. In China, the pest has only one generation in a year (Luh Nientsin, 1936).

Leptoglossus australis causes considerable damage to citrus fruits specially in South India, South Africa and Malaysia. Adults are more destructive than the nymphs. They congregate on fruits, pierce the rind and suck the juice. In case of severe infestation, as many as six bugs may be found on a single fruit. Such fruits do

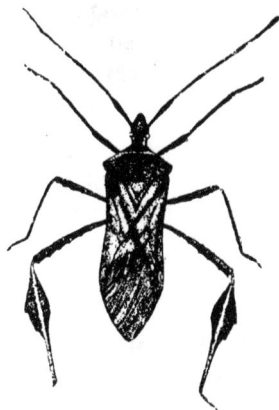

Leptoglossus australis (Fabricius)

not increase in size and subsequently dry and fall prematurely. Besides citrus, these bugs are reported to attack pomegranate in USA (Quayle, 1938) and cucurbit fruits including melons in India.

Cappoea taprobanenesis has been reported from India, Sri Lanka and Indonesia. It is a minor pest of oranges occurring regularly in the hills of North India. Nymphs and adults may often be seen crowded on trunks of the trees.

Spilostethus pandurus – the brightly coloured bug, appears in large number but its status as pest is doubtful. These bugs have been reported feeding on mango and peach fruits also. As a result of their feeding, the fruits start fermenting and fall down in a day or two.

FRUIT FLIES

Atleast half a dozen *Dacus* species have been recorded damaging citrus fruits. These are *Dacus correctus* (Bezzi), *D. cucurbitae* (Coquillett), *D. diversus* (Coquillett), *D. dorsalis* (Hendel), *D. tau* (Walker) and *D. zonatus* (Saunders). All these are polyphagous, attacking a wide range of fruits but none cause any major loss to *Citrus* species though most of these are major pests of one or the other host plant. Grubs of all these species look alike and are hard to differentiate whereas adults can be distinguished from one another. Apart from colour variation, *D.cucurbitae* and *D. tau* flies are slightly bigger in size with costal band broad and anal stripes well developed; *D. correctus* and *D. zonatus* are smaller in size with costal band incomplete and no anal stripes whereas *D.diversus* and *D. dorsalis* are medium sized with narrow costal bands and anal stripes (Kapoor, 1972).

As these are minor pests of citrus fruits, no control measures are adopted to check the infestation of these flies. Nevertheless, it is advisable that all the infested fruits be removed and destroyed immediately to prevent the population build up of the pest.

FRUIT SUCKING MOTHS

A large number of noctuid moths have been recorded attacking citus fruits. The caterpillars of these are leaf defoliators and are generally found on other host plants usually the wild ones (menispermaceous and anacardiaceous) though sometimes economic crops like castor, *ber*, pomegranate are also attacked and defoliated.

It is only the moths that are destructive to citrus fruits. The moths are distinguished by having particularly well developed proboscis with dentate tips with which they are able to pierce the ripening fruits. The moths are nocturnal in habit and may be seen flying about in orchards after dusk specially during rainy season. The damaged fruits soon start rotting as the punctured regions are easily infected with bacteria and fungi and ultimately the fruits drop prematurely. Of the various species *Othreis* spp. are comparatively more harmful. On an average these moths damage 3 to 5 per cent fruits every year.

Othreis fullonia (Clerck) is widely distributed throughout Orient extending from Africa to New Guinea and Australia. The head and thorax of the moths are reddish-brown and abdomen orange with greenish tinge. Forewings are variegated and striated with dark reddish-brown, a triangular white mark is usually present; hind wings are orange in colour with kidney-shaped black blotch in the centre. Wing expanse is 80 to 94 in males and 90 mm to 110 mm in females. Incubation, larval and pupal periods last for 3 to 4, 13 to 17 and 12 to 18 days respectively (Chowdhury and Majid, 1954) with two to three broods in a year.

Othreis materna (Linnaeus) has been reported from Indian subcontinent including Sri Lanka and Burma and also from Indonesia. Eggs are oval in shape and shining pale green in colour. Freshly hatched caterpillars are slender, thread like, 3 to 4 mm long and pale greenish in colour. Full grown caterpillars are 50 to 60 mm

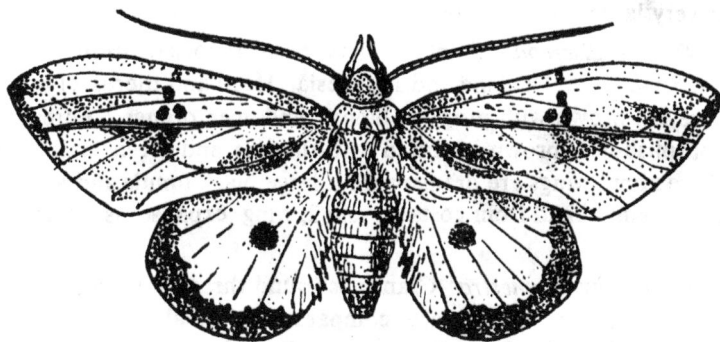

Othreis materna (L'nnaeus)

long, stout, velvety-blue with yellow patterns on dorsal and lateral sides and having a hump at anal end. Pupation takes place in their transparent pale whitish silken cover enclosed in leaf fold. Pupae are smooth, stout, 18 to 22 mm long, reddish-brown in colour with anal cremaster conical having two small hooks. Head and thorax of the moths are greenish-grey and abdomen orange. Forewings are greenish-grey with numerous faint striated reddish lines and three rufous spots; hind wings have apical area blotched with rufous, a round black spot in the centre and a marginal black band. Wing expanse is 90 to 110 mm. Incubation, larval and pupal periods are, 8 to 10, 28 to 35 and 14 to 18 days respectively.

Other species of *Othreis* are comparatively less common. Description of these moths as also other fruit sucking moths found damaging citrus fruits have been given in detail by Hampson (1896).

Othreis aurantia Moore has been reported from Indian subcontinent and Indonesia. Head and throax of the moths are ferruginous suffused with plum colour and slight purple bloom and sometimes having green patches; hind wings are orange in colour with large black Iunule and a submarginal patch. These moths are comparatively bigger as compared to other *Othreis* spp., having wing expanse of 110 to 120 mm

Othreis tyrannus Guenée is found along the entire Himalayan region including, Nepal, Sikkim as also China and Japan. The moths are nearly as big as those of *O. aurantia*. Head and thorax are dark reddish-brown and abdomen orange coloured. Forewings are chestnut red or green suffused and striated with rufous; hind wings have very large black lunule.

Othreis discrepans (Walker) has been reported from eastern India, Singapore, Thailand and Indonesia. Head and thorax of adults are greyish-brown with purplish bloom, abdomen orange. Forewings are purplish-grey-brown irrorated with fuscous and green patches at the base ; hind wings are orange with a large black lunule and a broad black marginal band from costa to vein 2, wing expanse is 90 to 110 mm.

Othreis hypermnestra Cramer is found throughout the Indian subcontinent. Its moths are comparatively smaller than those described above. Head and throax are yellowish-green and abdomen orange. Forewings are yellowish-green with dark striae and some grey patches; hind wings are orange coloured with a black spot and marginal black band from apex to vein 2. Wing expanse is 86 to 92 mm.

Othreis cocalus Cramer has been recorded from India, Bangladesh and Indonesia. The moths are slightly smaller than those of *O. hypermnestra* and further differ from those in having no black spots on hind wings and marginal band extending up to anal angle. Wing expanse is 82 to 88 mm.

Othreis ancilla Cramer is widely distributed in Indian subcontinent including Sri Lanka and Burma. Among *Othreis* spp. recorded on citrus in India, these moths are the smallest (average wing expanse 80 mm). Head and thorax are fulvous brown and abdomen

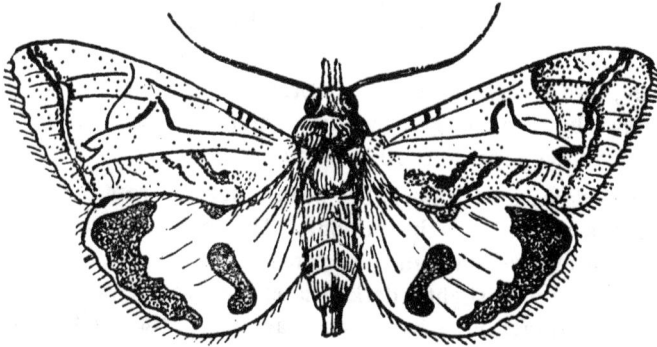

Othreis ancilla Cramer

orange. Forewings are olive-green suffused with purplish-reddish-brown and striated with rufous hind wings with large black lunule in the centre and a narrow submarginal band with wavy edges. Wing expanse 68 to 75 mm.

Othreis slaminia Fabricius is widely distributed in entire Oriental region from Malagasy to Australia. The moths have plum coloured head, green thorax and orange abdomen. Forewings are yellow and hind wings orange with a large black lunule. Wing expanse is to 80 to 92 mm in males and 90 to 194 mm in females.

Achaea janata (Linnaeus) is found in Ethiopian, Oriental and Australian regions. These are pale reddish-brown moths. Forewings cryptic coloured with diffused rufous band beyond post medial line; hind wings black with medial white band and 3 to 4 white spots on outer margin. Wing expanse is 60 to 70 mm.

Anomis fulvida (Guenée) is found throughout India, Burma, Sri

Lanka, China, Indonesia, Pacific Islands, Australia, etc. Head and thorax of the moths are ferruginous reddish-brown, abdomen reddish-fuscous, forewings dark ferruginous with 2 to 3 white spots in the middle, hind wings vinous red to dark brown in colour. Wing expanse is 50 to 58 mm.

Anua coronata (Fabricius) is found throughout Indian sub-continent as also Indonesia and Australia. Head and thorax of moths are pale reddish-brown and abdomen orange. Forewings are irrorated with dark specks and having a black patch near outer angle. Hind wings are orange coloured with broad medial and submarginal fuscous black bands not reaching inner margins. Wing expanse is 82 to 96 mm.

Anua mejanesi (Guenée) is widely distributed throughout India and West Africa. Head and thorax of moths are reddish-chocolate. Forewings are reddish-chocolate irrorated with dark specks while hind wings have basal area whitish, more or less suffused with fuscous. Wings expanse is 50 to 58 mm.

Anua tirrhaca (Cramer) has been reported from South Africa, Malagasy and Indian subcontinent. The moths have head and thorax greenish-yellow and abdomen orange. Forewings are greenish-yellow with slightly dark striate; hind wings are orange with broad submarginal black band not reaching costa. Wing expanse is 64 to 80 mm.

Calpe emarginata (Fabricius) is very common all over India, Sri Lanka and China. Head and collar region of the moths are fiery orange, thorax reddish-brown and abdomen fuscous. Forewings are reddish-brown suffused with purplish tinge and hind wings are ochreous-white suffused with fuscous towards outer margin. Wing expanse is 38 to 46 mm.

Calpe ophideroides Guenée is confined to foot-hills of Himalayas including Nepal as also Singapore and Malaysia. Head and thorax of the moths are reddish-brown, irrorated with grey and abdomen orange-yellow. Forewings are reddish-brown suffused with grey and having numerous fine pale striae; hind wings are orange-yellow. Wing expanse is 66 to 74 mm.

Calpe bicolor Moore has been reported from Kangra valley in India. The moths are similar to those of *C. ophideroides* except that the forewings have silvery sheen and are smaller in size; underside of wings being orange-yellow and wing expanse 52 to 60 mm.

Calpe fasciata Moore has been recorded from Kangra valley and Sikkim. The moths differ from *C. bicolor* in being without reddish-tinge, slightly smaller in size and having underside of wings brown. Wing expanse is 46 to 54 mm

Ercheia cyllaria Cramer has been recorded from India, Sri Lanka, Burma, Indonesia, etc. Head and thorax of the moths are pale reddish-brown, abdomen fuscous black; forewings suffused with fuscous and streaked with black and hind wings fuscous black with three medial white spots. Wing expanse is 50 to 60 mm.

Erebus hieroglyphica (Drury) has been reported from Malagasy India, Sri Lanka, Burma, Malaysia, Philippines and Indonesia. These are big blackish-brown moths; forewings blackish having a conspicuous whorl-shaped jet black mark. Wing expanse is 85 to 95 mm. The moths usually swarm on fallen fruits and seldom damage the fruits on trees, but this is for the sake of sucking the oozing sap and they do not make any fresh punctures on these fruits. The moths are also attracted to syrupy solution split on ground or trunks.

Lagoptera dotata (Fabricius) is widely distributed in Indian subcontinent including Sri Lanka and Burma. These are bronze brown moths having fore wings irrorated with white specks and prominent marginal grey band with a wavy line on it; hind wings are dark fuscous with margin and cilia whitish. Wing expanse is 72 to 82 mm.

Lagoptera honesta (Hubner) is found throughout Indian subcontinent and Philippines. Moths have head and thorax reddish-chestnut and abdomen crimson. Fore wings are reddish-chestnut slightly irrorated with dark specks and hind wings are crimson red with black submarginal medial patch. Wing expanse is 84 to 94 mm.

Lagoptera submira (Walker) has been recorded from India, Bangladesh and Burma. The moths differ from those of *L. dotata* in being much darker reddish-brown; forewings with costal and medial areas suffused with bluish-white and hind wings dark reddish-fuscous without any white medial band. Wing expanse is 70 to 80 mm.

Pericyma glaucinans (Guenée) has been reported from Congo,

South Africa, Indian subcontinent and Indonesia. Moths are dark reddish-brown; both wings suffused and blotched with fuscous. Fore wings with sinuous subbasal black line and hind wings with numerous fine oblique lines. Wing expanse is 36 to 42 mm.

Pelochyta astrea Drury is found throughout India, Sri Lanka, Burma and Formosa. The moths are very conspicuous, having white fuscous coloured body with a number of small round black spots, two on head, two pairs on collar, one pair each on pro- meso- and metathorax and one pair at the base of each forewing. Forewings and hind wings are hyaline, the former having margins and apical areas pale fuscous while the latter have marginal fuscous band. Wing expanse is 50 to 54 mm and 64 to 72 mm in case of males and females respectively.

Polydesma quenavadi Guenée has been recorded from India, Sri Lanka and Burma. These are brownish-grey moths; forewings havings wavy lines and marginal series of black specks and hind wings with basal area whitish, outer brownish, having wavy lines. Wing expanse is 40 to 56 mm.

Remigia frugolis (Fabricius) is found in the entire Oriental region from Africa to Australia. These are greyish-brown moths; forewings have diffused dark markings and submarginal series of black specks; whereas hind wings are ochreous fuscous with diffused submarginal lines. Wing expanse is 36 to 50 mm.

Serrodes inara Cramer has been recorded from Africa, India, Sri Lanka, Burma, Indonesia and Australia. The moths are stout greyish-brown; forewings are pale olivaceous-grey and hind wings fuscous. Wing expanse is 52 to 74 mm.

Sphingomorpha chlorea Cramer has been reported from Africa, Arabia, India, Sri Lanka, Burma, etc. The moths have ochreous white head and brown thorax and abdomen with ochreous white spots. Forewings are reddish-brown with dark striae and hind wings fuscous brown having an ochreous patch with black striae at centre of outer margin. Wing expanse is 60 to 84 mm.

Control of these moths is rather difficult as the immature stages are invariably present on other host plants and the adults appear only after dusk and do the damage during night. Systematic destruction of alternate host plants in the vicinity of orchards helps to check the pest population. Bagging of fruits has been suggested.

This is no doubt effective but it is very laborious and expensive process. Creating smoke in the orchrads after sunset may keep the pest at bay but this method too is rather cumbersome and not practicable on large scale. Poison baiting (20 gm malathion 50% W.P. or 50 ml diazinon E.C.+200 gm gur or molasses in 2 litres of water) has been found quite affective and may be practised (Butani, 1973 d, e). This bait will also trap and kill the fruit fies.

FRUIT BORERS

Citrus fruits are occasionally bored by some lepidopterous larvae. Those recorded so far, include, *Dichocrocis punctiferalis* (Guenée), *Heliothis armigera* (Hübner). *Prays endocarpa* Meyrick and *Virachola isocrates* (Fabricius). *D.punctiferalis*, commonly called castor capsule borer, is a major pest of guava and *V.isocrates*, a polyphagous pest, is the worst enemy of pomegranate.

Heliothis armigera, the gram pod borer, is widely distributed in tropics, subtropics and warmer temperate regions of the old world (New world species is, *H. zea* Boddie — CIE map No. A-15). In India it is a major pest of pulses, sporadically serious on cotton and a minor pest of various wild and cultivated fruits. Young caterpillars bore clean circular holes in flower-buds, later they bore into the fruits and feed within. The affected buds do not open and are shed while the fruits start rotting and ultimately fall down. A female may lay 1000 or more eggs. Egge are spherical, about 0.5 mm in diameter and yellowish in colour. Caterpillars when full grown, are about 40 mm long, stout, greenish-brown in colour with dark and pale longitudinal bands. When fully fed, they burrow in the soil and pupate there. Pupae are shiny-brown and about 16 mm long. Adults are brown moths with wing expanse of about 40 mm. Incubation, larval and pupal periods last for 2 to 4 , 15 to 24 and 10 to 14 days respectively. There are on an average eight generations in a year.

Prays endocarpa – the citrus rind borer, is closely related to citrus flower moth, *P.citri*. But caterpillars of *P.endocarpa* feed strictly on citrus fruits and do not attack the blossoms like *P.citri*. Eggs are laid singly on the fruits. On hatching the larvae bore into the rind of those fruits. Gall-like swellings are caused by the feeding and lignified tissue may extend down the fruit pulp and thereby make the fruits unfit for human consumption The caterpillars, however, remain in the rind. As many as 84 moths have been found to emerge from a single small lemon (Pagden, 1931).

Prays cndocarpa Meyrick

Generally no control measures are required against these borers However, the infested fruits must be removed and destroyed promptly. If need be, spray with 0.05% endosulfan or monocrotophos.

TERMITES

These are social insects, polymorphic, living in large colonies comprising of reproductive forms, soldiers and workers in elaborate nests – termitaria. The nests are constructed both above and below the ground. The species damaging citrus in India is mostly, *Odontoterms obesus* (Rambur). This is a soil inhabiting species and a major pest of a number of crops but minor pest of citrus. The damage is more pronounced in nurseries and in orchards having sandy-loam soils. It is the worker caste that does all the damage by feeding on roots and killing the seedlings outright. In case of established trees, the workers gnaw the roots and tunnel upwards in the main trunk, which may be completely hollowed if the pest is not checked in time. In case of severe infestation, the termites also attack the trunk and main branches from outside and in such cases they build the mud galleries on the host trees and feed beneath these galleries which are connected with the under ground termitarium. But this sort of damage in case of citrus is rather rare.

Soil application with 5% aldrin, chlordane or heptachlor dust @ 22 to 25 kg per hectare checks the pest effectively (Butani, 1973 a). Aldrin emulsion @ 10 kg a.i. per hectare added in the irrigational water is also equally effective. The treatment may be repeated annually. This will take care of other soil borne insect pests as well. In case of mud galleries on tree trunks and main branches, break and scrape the same and sponge the affected parts with aldrin solution or even with kerosene oil.

GRAPEVINE

GRAPEVINE, *Vitis vinifera* Linnaeus is one of the oldest fruits grown in the world. Its cultivation started in Transcaucasia (Caspian Sea area) about 6000 B. C. and gradually spread around the Mediterranean coast and to western European countries. Today, the grapes are grown throughout the subtropical and warm temperate regions of the world. The world's vineyards cover 10 million hectares (Bournier, 1977). Of the total production, 80% is used for wine making (about 300 million hectolitres per year), 10% for making raisins (5 million quintals annually) and only 10% is eaten as fresh fruit (40 million quintals) including juice and syrup making. In India though there is mention of grapevine cultivation in old Sanskrit literature like *Arthashastra* written in 4th century B.C. and medical treatises, *Charaka Samhuta* and *Sushruta Samhuta,* written in first century A.D., it is not clear if these records relate to India that's Bharat or to the Indian subcontinent, including Burma, Pakistan and even parts of Afghanistan. Yuan-Chwang, a Chinese Buddhist pilgrim passed through India (629 to 645 A.D.), and reported grape growing 'from Kashmir on'. It has not been specified if he found grapes in areas South and East of Kashmir or North and West. The vines were introduced from Afghanistan and established into India by the Moguls also. Till recently, grapes were mainly grown in southern parts of India (Maharashtra, Karnataka, Andhra Pradesh and Tamil Nadu), but of late viticulture has been spreading to North India as well, specially in the Punjab, North Rajasthan and parts of Madhya Pradesh and Uttar Pradesh. The area under this crop has risen from 648 hectares in 1955 to 3494 in 1963-64 and 8027 in 1967-68 (FAO report, 1970). At present there are over 10 million hectares under grapevines in India. The fruits are mostly used for table purpose though some quantity is used for wine making as well. Besides, a small quantity is also utilized for making juices, soft drinks and canning purpose. The sugar content of grapes grown in India varies between 13 and 22 per cent. Therefore, Indian grapes are not so good for making raisins. The raisin-grapes generally have 24 to 28 per cent sugars.

Over 85 species of insects have been reported from grapevines in India. Those of major importance are, flea beetles, cockchafer beetles, thrips and scale insects; the lesser important ones include, girdler beetle, stem boring beetle, leaf roller and termites; while *ak* grasshopper, whitefly, leaf sucking bags, leaf eating caterpillars, fruit sucking moths, wasps, etc. are of minor importance.

FLEA BEETLES

Scelodonta strigicollis (Motschulsky) is a serious pest, specific to grapevine. It is widely distributed all over India though it causes comparatively more loss in South India, Maharashtra, Gujarat and Rajasthan. In South India, it is active during September to December and again from March to May whereas in Haryana, Punjab and Himachal Pradesh the pest appears from March to November (Vevai, 1969 a). The grubs feed on cortical portion of roots and pupate in the soil, 60 to 80 mm deep in earthen cells. Adults are nocturnal in habit and comparatively more destructive than the grubs. They feed

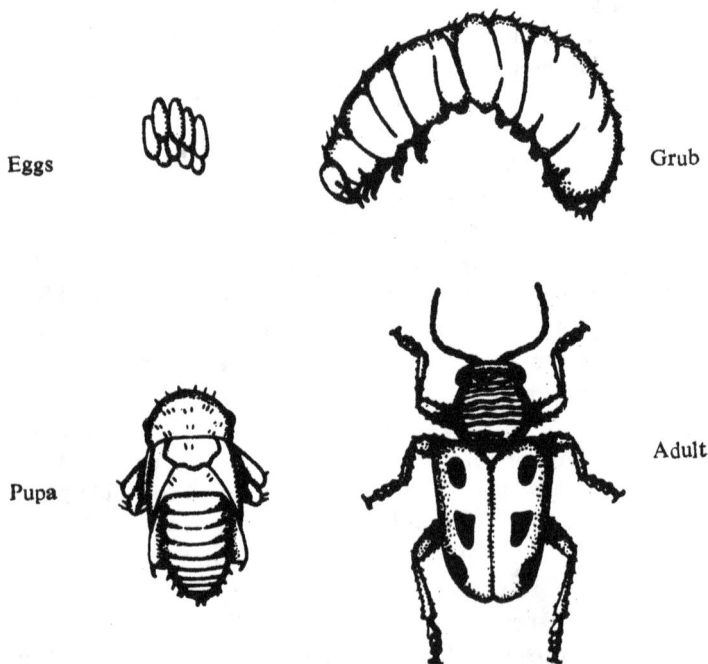

Eggs   Grub

Pupa   Adult

Scelodonta strigicollis (Motschulsky)

voraciously on swelling buds and tender shoots soon after pruning of the vines. During day they hide under the loose bark or clods. The affected buds or sprouts soon dry up.

Eggs are cigar-shaped, creamy-white and about one mm long. Adults are small, 4 to 5 mm long, copper coloured beetles with 3 black spots on each elytra. Mating takes place a month after emergence and the eggs are laid 4 to 7 days there after beneath the bark or in crevices in the stems. A female lays 200 to 600 eggs in her life time of about 8 months (March to October), in clusters of 20 to 40 eggs. Incubation period is 4 to 8 days, grub stage 34 to 45 days and pupal period 7 to 11 days (Pruthi and Batra, 1960) There are four overlapping generations in a year (Trehan *et al.*, 1947); hibernation is in adult stage during November-December to March.

To control this beetle, remove the loose bark and spray with 0.05% phosphamidon or 0.075% dichlorvos @ 1000 litres per hectare, immediately after prunning and again after 10 days (Butani, 1974 g); two to three sprayings are enough to check this pest.

Other flea beetles (Chrysomelids) found feeding on grapevine leaves are, *Monolepta erthrycephala* Baly, *M.signata* Olivier, *Mimastra cyanura* Hope, *Nodostoma subcastatum* Jacoby, *N. viridipennis* Motschulsky and *Oides scutellata* Hope. These are minor pests of grapevine. *M. erthrycephala* is a major pest of walnut, *M. signata* of various vegetable crops and *Nodostoma* spp. cause severe damage to banana.

Monolepta signata, commonly called white spotted flea beetle, is 3 to 4 mm long, having reddish-brown body and pale brown elytra with two big white spots on each.

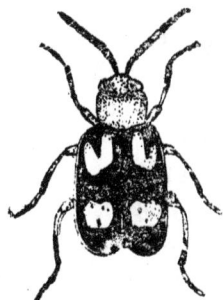

Monolepta signata Olivier

Eggs of *Oides scutellata* are brownish-yellow and round in shape; grubs are 6 to 9 mm long, hemispherical in shape and bluish in colour with green spots; pupae are 8 to 12 mm long and shinning yellow in colour while the adult beetles are 9 to 13 mm long, hemispherical with yellowish-green elytra having a brick red longitudinal stripe on each elytron. Egg, grub, pupal and adult stages last for 2 to 4, 3 to 5, 2 to 3 and 39 to 50 weeks respectively; hibernation takes place in adult stage under the soil.

LEAF EATING BEETLES

Cockchafer beetles or white grubs are widely distributed all over India damaging a variety of economic crops including various fruit trees. Those recorded damaging grapevine are, *Apogonia* species, *Brahmina coriacea* (Hope), *Adoretus bengalensis* Blanchard, *A. brachypygus* Burmeister, *A. duvauceli* (Blanchard), *A. horticola* Arrow, *A. lasiopygus* Burmeister (*ovalis* Blanchard), *A. versutus* Harold, *Anomala bengalensis* Blanchard, *A. dimidiata* Hope, *A. dorsalis* (Fabricius), *A. ruficapilla* Burmeister, *Pachyrrhinadoretus frontatus* (Burmeister), *Schizonycha ruficollis* Blanchard and others (Batra *et al.*, 1973; Bindra *et al.*, 1973).

Eggs are laid in soil. On hatching, the grubs feed on rootlets of various grasses and even economic crops including grapevines. Adult beetles are nocturnal, these come out of the soil after dusk and feed on leaves making innumerable holes in leaf lamina. In case of severe infestation, the entire leaf lamina may be eaten away. The adult beetles appear soon after the first heavy shower (May-June) and are active for about two months.

Besides the above mentioned pests, ground beetles *Dasus* (*Gonocephalum*) *depressum* Fabricius and *D. hoffmannseggi* Steven have also been recorded from South India damaging grapevines. Eggs are laid loose in the soil. On hatching, the grubs gnaw the hair of roots and rootlets. Pupation takes place in the soil. Grubs of *D. depressum* are cylindrical, shining and yellowish-brown in colour. Adults are dull black beetles about 10 mm long usually covered with crust of sand particles. Adults of *D. depressum* are seen swarming on leaves and nibbling the same as also scrapping the berries, whereas those of *D. hoffmannseggi* feed on tender vines.

Dasus hoffmannseggi Steven

To control the beetle grubs, soil application with 5% aldrin or chlordane dust @ 22 to 25 kg per hectare is recommended. This will also combat termites and other soil inhabiting insect pests. For adults beetles spraying with 0.15% carbaryl or 0.04% endosulfan has been found very effective (Bindra *et al.*, 1973). A waiting period of 5 to 7 days may be necessary, so that the residues may dissipate to nontoxic levels or safe limits.

THRIPS

Grapevine thrip, *Rhipiphororthrips cruentatus* Hood, is a major pest of grapevine. It is highly polyphagous and appears to have a particular preference for anacardiaceous and myrtaceous plants such as country almond, cashewnut, guava, *jamun* mango, etc. The nymphs and adults rasp the ventral surface of tender leaves and flower-stalks and suck the oozing cell sap. A single individual damage on an average 0.1 sq cm area of leaf lamina per day. As a result of their intensive feeding the leaves develo silvery white scorchy patches with curly tips, gradually get deformed and ultimately fall down. Attack on flower-stalks results in shedding of flowers. Sometimes the fruits are also attacked resulting in scab formation on the berries. The peak infestation is during hot weather – May to July in South India, March to April and August to October in Haryana and Punjab, whereas in Himachal Pradesh it is March to October. Varieties of grapevine with leaves thick and hairy underside are known to be tolerant to the attack.

Nymphs are reddish in colour and adults bright yellow (♂) to dark brown (♀). Both sexual and parthenogenetic reproduction occur side by side; the progeny of the latter being always males. Eggs are laid @ 2 to 6 per day. Incubation, pre-imaginal, pupal and total life-cycle durations occupy, 3 to 8, 11 to 22, 2 to 5 and 14 to 33 days respectively (Rahman and Bhardwaj, 1937). Males die after 2 to 6 days of mating while the females live up to 18 to 20 days. There are five to eight generations in a year and overwintering is in pupal stage, 80 to 150 mm deep in the soil.

Retithrips syriacus (Mayet) is the typical grapevine thrip of Egypt. It is reported from India by Seshadri and Ananthakrishnan (1954). This is also polyphagous and amongst fruit trees, beside grapevine, it is found on leaves of custard apple and pomegranate. It has flattened body which is uniformily dark brown and has dark grey forewings. Life-cycle is completed in 9 to 13 days.

Taeniothrips traegardhi (Trybom) is mainly a pest of maize and other graminaceous plants. Swarms of this thrip may occasionally be found on grapevin e, causing bleaching of leaves. It is however, a minor pest reported only from Tamil Nadu.

Blossom thrips, *Thrips florum* Schmutz (*hawaiiensis* Morgan) and *Frankliniella dampfi* Priesner (*sulphurea* Schmutz) are the common thrips found in various ornamental plants specially roses. These are minor pests of grapevine. Besides grapevine, the former is found on apple and banana while the latter has been recorded on apple, citrus, melons and sapota. The affected flowers fade and shrink, ultimately affecting adversely the fruit setting capacity.

Scirtothrips dorsalis Hood – the chilli thrip, has been reported damaging grapevine in Tamil Nadu, Andhra Pradesh and Maharashtra. This too is polyphagous having a very wide range of host plants (Ananthakrishnan, 1971). It has been observed to infest flower bunches and young fruits of grape, causing heavy damage by reducing the fruit setting capacity of vines and causing the development of corky layer with cracks cn surface of fruits. These thrips reproduce sexually as well as parthenogenetically. They oviposit within tissues of tender leaves. A single female lays 50 to 100 eggs. Life-cycle on tea is completed in 13 to 18 days (Dev, 1964) whereas Raizada (1965) observed the life-cycle to cccupy 15 to 20 days.

To control these thrips, spray with 0.03% dimethoate, oxydemeton methyl or phosphamidon, twice at an interval of 10 days.

COCCOIDS

More than 20 species of mealy bugs and scale insects have been reported damaging grapevine from various parts of India. These include, *Drosicha mangiferae* (Green), *Icerya aegyptiaca* (Douglas), *I. purchasi* Maskell, *Ferrisia virgata* (Cockerell), *Maconellicoccus hirsutus* (Green), *Nipaecoccus viridis* (Newstead), *Kerria communis* (Mahdihassan), *K. lacca* (Kerr), *Aspidiotus destructor* Signoret, *Chrysomphalus ficus* Ashmead, *Duplaspidiotus terreratus* (de Charmoy), *Hemiberlesia lataniae* (Signoret), *Parlatoria camelliae* Comstock, *P. oleae* (Colvée), *Phenacaspis vitis* (Green), *Ceroplastes actiniformis* Green, *Chloropulvinaria polygonata* (Cockerell), *Coccus discrepans* (Green), *C. elongatus* (Signoret), *Parasaissetia nigra* (Nietner), *Pulvinaria maxima* Green, etc. These are polyphagous pests having a wide range of host plants. The loss caused to grapevine by any one species may be of minor importance, but usually more than one species are found damaging the vines and their collective damage invariably is substantial.

In the initial stage of attack, remove the infested leaves and twigs and burn the same immediately. This will prevent the pest from spreading. In case of severe infestation, spary with 0.1% BHC+0.1% DDT or 0.05% carbophenothion or fenitrothion.

GIRDLER BEETLE

Sthenias grisator Fabricius has been reported as a serious pest from South India and Sri Lanka. Besides grapevine, the beetles also damage almond, apple, citrus, jack-fruit, mango, mulberry, ornamentals, etc. The adult beetles emerge usually around July-August and maximum damage is caused during August to October. The beetles with the strong mandibles, ring or girdle the young green branches of the host plant. This is a unique device that invariably preceeds egg-laying and allows the girdled branch to dry up so as to enable the grubs to bore and tunnel the dry wood. For egg-laying, the females cut transverse slits under the bark of girdled branch and thrust inbetweeen the bast and sap wood, one to four eggs. On hatching, the grubs straight away tunnel into the wood. Pupation takes place in an elongate chamber inside the tunnel and the beetles emerge out by circular exit-holes.

Eggs are elongated oval in shape, about 4 mm long and whitish in colour. Freshly formed grubs are 2 to 3 mm long, having dark brown head, globular thorax and cylindrical abdomen sparsely covered with hair. When full grown, grubs are 11 to 14 mm long. Adult beetles are about 24 mm long, greyish-brown with white and brown irregular markings resembling the bark colour; elytra have an elliptical greyish median spot and an eye-shaped patch.

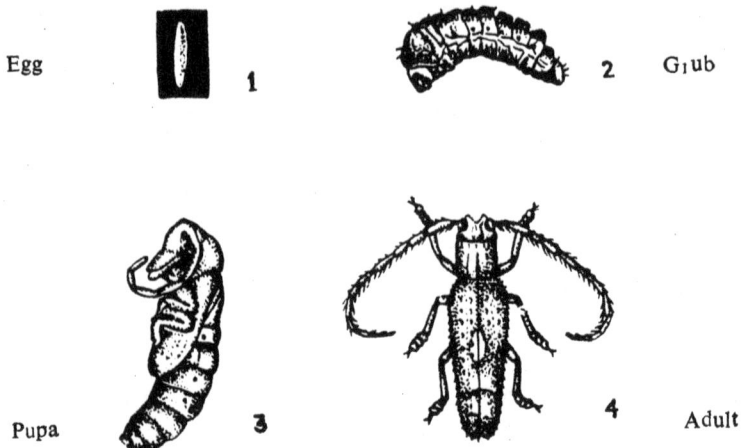

Egg   1   2   Grub

Pupa   3   4   Adult

Sthenias grisator Fabricius

Incubation period is 8 to 10 days, grub stage varies considerably according to dryness and nutritive value of food material and competition; when the infestation is severe; usually it is 7 to 8 months with only one generation in a year.

To control this longicorn beetle, cut the attacked branches below girdling point and burn the same. Hand collection and destruction of beetles also helps in mitigating this pest.

STEM BORING BEETLE

Sinoxylon anale Lesne is an important pest reported from Punjab (Atwal, 1976). It has also been recorded in France, Italy, USSR, China and Japan. In addition to grapevines, the beetle also attacks a number of forest trees in India. The adults appear with the onset of Winter, bore inside the stems and feed within, making longitudinal tunnels and a number of exit holes. Chewed up wood particles are thrown out of these holes. The eggs are laid in these tunnels and on hatching, the grubs feed therein causing further damage. As a result of the attack the plant portion above the point of attack gradually dries and dies away. The pest is active only in dormant vines and perfers late-sprouting varieties like, *Anab-e-shahi*. The breeding of the pest continues in the dead vines throughout the year.

Grubs are thick, slightly curved and yellowish-white in colour. Adults are dark brown, stout beetles, 3 to 5 mm long having three-bladed antennae and a pair of spines on the posterior elytral extremity.

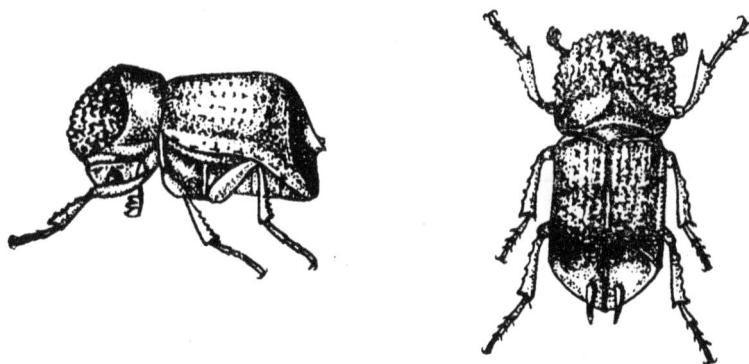

Sinoxylon anale Lesne

To prevent the infestation, clean cultivation including removal of loose bark, coupled with careful prunning and destruction of infested parts is very helpful. During the dormant period of vines, give a cover spray with 0.1% BHC + 0.1% DDT.

LEAF ROLLER

Sylepta lunalis Guenée is found all over India and though a specific pest of grapevine, it has been found attacking eleven species of Vitaceae in South India (David, 1961).

A famale lays 100 to 150 eggs, usually on ventral surface of leaves. On hatching, the young caterpillars feed on epidermis of leaves and skeletonize the same. Later, these caterpillars roll the leaves and feed within. Pupation takes place within rolled leaves. Generally each leaf-roll contains only one caterpillar or pupa. Sometimes the pupae may fall down from the rolled leaves and are seen lying on the ground on fallen leaves and debris. The activity of the pest is accelerated with the onset of monsoon (July-August) and continues till October.

Eggs are oval, about 1.5 mm long and creamy-white in colour. Caterpillars are hairy, cylindrical, 18 to 24 mm long when full grown and pale green in colour. Pupae are 8 to 12 mm long and brown in colour. Adults are dirty brown moths with white spots and wavy patterns on fore- and hind wings. Wing expanse is 20 to 24 mm($\male$) and 24 to 28 mm ($\female$). Egg, larval and pupal stages last for 2 to 3, 14 to 20 and 5 to 7 days respectively. The total life-cycle during the active period lasts for 3 to 4 weeks.

In the initial stage of attack, remove the rolled up leaves and destroy the same with larvae/pupae within. In case of severe infestation spray with 0.05% dimethoate, endosulfan, malathion or phosphamidon (Sohi and Bindra, 1974).

TERMITE

Odontotermes obesus (Rambur) is polyphagous subterranean pest that attacks the root zone. Though the pest is present in the soil throughout the year, it is more active during April to June in North India and from January to May in South India. Heavy rains or flood irrigation drives away the pest deep down in the soil. The damage is caused only by workers who gnaw the roots, with the result the affected vines turn pale, wither and die away.

To safegaurd against the ravages of termites, treat the upper 500 mm layer of soil with 5% aldrin or chlordane dust @ 22 to 25 kg per hectare before planting the vines. Repeat the treatment once every year specially if the soil is sandy or sandyloam. Addition of aldrin emulsion in irrigational water also helps to keep the pests away.

SAP SUCKING INSECTS

Leaf hoppers, *Empoasca minor* Pruthi, *Arboridia viniferata*

Sohi and Sandhu *Erythroneura* species and *Typhlocyba* species have been reported damaging grapevines but these are minor sporadic pests. Nymphs and adults of these hoppers usually suck cell sap from the ventral surface of leaves. The feeding spots on the leaves become pale. In case of severe infestation, the affected leaves turn yellow, gradually start curling; become brown and ultimately fall down. The hoppers are active during Spring and the population declines after heavy showers. Overwintering is in adult stage, generally under debris or plant remnants on the ground. Incubation period of *Erythroneura* species is about 14 days and its nymphal duration lasts for 2 to 3 weeks with two to three generations in a season (Atwal 1976),

Peach mealy aphid, *Hyalopterus pruni* (Geoffrey) (*Brachycaudus arundinis* Fabricius) – a major pest of peach has also been recorded feeding on grapevine foliage, but the damage caused by desapping the foliage is of minor importance. Fortunately, *Phylloxera vitifoliae* Fitch – the notorious pest of grapevines in Europe which once threatened the entire wine industry of Europe, has not yet been recorded from India (Pruthi, 1969).

Whitefly, *Aleurocanthus spiniferus* (Quaintance) – a major pest of citrus, has also been recorded sucking the cell sap of grapevine foliage though not causing any severe loss.

These sucking pests are also responsible for an indirect loss by producing honeydew. This sweet secretion serves as a substrate for the growth of sooty mould fungus on foliage and fruits which in turn affects the production of fruit and also depreciate the quality of grapes.

To control these aphids, leaf hoppers and whitefly, spray with 0.03% dimethoate, oxydemeton methyl or phosphamidon.

Besides these sucking pests, a pentatomid bug, *Scutellera nobilis* (Fabricius); a coreid bug, *Anoplocnemis phasiana* (Fabricius) and tea mosquito bug, *Helopeltis antonii* Signoret have also been occasionally found sucking the cell sap from grapevine foliage. These are minor pests and require no control measures.

LEAF EATING INSECTS

Grapevine leaves are occasionally eaten away by the grasshoppers:

Poekilocerus pictus (Fabricius) and   *Tetratodus monticollis* Grey;
hairy caterpillars :   *Euproctis flava*   (Bremer),   *E. freterna* (Moore),
E. lunata (Walker); sphingid caterpillars :   *Hippotion celerio* (Linnaeus)
Rhyncholoba acteus Cramer,   *Thereta alecto*   Linnaeus,   *T.gnoma*
Fabricius and *T.pallicocta* Walker. All these are minor pests. If and
when necessary, dust with 10% BHC or spray with 0.1%
BHC + 0.1% DDT.

Bagworm, *Clania crameri* Westwood – a polyphagous pest, has
been reported from Karnataka (Veeresh and Rajagopal, 1977). The
caterpillars bind together pieces of twigs, grasses, wood particles,
etc. by means of silken strands and make small bags over their body.
Only the head and part of thorax are protruded outside the bag
for feeding purpose.

Archips miaceana (Walker) is   another minor pest reported
from Karnataka. The caterpillars web together the leaves, main
stalks of fruits and even a few grape berries and feed within on
epidermis of leaves. In addition, *Phyllocnistis toparcha* Meyrick
and *Antispila argostomia* Meyrick have been recorded as leaf miners.
The loss caused by these pests is however, negligible.

FRUIT SUCKING MOTHS

Achaea janata (Linnaeus), *Anomis flava* Fabricius, *Grammodes
stolida* Fabricius, *Hulodes caranea* Cramer, *Lagoptera dotata* Fabricius,
Othreis ancilla Cramer, *O. cojeta* (Cramar), *O. materna* (Linnaeus)
Parallelia algira (Linnaeus), *Pericyma glaucinans* Guenée, *Perigea
capensis* Guenée, *Polydesma umbricola* Boisduval, *Remigia archesia*
Cramer, *R. frugalis* Fabricius, *Serrodes inara* Cramer, *Sphingomorpha
chlorea* Cramer, *Spodoptera litura* (Fabricius) and *Aganais ficus*
(Fabricius) have been reported damaging the grapes (Sundra Babu and
David, 1973). These are minor pests. Their caterpillars feed on
leaves of other host plants, mostly those growing wild and only the
moths are harmful to economic crops including grapes. These moths
have saw-and-file-like proboscis with which they pierce the ripe berries
and suck the juice. Subsequently, the punctures made by these
moths in the fruits, also permit the entry of various bacteria and
fungal organisms and the affected fruts start rotting and emitting
offensive smell. Even though only a few berries may be damaged in
each bunch, the entire bunch may lose its market value, unless the

affected berries are hand-picked and removed. This however, will be uneconomical on large scale.

Hawk moth, *Agrius convolvuli* (Linnaeus) is another minor pest reported from Haryana, Punjab and Himachal Pradesh. The caterpillars are voracious feeders of foliage while the moths, which are crepuscular, attack the fruits and suck the juice. Full grown caterpillars are about 100 mm long, dark brownish-green in colour with reddish-yellow lateral stripes and a pair of conspicuous horns. Pupae are easily recognised by the long, free proboscis sheath. Moths have exceptionally long proboscis (60 to 100 mm) and violet bands on abdomen; wings are pale grey and wing expanse is 55 to 70 mm.

Agrius convolvuli (Linnaeus)

As these are minor pests, generally no control measures are adopted against these pests in vineyards. However, bagging the grape bunches or creating smoke, after sunset, in the vineyards help in preventing the attack by these moths.

FRUiT BCRERS

Oxyptilus regulus Meyrick – grape boring plume moth, has been reported from India, Sri Lanka and Australia. In India, it is more common in Karnataka and other southern states than in the North. Eggs are laid singly at night on fruits, around fruit stalks or peduncles. On hatching, the caterpillars bore into the ripening berries and tunnel through the pulp upto the seeds. A single caterpillar may damage number of berries. Pupation generally takes place on peduncle of healthy berries, some of the larvae that fall down pupate on weeds or dry leaves lying on the ground.

WASPS

The hornet, *Vespa orientalis* Linnaeus and house-wasp, *Polistes hebraeus* (Fabricius) are often seen hovering and damaging ripe fruits not only in grapevine orchards but also at places where grapes are kept for drying purpose. These wasps are also a great nuisance to the pickers and other workers. *V. orientalis* wasps are mainly predaceous on various insects including, honeybees, lepidopterous larvae, etc. In addition, these wasps are avid feeders on any sweet material and besides grapes, these wasps also damage the fruits of citrus, mulberry, peach, pear, plum, pomegranate, etc. *P. hebraeus* wasps are generally found in houses and offices, damaging the books and files.

Adults of *Vespa orientalis* are light chestnut-red with 3rd and 4th abdominal segments pale-sulphur-yellow; wings are flavo-hyaline and wing expanse is 48-58 mm. *Polisetes hebraeus* wasps are either bright yellow or fulvous-brown; wings in both forms are ferruginous and wings expanse is 36 to 46 mm.

Vespa orientalis Linnaeus

The wasps being minor pests, no control measures are adopted against these.

BER, *Zizyphus* spp., is said to be indigenous to south-eastern China, India and Malaysia. It is grown all over India, cultivated as well as wild and occupy an area of about 10,000 hectares; half of which is in Madhya Pradesh and over 2000 hectares in Bihar (Vevai, 1971 b). It is most hardy fruit tree and can grow successfully on all types of soils and even under prolonged drought and water-logged conditions. The fruits of *Z. mauritiana* Lamark – the variety commonly grown in India, contains 50 to 150 mg of ascorbic acid (vitamin C) per 100 gm of fruit and 12.0 to 18.7% sugar. It is often called poor man's fruit and that's why it was hitherto considered uneconomical to adopt any plant protection measures to safeguard these trees from the ravages of insect pests. But now that consumers are becoming quality conscious, it is time to know about the activities of pests on these trees as well and methods to prevent the same.

As many as 80 insect species have been found feeding on *ber* trees in India. Those of major importance are *ber* fruit fly, *ber* mealy bug, beetles and weevils. Besides, there are various bugs, scale insects and thrips that suck the cell sap from leaves and a large number of lepidopterous pests including bark eating caterpillars, leaf eating caterpillars, leaf rollers, leaf webbers, shoot and fruit borers that occasionally damage the *ber* trees.

FRUIT FLIES

Ber fruit fly, *Carpomyia vesuviana* Costa, was originally recorded from Italy on the slopes of Mount Vesuvius – hence the name. It is widely distributed and has been recorded from India, Pakistan and Mediterranean countries. It is a specific pest of *ber* (*Zizyphus* spp.), both wild and cultivated varieties. Fleshy and late maturing varieties are comparatively more susceptible. Upto 77% damage to fruits has been reported from Andhra Pradesh (Nair, 1975). Unlike other fruit flies (*Dacus* spp.), this fly becomes active in Autumn and the activity continues throughout Winter and Spring, though maximum damage is caused during February-March. The female flies puncture the ripening fruits and lay their eggs inside the epidermis. The development around the punctures is arrested and this causes the deformation of fruits. On hatching, maggots feed on fleshy and juicy pulp. As many as 18 maggots have been found in a single fruit. The infested fruits turn dark brown, rot and smell offensively. One maggot is sufficient to destroy the entire fruit. When full fed the maggots drop down from the fruits to pupate in the soil, 50 to 75 mm deep.

Eggs are dull-creamy-white, opaque and 0.9 to 1.0 mm long. Full grown maggots are amphineustic, cretaceous or creamy-white and about 6 mm long. Pupae are barrel-shaped, 4 to 5 mm long and light brown when freshly formed becoming ochraceous in 3 to 4 days. Adult flies are small, brownish-yellow with brown longitudinal stripes on the thorax surrounded by black spots; wings are hyaline and transparent with four yellowish cross bands.

Carpomyia vesuviana Costa

A single female lays 12 to 18 eggs (maximum 22) singly or in groups of 2 to 4 eggs. These hatch in 1 to 4 days. Maggot stage lasts for 9 to 12 days in March extending up to 22 days in November. Pupal stage lasts for 11 to 31 days in Summer and 45 to 87 days during Winter except during aestivation when it may last for 133 to 187 days (Batra, 1953). There are three to four generations in a year.

Beside this fruit fly, *Dacus dorsalis* (Hendel), *D. correctus* (Bezzi) and *D. zonatus* (Saunders) have also been reported damaging *ber* fruits but these are of minor importance.

To suppress the population of these flies remove and destroy all fruits; remove all the wild *ber* bushes from vicinity of grafted *ber* orchards; rake the soil around the trees frequently, specially during Summer, to expose and kill the pupae. In case of severe infestation, spray 3 to 4 times at triweekly interval commencing from last week of November – first two sprayings with 0.1% lindane and last one or two with 0.06% malathion (Nayar *et al.*, 1976).

MEALY BUGS

Ber mealy bug, *Perissopneumon tamarindus* (Green) has been reported from India and Pakistan. It is a polyphagous pest found damaging apple, banana, *ber*, citrus, fig, mulberry, tamarind, etc. In India, the pest is more common and destructive in Punjab, Haryana and Rajasthan.

Innumerable nymphs and adult females may be seen congregating on shoots, twigs and tender leaves during Summer, sucking the cell sap. As a result of desappings there is loss of vitality and in case of severe infestation, the growth of tree and its fruiting capacity are adversely affected. The bugs also exude honeydew which falls on dorsal surfaces of lower leaves, stems and even fruits. Sooty mould grows rapidly on honeydew, covering the affected areas with a superficial black coating.

Eggs are oval, shiny yellow and less than one mm long. Grubs are flattened-oval, dull, ash-brown in colour with a slight mealy covering. Adult females are wrinkled, apterous, 10 to 14 mm long and completely covered with waxy secretion. Males are alate with only one pair of wings well developed. The gravid females descend from the trees in September-October and lay eggs in the soil about 50 to 150 mm deep. A single female lays 100 to 300 eggs in 50 to 55 days. The eggs overwinter and hatch in Spring (April-May). Nymphal periods are 40 to 62 and 42 to 65 days in case of males and females respectively. There are three nymphal instars in case of females while in males there are only two nymphal instars followed by prepupal and pupal stages. Prepupal period lasts for 1 to 3 days and pupal stage varies between 10 and 14 days (June-July). Males are on wings about a month earlier, fertilize the females and die. Longevity of females is 59 to 81 days and there is only one generation in a year.

Drosicha mangiferae (Green), *Nipaecoccus viridis* (Newstead) *(filamentosus* Cockerell, *vestator* Maskell) are other mealy bugs reported on *ber* trees in India. These are of minor importance.

Frequent raking of the soil under the trees during Winter helps in destroying the eggs and thereby reducing the pest population. Soil treatment with 5% aldrin, chlordane or heptachlor dust @ 300 to 500 gm per tree reduces the population of newly hatched nymphs. Tying around the tree trunks 300 mm wide alkathene sheet (400 gauge) prevents the nymphs from ascending the trees. In case of severe infestation, spray with 0.1% diazinon or moncrotophos.

LEAF DEFOLIATING BEETLES

A few species of cockchafer beetles have been reported damaging *ber* trees. These include, *Holotrichia insularis* Brenske, *Adorestus nitidus* Arrow, *A. pallens* Arrow, *A. versutus* Harold, *Anomala bengalensis* Blanchard, *A. dimidiata* Hope, *A. dorsalis* Fabricius, *A. ruficapilla* Burmeister. Of these *A. nitidus* and *A. pallens* are comparatively more destructive to *ber* trees, specially in North India and *A. versutus* in South India. Though polyphagous, these beetles prefer *ber* and grapevine.

Only adult beetles are destructive to *ber* trees. The grubs remain in the soil and feed on soil humus, roots of various grasses and other vegetation growing around.

Adults are nocturnal, their feeding and mating activities start at dusk and before dawn they go down in the soil to hide during day. The feeding pattern of these beetles is characteristic; the leaf laminae are perforated by numerous small irregular holes, which are bitten out by the strongly toothed maxilla and not by mandibles. The damage is more pronounced in neglected orchards or in orchards grown in sandy or sandy-loam soils; the pest cannot thrive in stagnated water or in soils with poor drainage. Swarms of adults appear early in Spring, their activity is maximum during Summer and declines with heavy showers of monsoon.

Eggs of *A. pallens* are smooth, white and elongated. Grubs are creamy-white and wrinkled. Adults are stout, hardy, convex, dirty-yellowish beetles, about 8 mm long. Eggs are laid singly in soil during May to August. These hatch in 6 to 9 days. Grubs remain active throughout the Summer and make earthen cells in

Autumn to hibernate in these during Winter. Pupation takes place
in the soil during next Spring. Pupal period is 11 to 12 days – thus
there is only one generation in a year.

Crop sanitation, like keeping the orchards free of weeds, etc.,
help in mitigating the pest problem. Light-traps have been found
very useful for trapping the adult beetles (Kalra and Kulshreshtha,
1961). Ploughing around the trees is useful in exposing the
hibernating grubs and killing the same. Flooding the fields prevents
the beetles from egg-laying as also kills some of the grubs and drives
away the other grubs deep down in the soil. In case of heavy
infestation, dusting may be done with 10% BHC, but then the fruits
should not be harvested for atleast two weeks after the insecticidal
treatment.

LEAF EATING WEEVILS

A large number of grey and black weevils have been reported
from different parts of India as foliage feeders of *ber* trees. These
include, *Amblyrrhinus poricollis* Boheman, *Atmetonychus pergrinus*
Olivier, *Dereodus pollinosus* Redtenbecher, *Hypolixus pica*
Fabricius, *Myllocerus discolor* Boheman, *M. sabulosus* Marshall, *M.
spinicollis* Marshall, *M. transmarinus* Herbest, *M. unidecimpustuletus*
Faust, *Peltotrachelus pubes* Faust, *Phytoscaphus triangularis* Olivier,
Platymycterus sjostedti Marshall, *Sitones crinitus* Olivier, *Tanymecus
circumdatus* Wiedemann, *T. hispidus* Marshall, *Xanthochelus faunus*
Olivier and *X. superciliosus* Gyllenhal. Of these only *X. superciliosus* is
of regular occurrence, often causing severe damage; all others
are sporadic pests, causing hardly any significant loss. Both grubs
and adults of these weevils do the damage. Grubs usually feed on
roots specially during Summer. The adults prefer leaves though
sometimes flower-buds are also attacked.

Eggs of *X. superciliosus* are ovoid and whitish-yellow in colour.
Grubs are fleshy, wrinkled and curved in shape, about 10 mm long
and white in colour. Adults are large grey weevils about 15 mm long
with a conspicuous snout. Eggs are laid in soil. These hatch in 4 to
6 days, while the grubs mature in 4 to 10 weeks. There are five to
six overlapping generations in a year (Butani, 1973 a).

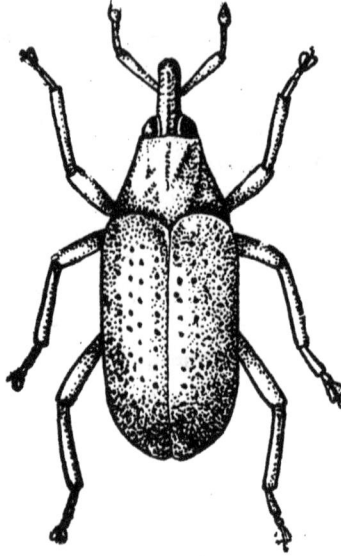

Xanthochelus supercilious Gyllenhal

To control these weevils, dust the trees with 10% BHC or spray with 0.04% phosphamidon or quinalphos. A gap of atleast two weeks should be kept between insecticidal treatment and harvest of the fruits.

Other beetles found damaging *ber* leaves, include, spiny beetle *Platypria andrewesi* Weise and tortoise beetles, *Oocassida cruentata* (Fabricius), *O. obscura* (Fabricius) and *O. pudibunda* Boheman. However, these are all minor pests.

Occassida obscura has been recorded from South India. Grubs and beetles scrape and bite holes in leaf lamina. Grubs are flattened elongate oval in shape and spiny. They have a forked process at their posterior end, which is bent upward and carry cast off skins forming a shield-like covering over the body. Adults are brightly coloured beetles, 5 to 6 mm long, broadly oval in shape with head concealed under prothorax and abdomen and elytra broadly widened.

Platypria andrewesi has been reported from Tamil Nadu, Maharashtra, Madhya Pradesh and Punjab on *ber* leaves and from Gujarat on sugarcane leaves (Maulik, 1919). Eggs are laid on tender leaves. On hatching the grubs mine the leaves eating out small cavities or pockets. The grubs do not remain in one place or cavity but move about making more and more cavities. Pupation takes place on the leaves in one of these cavities. Grubs are flat with large and hard head having short antennae; legs are well developed that enable the grubs to run actively. Adult beetles are small, 4 to 5 mm long, reddish-brown in colour and covered with whitish hair. Elytra broader at the base than near the prothorax and full of black spines.

Platypria andrewesi Wiese

LEAF HOPPERS AND BUGS

Eurybrachys tomentosa Fabricius, commonly known as green stripped leaf hopper, sucks the sap from leaves. It is a polyphagous pest found throughout India. Adult females are greenish-olivaceous in colour while males are brownish in colour and smaller than females. Eggs are laid in clusters on leaves and shoots and are covered with a thick mat of white flocculent efflorescence by the females. A female lays on an average six egg-clusters of 30 to 42 eggs each. Incubation period is 2 to 3 weeks and nymphal 13 to 19 weeks with three overlapping generations in a year.

Zygnidia pakistanica (Ahmed) is another leaf hopper reported from Pakistan and India (Punjab). It is active during July to April, when nymphs and adults suck the sap from ventral surface of leaves producing whitish stippling on dorsal surface of leaves.

Cotton jassid, *Amrasca lybica* (de Berg.) has also been reported feeding on leaves of *ber* trees. This is a major pest of cotton and number of vegetables in various African countries, Mediterranean region and Middle East (CIE map No. A–223). In India, there are only stray records of this jassid occurring mostly in association with *Amrasca biguttula biguttula* (Ishida).

Machaerota planitiae Distant, the spittle bug is a minor sporadic pest recorded from Bihar. Eggs are laid singly in tissues of leaves, shoots and flower-buds. On hatching, the nymphs feed on leaves and shoots and get covered with froth which hardens and forms a white calcareous tube. Number of these tubes may be seen attached to stems and shoots of the affected trees.

Besides, there are various other bugs reported time and again from various parts of India. These include, *Agonoscelis nubila* Fabricius, *Tessaratoma javanica* Thunberg, *Leptocentrus obliquis* Walker, *L. taurus* (Fabricius), *Otinotus oneratus* Walker, *Iricentrus bicolor* Distant, *Helopeltis febriculosa*, *Monosteira minutula* Montandon, *M. edeia* Drake and Livingstone and *Urentius ziziphifolius* Menon and Hakk. None of these cause any major loss to *ber* trees.

Spraying with 0.03% dimethoate, phosphamidon, oxydemeton methyl, diazinon or quinalphos is quite effective in controlling these bugs.

SCALE INSECTS

A number of hard and soft scale insects have been recorded almost on all fruit trees including *ber*. Those found damaging *ber* trees are, *Aonidia zizyphi* Rahman and Ansari, *Aonidiella aurantii* (Maskell), *A. orientalis* (Newstead), *Aspidiotus destructor* Signoret, *Chrysomphalus ficus* Ashmead, *Hemiberlesia lataniae* (Signoret), *Howardia biclavis* (Comstock), *Phenacaspis megaloba* (Green), *Parlatoria oleae* (Colvée), *P. pseudopyri* Kuwana, *Ceroplastodes cajani* (Maskell), *Coccus acutissimus* (Green), *C. discrepans* (Green), *Pulvinaria burkilli* Green and *P. maxima* Green. Besides, lac insects, namely, *Kerria communis* (Mahdihassan), *K. fici* (Green), *K. indicola* (Kapur), *K. jhansiensis* (Misra), *K. lacca* (Kerr) *K. mysorensis* (Mahdihassan) and *K. pusana* (Misra) have also been recorded on *ber* trees. Until unless these species are reared commercially for lac production, their presence is considered as harmful, because they do devitalize the trees and affect adversely the yield of fruit.

Most of these insects are polyphagous and some are even major pests of various fruit crops. However, on *ber* trees these are generally of minor importance.

Remove and destroy promptly the twigs bearing infested leaves to prevent the population build-up. If necessary, spray with 0.05% carbophenothion, dichlorvos or fenitrothion.

THRIPS

Dolichothrips indicus (Hood), *Scirtothrips dorsalis* (Hood) and *Scolothrips asura* Ramakrishna and Margabandhu have been reported occurring occasionally on *ber* leaves the former is also found on *jamun* and litchi whereas the latter has been recorded on banana leaves. The damage caused by these thrips to *ber* leaves is seldom severe.

Scirtothrips dorsalis, commonly known as chilli thrip, has been reported infesting the flowers of *ber*, citrus, grapevine (also fruits), mango, pomegranate, tamarind, etc. The infested flowers gradually fade, wilt and naturally bear no fruit.

Spraying with 0.03% dimethoate, diazinon or phosphamindon is quite effective in controlling these thrips.

BARK EATING CATERPILLARS

Indarbela quadrinotata (Walker) and *I.tetraonis* (Moore) have been reported feeding on tree trunks of various fruit trees throughout Indian subcontinent. In India, *I. quadrinotata* is more destructive to mango trees and *I.tetraonis* to guava trees. The caterpillars of both species have been recorded feeding on trunks and main branches of *ber* trees but not as major pests. These caterpillars are dirty brown in colour. Adult moths of both species are pale brown having no proboscis and the antennae of male moths are bipectinate to the apex (Butani, 1977 a). Moths of *I. tetraonis* are slightly bigger in size and paler in colour than those of *I. quadrinotata*. There is only one generation in a year, the feeding period of caterpillars being 10 to 11 months.

LEAF EATING CATERPILLARS

Ber butterfly, *Tarucus theophrastus* (Fabricius) is widely distributed all over Indian subcontinent. Besides, it has also been reported from UAR, Iran, northern and western Africa, Arabia and Baluchistan. It is a major pest of various *Zizyphus* species and a minor pest of *Citrus* spp. Eggs are laid on tender leaves. On hatching, the caterpillars feed on leaves and flower-buds. The maximum activity of the pest has been observed during May-June and September-October. Eggs are small, flat, round and pale green in colour. Full grown caterpillars are 11 to 15 mm long, fleshy and flattened with pale ochreous head and pale green body. Pupae are smooth and green

in colour, head, thorax and wing-pads speckled thickly with black.
Adult butterflies have dark brown to black body and beautifully
patterned bluish wings; wing expanse is 22 to 31 mm (Bingham, 1907).

Tarucus theophrastus (Fabricius)

Hairy caterpillars, *Thiacidas postica* Walker, *Euproctis fraternæ*
(Moore), *E. lunata* (Walker), *E.flava* (Bremer) and *Beara dichromella*
Walker have been reported feeding on *ber* leaves (Pruthi and Batra,
1960). Of these only *T. postica* is commonly met with, specially in
South India. The young caterpillars feed gregariously but later they
disperse and feed voraciously defoliating the entire twig. Eggs are
spherical. Full grown caterpillars are 18 to 25 mm long, yellowish-
brown in colour with plenty of hair all over the body. Adult males
are stout, medium sized, dull coloured moths that fly by night and
hide during day. Females are apterous and sluggish.

Thiacidas postica Walker

Eggs are laid in batches and normally all the eggs in a batch
hatch simultaneously. Incubation, larval and pupal periods last for,
on an average, 8, 25 and 8 days respectively.

Streblote siva (Lefevre) is another hairy caterpillar found feeding on *ber* and guava leaves specially in Bengal. Its main host is rose. Caterpillars are pale-ochreous-brown with small black spots and long lateral tufts of ochreous hair. Cocoons are formed of pale hair on the twigs of the trees. Adults are beautifully coloured moths having head and thorax greyish-white, abdomen white, forewings with conspicuous reddish-brown spot ringed with white, hind wings white with slight fuscous suffusion on outer area and a dark patch on outer margin (Hampson, 1896). Wing expanse is 42 to 52 mm and 70 to 80 mm in case of males and females respectively.

1. Eggs

2. Caterpillar

3. ⎫ Adults
4. ⎭ ♀ ♀

5. ⎫ Adults
6. ⎭ ♂ ♂

Streblote siva (Lefevre)

Porthmologa parachina Meyrick is confined to eastern India and has been reported causing considerable damage in Assam (Chowdhury and Majid, 1954) and Bihar. The caterpillars are dull green with black spots on their bodies. They web silken covers on *ber* leaves and feed within on chlorophyll. Occasionally these caterpillars bore into the shoots as well.

Wild-silk moth, *Cricula trifenestrata* Helfer is widely distributed in India, Bangladesh, Burma, Sri Lanka and Indonesia. In India, it appears sporadically in large number in some seasons defoliating the leaves of *ber,* cashew nut, mango, etc. Full grown caterpillars are stout, dark brown and about 50 mm long. These pupate in golden-yellow silken cocoons. Adults are reddish-brown; females having three clear hyaline spots on forewings and one on hind wing. Wing expanse is 60 to 82 mm in case of males and 70 to 92 mm in case of females.

Cricula trifenestrata Helfer

Caterpillars of Tassar silk moth, *Amthearea paphia* ·Linnaeus have also been reported occasionally defoliating *ber* trees.

To control the damage caused by leaf eating caterpillars hand-picking and prompt destruction of caterpillars in the initial stage of attack is the most economical and effective method. If necessary, dust with 10% BHC. These control measures are also effective against leaf rollers and webbers.

LEAF ROLLERS AND WEBBERS

Ancylis spp. *Porthmologa paraclina* Meyrick, *Psorosticha zizyphi* (Stainton) and *Synclera univocalis* (Walker) are some of the common leaf rollers recorded damaging *ber* trees in India.

Ancylis lutescens Meyrick has been reported from Assam, Bihar, and Madhya Pradesh. The caterpillars roll up the leaves, feed on epidermis of dorsal surface of rolled leaves (external surface of rolled portion is not eaten). Full grown caterpillars are 30 mm long, cylindrical in shape, yellowish-green with alternate brownish and light green longitudinal stripes on dorsal and lateral sides. Pupation takes place in the rolled leaves. Pupae are 7 to 10 mm long, cylindrical in shape and brown in colour. Adults are medium sized moths, fore-wings ferruginous, darker anteriorly, hind wings grey and wing expanse 18 to 22 mm.

Ancylis aromatias Meyrick and *A.cyanostoma* Meyrick have been reported from Bihar and Karnataka respectively as minor pests.

Psorostica zizyphi is found all over India, Pakistan and Śri Lanka. *Ber* and citrus are its main hosts. Eggs are laid along the midribs of leaves. On hatching, the tiny caterpillars mine the leaves, later they roll up longitudinally and web together several apical leaves and feed within; first on epidermis and later on leaf lamina. In case of older leaves only chlorophyll is eaten away. Pupation takes place in transparent white silken cocoons spun on folded leaves. Full grown caterpillars are about 15 mm long, slender and yellowish-green in colour. Pupae are 6 to 8 mm long and reddish-brown. Adults are small moths, brownish in colour with specially rounded broad wings.

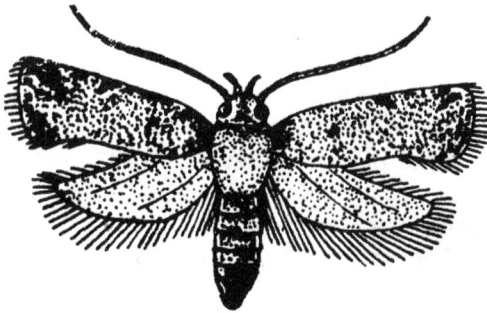

Psorostica zizyphi (Stainton)

Egg, larval and pupal stages last for 3 to 5, 9 to 11 and 5 to 10 days respectively (Nair, 1975). Adult longevity is 15 to 16 days.

To control these leaf rollers, hand-picking and mechanical destruction of caterpillars is suggested. If need be, dust with 10 % BHC.

STEM BORING BEETLES

Longicorn beetles, *Celosterna scabrator* Fabricius and *C.spinator* Fabricius have been reported boring the stems of *ber* trees. Both are minor pests. The grubs tunnel into hard woody tissues, causing hinderance in translocation of cell sap. Grubs are legless, head small with powerful mandibles, prothorax slightly swollen with a dorsal broad plate and abdomen having dorsal plate on each segment.

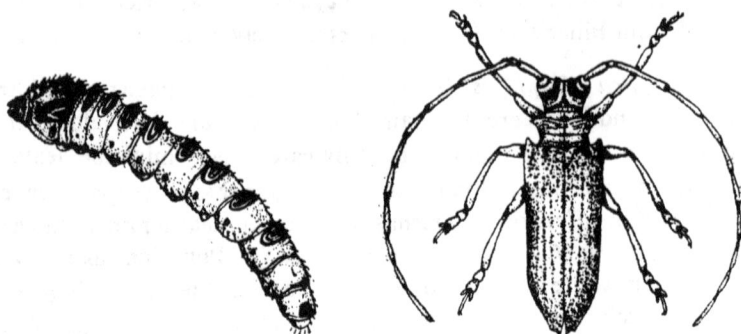

Celosterna scabrator Fabricius

Adults are mottled grey in colour, 22 to 28 mm long; antennae of males are as long as their body and those of females twice the body length.

FRUIT BORERS

Ber fruit borer, *Meridarchis scyrodes* Meyrick is commonly found in South India though it has also been reported from Bihar (Pusa) and Maharashtra (Nagpur). Besides *ber*, the pest has also been recorded on *jamun* and olive fruits. On hatching the caterpillars bore into the fruits, feed on pulp and fill the cavities with faecal matter. Caterpillars are pale yellow during earlier stages, later these become pinkish to reddish in colour while adult moths are dark brown (Basha, 1952).

Pomegranate butterfly, *Virachola isocrates* (Fabricius) – a major pest of pomegranate, is sporadic minor pest of *ber* fruits. It breeds throughout the year migrating from one variety of fruit to another. Eggs are laid on flower-buds and fruits. On hatching, the tiny caterpillars bore inside the fruits and feed within. Pupation also takes place inside the fruits. The infested fruits are subsequently attacked by various bacteria and fungi. As a result the fruits rot and emit offensive smell.

To check the infestation of these fruit borers, remove and destory promptly the infested fruits.

L ITCHI, *Litchi chinensis* Sonnerat (nec. *Nephelium litchi* Cambess) –
a delicious fruit, is indigenous to southern parts of China. It has
been introduced in India, Bangladesh and Burma via Japan during
the end of 17th century. A monograph on litchi by Ts'ai Hsiang in
1059 is considered to be the oldest publication on fruits. Today,
India is the second largest producer of litchi with more than
10,000 hectares under its cultivation. Over 8000 hectares are
in North Bihar and remaining 2000 hectares are spread over Assam,
the submontane area of Uttar Pradesh, Himachal Pradesh and
Punjab. A few trees are also found on slopes in the Nilgiris. There
is ample scope of increasing the area under this fruit, specially along
the entire foot-hill region of Himalayas in Uttar Pradesh, Himachal
Pradesh and Punjab. Litchi is rather exacting in its climatic
requirements. It requires subtropical climate without heavy frost in
Winter and dry heat it Summer. Litchi fruits contain 15% soluble
sugars and 1.15% protein. These are rich in vitamin C and also
contain fair amount of phosphorus, calcium and iron (Singh, 1958).

Litchi trees are attacked by about 40 insect species in nature
(Singh and Singh, 1954; Wadhi and Batra, 1964; Vevai, 1971 b). Tree
trunks and main branches are damaged by bark eating caterpillars;
leaves are attacked by leaf curl mite, aphids, scale insects, bugs,
whiteflies, thrips, leaf miners, leaf eating caterpillars, leaf rollers,
beetles, weevils, stem borer and red ants while fruits are spoiled
by fruit and seed borers, including pomegranate butterfly. Fortunately,
most of these are either sporadic or of minor importance; only bark
eating caterpillars and leaf curl mite cause serious damage.

BARK EATING CATERPILLARS

Indarbela spp. are the most common pests of fruit trees found
in almost all orchards specially in those, that are not well maintained
or are fairly old. The species found damaging litchi trees are,
I. quadrinotata (Walker) and *I. tetroanis* (Moore). Both the species are
widely distributed all over the Indian subcontinent and are

polyphagous with a wide range of host plants including, *aonla, ber,*
citrus, *falsa*, guava, jack-fruit, *jamun*, litchi, loquat, mango, mulberry
and pomegranate (Verma and Khurana, 1974).

Moths and caterpillars are nocturnal in habit. Caterpillars bore
inside the trunk or main stems about 150 to 250 mm deep. At
night the caterpillars come out and feed on the bark protected by
large silken webs that cover the entire affected portion. Later, the
caterpillars eat through the bark into the wood and in case of
severe infestation, sap movement is interfered with and the trees
cease to flush.

To control these borers, clean the affected portions of the
tree by removing all the webs, etc. and insert into the holes cotton-
wool soaked in carbon bisulphide, chloroform, formalin or petrol
and then seal the holes with mud. Shah (1946) considered use of
these chemicals to be · expensive and suggested introduction of hot
water into the holes by means of a cheap syringe. This method is
rather cumbersome and non-practicable, specially in case of big
orchards. Khurana and Gupta (1972) have suggested injecting in
the holes 0.013 % dichlorvos, 0.05 % trichlorfon or 0.05 %
endosulfan.

In addition to *Indarbela* spp., *Lymantria mathura* Moore has been
recorded from Uttar Pradesh damaging inflorescences and bark of
litchi and mango trees (Singh, 1954 a). The pest has also been
reported as leaf defoliator of various forest trees and *jamun*. It is
found throughout the year but is more active during hot weather and
early rains. Caterpillars are 50 mm (♂) to 85 mm (♀) in length, ash-
grey in colour with yellow bands across the thorax and abdomen having
rows of papules bearing tufts of long hair. Pupae are found on the
leaves fastened with a few silken strands. Pupae are 15 to 25 mm long,
smooth except for a few groups of short bristles. Moths exhibit
dimorphism – forewings of both the sexes are white with brown

♂   ♀

Lymantria mathura Moore

patches but those of females have pinkish edges; hind wings of males are light orange with brown patches while those of females are white with brown margins. Wing expanse of males is 40 to 50 mm and that of females, 80 to 100 mm.

Egg, larval and pupal periods last for 8 to 10, 20 to 22 and 8 to 10 days respectively.

The red borer, *Zeuzera coffeae* (Nietner) has been reported as a major pest of litchi in Bangladesh boring the bark of the trees (Alam *et al.*, 1964). In India, this is a major pest of coffee and tea and minor pest of various fruit trees including litchi.

LEAF CURL MITE

Eriophyid mite, *Aceria* (*Eriophyes*) *litchi* (Keifer, 1943) is the most destructive pest of litchi and has been reported from almost all the litchi growing countries of the world. In India, it has been reported from Assam, Bengal, North Bihar, parts of Uttar Pradesh and South India (Misra, 1912; Puttarudriah and Channa Basavana, 1959). It is specific pest of litchi. Young plants and seedlings in nursery are more liable to attack. Nymphs and adults are usually found near the base of hair on ventral surface of leaves. The mites puncture and lacerate the tissues of leaf with their stout rostrum and suck the cell sap. Chocolate-brown-velvety growth (erinose) on the ventral surface of infested leaves is the characteristic symptom of attack by this pest. In the beginning, small deep excavated pits may be found lined throughout with brownish-velvety pubescence and when these coalesce, the leaves curl up apically or double over vertically forming hollow cylinders; ultimately the attacked leaves wither and fall down. No sooner the withering of leaves starts, most of the mites move upwards to infest the fresh succulent leaves. The attack by this mite generally begins from lower portion of the tree and gradually extends upwards. Besides the leaves, sometimes flower-buds and immature fruits are also attacked (Alam and Wadud, 1964). The mite also causes the leaf curl disease of litchi (Ghai, 1976).

Eggs are extremely small, 0.04 mm in diameter, round in shape and whitish in colour. Nymphs and adults are also whitish in colour and similar in appearance, only the nymphs are smaller in size and have lesser number of lateral setae than adults. Both are minute vermiform four legged acarines – a unique feature, separating

eriophyid mites from all other acarina. Both pairs of legs are
situated at anterior end. Adults are 0.15 to 0.2 mm long, abdomen
greatly enlarged having about 55 ring-like segments.

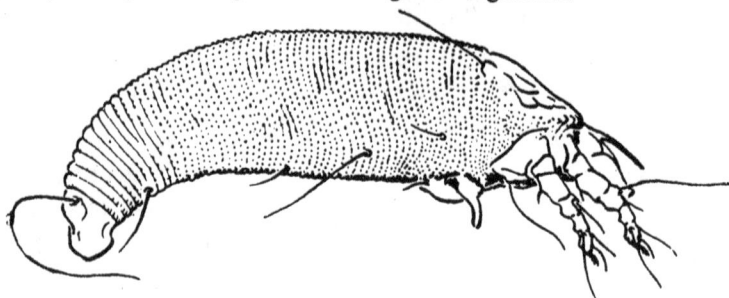

Aceria litchi (Keifer)

Eggs are laid singly on the ventral surface of leaves at the base
of hair. Incubation period is 2 to 3 days, nymphal 8 to 12 days and
adult longevity is 2 to 3 days. Sexual dimorphism is evident only in
adult stage. Overwintering is in adult stage.

To check the infestation of mites, collect and burn all the
infested leaves. Misra (1912) as also Nishida and Holdway (1955)
suggested spraying with sulphur. Use wettable sulphur @ 100 gm in
20 litres of water. Give one spraying in May-June (after harvest of
fruits) and one or two more sprayings in Winter (December-January),
when the trees are dormant and mites are torpid with cold. These
treatments have been found to be very effective.

SAP SUCKING INSECTS

Drosicha dalbergiae (Green), *D. mangiferae* (Green), *Fiorinia
nephelii* Makell, *Parlatoria pseudopyri* Kuwana, (*cinerea* Hadden),
Chloropulvinaria psidii (Maskell), *Geococcus radicum, Parasaissestia
coffeae* (Walker) and *Kerria albizziae* (Green) have been recorded
feeding on litchi leaves and twigs. All these are of minor
importance, causing negligible loss. Nevertheless, if and when these
coccoids appear, remove and destroy the infested leaves and twigs.

Black citrus aphid, *Toxoptera aurantii* (Fonscolombe) – a major
pest of citrus spp., is a minor pest of various fruit trees, including,
custard apple, jack-fruit, litchi, loquat, mango, sapota and tamarind. It
is also found in greenhouses on a variety of plants. Clusters of blackish-
brown individuals (nymphs and adults) may be seen on flush foliage

sucking the cell sap. As a result, there is discolouration of young leaves. The insects also exude honeydew on which sooty mould develops. This black coating interferes with the photosynthetic activity affecting adversely the growth and fruiting capacity of the tree. Dry weather followed by rainy season is most favourable for the multiplication of this aphid (Butani, 1977 c).

Whiteflies, *Aleurocanthus husaini* Corbett and *Dialeurolonga elongata* (Dozier) have been reported sucking the sap from litchi leaves. The main host of these species is citrus. In addition, *A. husaini* has also been recorded on peach, pear and plum trees. The affected leaves turn pale and get deformed.

Dolichothrips indicus (Hood) has been reported lacerating and rasping on the leaves while *Megalurothrips distalis* (Karny) (*nigricornis* Schmutz) and *M. usitatus* (Bagnall) infest the flowers, feeding on pedicels, sepals, petals and even on stigma. If the attack is severe, which is rare, the leaves curl and dry away whereas the flowers get devitalized,

To control the aphid, whiteflies, and thrips spray with 0.03% dimethoate, monocrotophos or phosphamidon. One to two sprayings at an interval of 8 to 10 days are good enough to check these pests effectively.

A membracid bug, *Gargara varicolor* Stal, has been reported from Uttar Pradesh, appearing in large number at the time of emergence of new leaves, sucking the sap from young twigs (Hukam Singh, 1978).

Pentatomid bugs, *Chrysocoris stolli* (Wolff), *Halys dentatus* Fabricius, *Tessaratoma javanica* Thunberg and *T. quadrata* Distant have been recorded feeding on litchi trees. These are all minor pests. *C. stolli*, an specific pest of litchi, is comparatively more common. This pest is active throughout the year except during severe cold (December- January). Incubation and nymphal periods last for 5 to 7 and 19 to 29 days respectively (Singh and Sharma, 1961).

Spilostethus pandurus (Scopoli) – a polyphagous pest, has been reported damaging litchi leaves (Bhattacharjee,1959). It has also been found feeding on apple, apricot, citrus, fig, *jamun*, pistachio (*Pistacia vera*), various vegetables, etc. Besides sucking the cell sap, these insects also inflict injury to flowers and tender fruits. The affected flowers dry and witheraway without forming fruits while the fruits when attacked drop down prematurely. Eggs hatch in 4 to 5 days and nymphal duration lasts for 25 to 33 and 28 to 38 days during Summer and Winter respectively. Adult longevity is about a month. When the attack is serious, spray with 0.05% diazinon or dichlorvos to control these bugs.

LEAF MINERS

Acrocercops heirocosma Meyrick has been reported from eastern India while *A. cramerella* Snellen has been recorded from Bihar and Uttar Pradesh. The caterpillars mine the leaves during August to October and feed within. Caterpillars of *A.cramerella* have also been observed damaging the ripe fruits while those of *A.heirocosma* bore into the midribs near the base of leaves and tunnel upwards, mining the leaf lamina on both sides of the midrib. The mined portion develops a rust-red colour and if the attack is severe the entire leaf dries up. Full grown caterpillars leave the mines, form transparent round scale-like cocoons on the leaves and pupate therein. Full grown caterpillars are green in colour and 4 to 6 mm long. Pupae are slender, cylindrical, 4 to 5 mm long and green in colour when freshly formed, turning yellowish-brown after a few hours.

As the damage caused by leaf miners is generally negligible, no control measures are normally warranted. However, Hukam Singh (1975) has suggested spraying with 0.05% quinalphos, fenitrothion or dichlorvos,

LEAF ROLLERS

Tortrix (*Cecoecia*) *epicyrta* (Meyrick), *Platypeplus aprobola* (Meyrick) and *Olethreutes* (*Argyroploce*) *leucaspis* (Meyrick) have been recorded on litchi (Wadhi and Batra 1964). *T. epicyrta* is found all over India and is a polyphagous pest damaging apple, apricot, citrus, guava, *jamun*, litchi, peach, etc., *P. aprobola* and *O. leucaspis* have been reported mainly from Bihar and Uttar Pradesh respectively. *P.aprobola* has also been found damaging *jamun* and mango trees whereas *O. leucaspis* is specific pest of litchi trees. The caterpillars web together tender leaves and adjacent flower-buds and feed within on chlorophyll and petals; occasionally they may bore into the buds and feed within. They also roll the leaves and feed inside the same. As a result, if the infestation is severe, the

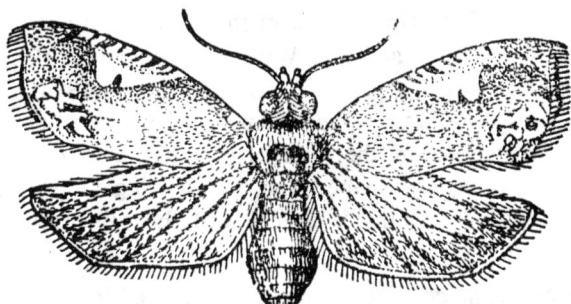

Olethreutes leucaspis (Meyrick)

fruiting capacity of the trees is adversely affected. Egg, larval and pupal durations of *O. leucaspis* last for 2 to 8, 12 to 53 and 7 to 24 days respectively (Misra and Pandey, 1965). Life-cycle is completed in 21 to 85 days with five to six generations in a year.

LEAF EATING CATERPILLARS

Procometis spoliatrix (Meyrick), *Selepa celtis* Moore, *Sympis rufibasis.* Guenée, *Oenospila quadraria* (Guenée) and *O. veraria* (Guenée) have been reported from various parts of India as defoliators of litchi trees. Of these, the last two are comparatively more common. Caterpillars of *Oenospila* spp. are the true loopers having elongated slender body with prolegs on sixth and tenth abdominal segments. The caterpillars are brown in colour with greenish tinge. This is a deceptive camouflage as it closely resembles the colour of twigs of the tree and often the larvae are mistaken for leaf petioles.

As these insects are not of any major importance, hardly any work has been done on biology, bionomics, etc. of these speciess Nevertheless, if and when these caterpillars appear in large number, spray with 0.05% endosulfan or quinalphos.

BEETLES AND WEEVILS

Diapromorpha melanopus Lacordaire *D. quadripunctata* Jacoby and *Cryptocephalus insubidus* Suffrain are the chrysomelid beetles found feeding on litchi leaves but not causing any severe damage.

Amblyrrhinus poricollis Boheman, *Myllocerus delecatulus* Boheman, *M. discolor* Boheman, *M. dorsatus* Fabricius, *M. undecimpustulatus* and *Ptechus* species are some of the weevils found nibbling the leaves starting from leaf margins. None of these cause any severe damage. These weevils are more active in Winter (November to February) than during Summer (April to July). Hukam Singh (1974) has reported that adults of *Apoderus blandus* Faust feed on chlorophyll of young leaves. For oviposition, the females roll the leaves transversely into a compact cylindrical shape and lay a single egg inside. On hatching, the grub feeds inside the roll and the roll subsequently dries up and falls down.

Normally, no control measures are required for these beetles and weevils. However, spraying with 0. 04% diazinon or dichlorvos or 0.05% fenitrothion is effective in checking the pest population.

SHOOT BORER

Chlumetia transversa Walker – a minor pest of mango and litchi, is widely distributed all over India and S. E. Asia. The caterpillars bore into tender shoots near the growing points and the affected shoots are often killed outright. To check the population build-up, remove and destroy the affected shoots as soon as the damage is observed.

FRUIT AND SEED BORERS

Pomegranate butterfly, *Virachola isocrates* (Fabricius) also attacks the litchi fruits occasionally. The infested fruits emit bad smell and drop down prematurely.

Singh and Singh (1954) reported pomegranate borer, *Deusorix epjarbas* (Moore) and Nayar *et al.* (1976) has mentioned *Rapala* species as fruit borers.

Cryptophlebia illepida (Butler) (*carpophaga* Walsingham) – a polyphagous pest, has been reported attacking number of fruits, including litchi. The caterpillars bore into the developing fruits and tunnelling through pulp they attack the seeds and feed inside the same. The tunnels get filled with the excreta of caterpillars and start rotting.

Cryptophlebia illepida (Butler)

All these are minor pests and require no control measures, However, remove and destroy the affected fruits to prevent the infestation from spreading.

RED ANT

Oecohpylla smaragdina (Fabricius) also build their nests on litchi trees. These ants are active throughout the year and are carnivorous in habit and ferocious by nature. As such, besides causing minor damage to litchi trees, they are a source of nuisance to the pickers, etc., who often get badly bitten by these ants. As soon as the ants' nests appear on the trees, these may be removed and destroyed mechanically.

Plus je connais les peuples
Plus j'aime les insectes

L OQUAT, *Eriobotry japonica* Lindley, also known as Japanese plum, is native of Central China. It is being cultivated in Japan since antiquity from where it spread throughout the subtropical regions of the world. At present, besides India, China and Japan, loquats are commercially grown in parts of Australia, South Africa and USA (Srivastava, 1957). In India, the fruit occupies an area of about 2000 hectares confined mainly to submontane area of Uttar Pradesh (1200 hectares), Himachal Pradesh and Punjab. A small area is scattered around Delhi, Haryana, parts of Gujarat and Maharashtra, Assam, Bengal as also hills of South India. The fruits are used for dessert purpose as well as for making jam and jelly.

Mango fruit fly and bark eating caterpillar are the only pests of major importance, while scale insects are sporadic pests causing severe damage only in certain pockets during certain years. The minor pests include, aphid, chafer beetles, grey weevils, carpenter bee and pomegranate butterfly.

MANGO FRUIT FLY

Dacus dorsalis (Hendel) is a polyphagous pest, mango being its most preferred host. It has been observed damaging loquat fruits during April-May and the late maturing varieties suffer comparatively more than the early ripening ones. Eggs are laid just beneath the epicarp of ripening fruits (1 to 2 mm deep), On hatching within 2 to 3 days, the maggots tunnel into the fruits and feed on pulp for 12 to 15 days. When full fed, the maggots come out of the fruits, fall on the ground and pupate in the soil. Pupal period is about one week in Summer and extends to over 6 weeks during Winter. The infested fruits do not fall down but remain on the tree with brownish syrupy juice oozing out from the punctures made by the female flies for ovipositing or by maggots for coming out. In South India the pest is active throughout the year migrating from host to host whereas in North India, it overwinters during November to March in pupal stage in the soil. It is because of the late emergence of adults, that the early ripening varieties of loquat escape the attack by this pest, while the fruits left over after harvest suffer the most.

To prevent the attack of fruit fly, harvest the fruits while still hard; do not leave any fruits, however small, on the trees after harvest, as these serve as breeding sources for the flies. Plough around the trees and rake up the soil specially during Winter to expose and kill the pupae. Spraying with yeast hydrolysate + sugar + malathion is also effective (Gupta, 1958).

BARK EATING CATERPILLAR

Indarbela quadrinotata (Walker) is a notorious pest, found all over India on a number of fruit trees, loquat being one of these. It is more common in Uttar Pradesh and is serious in neglected orchards.

Eggs are laid on the bark of the trees. Freshly hatched larvae nibble the bark for a couple of days then bore into the tree trunk and feed within. The damage is conspicuous by the presence of thick ribbon like webs comprising of chewed wood particles and excreta, hanging loosely on tree trunks and main branches specially near forking of the branches.

To control this pest, remove all the webs and insert into the holes, cotton wool soaked in carbon bisulphide, chloroform, 0.05% trichlorfon or endosulfan (Khurana and Gupta, 1972).

SCALE INSECTS

Coccus (*Lecanium*) *viridis* (Green), *Chloropulvinaria psidii* (Maskell) *Eulecanium tiliae* (Linnaeus), (*capreae* Linnaeus, *corylii* Linnaeus), *Saissetia coffeae* (Walker), *Parlatoria oleae* (Colvée) and *P. pseudopyri* Kuwana have been reported damaging loquat in India (Vevai, 1971b). These are all polyphagous pests having a wide range of host plants. On loquat, *C.viridis* is comparatively more common than the others.

Coccus viridis, popullary known as soft green scale, is a major pest of coffee all over the world. It is cosmopolitan in distribution in the tropics with the exception of Australia (CIE map No. A–305). In India specially in southern region, the pest is also found on *bael*, citrus, guava, loquat, mango and a number of other wild and cultivated hosts. Innumerable young ones and adults may be found crowding along the midribs and veins on the ventral surface of leaves. They also attack tender branches and developing fruits. As a result the affected parts become discoloured and malformed. Severely infested fruits fall off prematurely. These insects also cause further indirect loss by producing honeydew. This sweet secretion serves as a substrate for development of sooty mould and also attracts various ants like, *Oecophylla smaragdina* (Fabricius), *Camponotus compressas* Linnaeus, *Cremastogastor* species, etc. These ants protect the scale insects from their predators and also aid in dissemination from tree to tree.

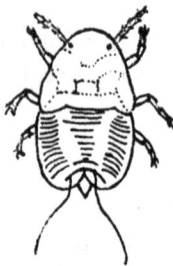

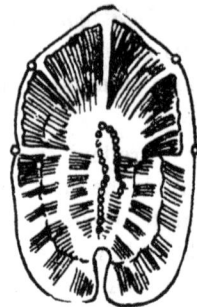

Coccus viridis (Green)

These insects reproduce ovo-viviparously, as males are very rare. A female lays 300 to 500 eggs in her life time of 2 to 5 months. These hatch within a few hours but the young ones remain under the body of the female for a few days then crawl about a little in search of a succulent spot and fix themselves there for feeding. The crawlers are flat, ovate, slightly convex, pale green to yellowish-green in colour; when mature these are 2 to 3 mm long.

Brown soft scale, *C. hesperidum* are also often found along with *C. viridis*. These have same feeding habits and similar life-history as *C. viridis*. A single generation may last for 30 to 70 days (maximum 180 days during Winter) and the longevity of adult females varies between 17 and 48 days.

Remove and destory all the infested leaves, twigs and fruits to prevent the pest population from spreading. Two sprayings with 0.05 % chlorfenvinphos or quinalphos at an interval of 7 days are quite effective in checking the pest for four weeks (Singh and Rao, 1977).

APHID

Black citrus aphid, *Toxoptera aurantii* (Fonscolombe) has been recorded damaging not only citrus but also custard apple, jack-fruit, litchi, loquat, mango, sapota, tamarind, etc. Both nymphs and adults suck the cell sap from tender leaves, twigs and shoots. They also secrete honeydew which encourages the development of sooty mould and very soon the affected parts get covered with a superficial black fungal growth. It is however, a pest of minor importance, can be controlled by spraying 0.03 % dimethoate, or phosphamidon or 0.05 % endosulfan monocrotophos or dichlorvos.

THRIPS

Haplothrips species and *Heliothrips* species have been reported infesting the flowers during Winter. These are sporadic pests and do not cause any appreciable damage. If and when necessary, spray with 0.03% dimethoate or phosphamidon or 0.05% endosulfan or dichlorvos.

CHAFER BEETLES

Adoretus duvauceli Blanchard. *A. lasiopygus* Burmeister, *A. horticola* Arrow and A. *versutus* Harold have been reported damaging loquat trees. These are polyphagous pests and loquat is not their primary or preferred host because the leaves of loquat trees are rather tough. It is only when other hosts are not easily available these beetles attack and feed on loquat leaves. The grubs remain in the soil feeding on roots of loquat and other grasses growing around while the adult beetles feed on leaf lamina making big irregular holes. The adults feed during night and hide in the soil or under big clods during day.

In case of severe attack, which is rather rare, dust the trees with 10% BHC or spray with 0.2% carbaryl; if necessary, repeat the treatment after 7-8 days (Butani 1974 j).

GREY WEEVILS

Apple weevil, *Myllocerus discolor* Boheman and almond weevil, *M. laetivirens* Marshall have been occasionally recorded damaging loquat trees. These are polyphagous pests and feed on a variety of host plants. The grubs live underground and feed on roots while the adult weevils nibble the leaves from margins and eat away small patches of leaf lamina.

Normally these weevils do not cause any severe damage to loquat trees and therefore, no control measures are required; but if there is a servere infestation, dust with 5% BHC or carbaryl or spray with 0.05% dichlorvos or fenitrothion.

CARPENTER BEE

Megachile anthracina Smith has been occasionally recorded damaging loquat leaves. These are stout bodied, densely haired solitary bees.

Megachile anthracina Smith

They build mud cells as their nests in racks and crevices in dead wood and hedges. For lining their nests the female bees cut and carry away large neat circular or oval patches of leaf lamina of various plants including loquat. This is usually done before and after monsoon. A series of earthen cells is constructed end to end and provisioned with honey and pollen. In each cell, an egg is laid on the stored food-mass and the cell is sealed. On hatching, the larvae feed on the available food-supply provided and pupate there in the empty cell. The defoliation in case of loquat trees is hardly of any economic importance. Nevertheless, dusting the leaves with pyrethrum or 5% BHC dust retards the attack of these bees (Butani, 1974 k).

POMEGRANATE BUTTERFLY

Virachola isocrates (Fabricius) though primarily a serious pest of pomegranate, has a wide range of host plants including loquat. Eggs are laid on the fruit panicles. On hatching, the caterpillars bore inside the fruits and feed within. The infested fruits give out offensive smell. The holes made by caterpillars also facilitate the entry of various bacteria and fungi. The affected fruits ultimately rot and fall down.

Bagging of fruits has been suggested but this is neither practicable nor economical specially in case of loquat. Removal of infested fruits and their prompt destruction prevents the population build-up and spread of this pest and should therefore be adopted.

Beneath these green mountains
where Spring rules the year,
The arbutus and LOQUAT
in the season appear;
And feasting on LYCHEE
three hundred a day
I shouldn't mind staying eternally here

SU SHIH (1904 AD)

3.6
DATE PALM

DATE PALM, *Phoenix dactylifera* Linnaeus is native of western Asia and the deserts of Middle East. As the legend goes, it was introduced in India by the army of Alexander the Great (Popenoe, 1913). Toddy palm, *P. sylvestris* Roxburghii, however, is said to be of Indian origin. Date palm is one of the earliest fruit, being cultivated for atleast 5000 years. It has been an important staple food for many people in the hot deserts South and East of Mediterranean. It can grow in arid areas better than any other crop. High temperature is necessary to ripen the fruits, but cold can also be tolerated by these trees. Low humidity is desirable. The fruit has very high food value with sugar content of about 54% and protein 7% (Hill, 1975). Besides, Iraq and Persian Gulf, dates are also grown in other Arabian countries, Pakistan, India, Spain, southern California, Texas, etc. In India date palms occupy an area of about 400 hactares but there are no commercial plantations of high quality dates. Dried dates worth rupees thirty millions are imported every year (Singh *et al.*, 1963). It is therefore necessary to plan and encourage the date cultivation on commercial scale in suitable agro-climatic regions. There is ample scope to extend date cultivation in Kutch, Rajasthan as also in arid districts of Haryana and Punjab.

Only 13 insect species have been reported damaging date palms in India. In addition, Bhaduri (1958) has appended six satyrid and 15 nymphalid butterflies, sipping juice from date palms tapped for toddy during November. Of all these, only balck palm beetle, red palm weevil and a few species of scale insects cause sufficient damage to warrant their inclusion in the category of major pests.

BLACK PALM BEETLE

Oryctes rhinoceros (Linnaeus) commonly known as Rhinoceros beetle was originally described by Linnaeus (1758) as *Scarabaeus rhinoceros*. It is indigenous to India and by now it has also been recorded from Sri Lanka, Bangladesh, Burma, Thailand, Malaysia,

South China, Taiwan, Indonesia and Philippines (CIE map No. A–54). Besides date palms, it is also found damaging coconut and various other palms as also occasionally, banana, pineapple, etc.

There are as many as 42 *Oryctes* species recorded from various parts of the world and most of these feed on various palm trees. Both grubs and pupae are usually found in undisturbed heaps of decaying organic debris, cowdung, manure pits or dead and decaying palm logs or trees. Beetles are the destructive stage of this pest. They are nocturnal in habit and not capable of long flights. At night these beetles feed on the crowns of palms by boring into the unopened tender fronds and tunnel downwards feeding on soft tissues of the growing plant. As a result, the growth of the trees is arrested. Young trees are more prone to attack by this pest and these trees, when attacked, ultimately wither and die away. The maximum damage by this pest is done during the monsoon. Incidentally, it may be mentioned that Ghosh (1924) has reported that the grubs are relished as a delicacy hy the Burmese.

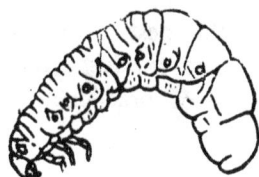

Grub

Adult

Pupa

Oryctes rhinocerus (Linnaeus)

Eggs are oval in shape, 3 to 4 mm long and white in colour. Grubs are 80 to 100 mm long, soft, fleshy and wrinkled. Pupae are 40 to 50 mm long and uniformly yellowish-brown. Beetles are robust, 35 to 50 mm long, glossy-dark-brown to black dorsally and reddish-brown ventrally. The pygidium in males is rather long, smooth and rounded while in females, it is short, hairy and relatively pointed. Nirula (1955) studied its bionomics. Eggs are laid in rotting vegetation, manure or compost heaps. These hatch in 8 to 14 days. Grub duration is 47 to 191 days and pupation takes place in the soil, 150 to 600 mm deep. Pupal period is 14 to 29 days and the adults remain in cocoons for 5 to 26 days till they become sexually mature. Total life-cycle is completed in 100 to 260 days and the adult longevity extends from 76 to 219 days.

To check the pest population effectively, keep the orchards clean and adopt all sort of sanitary precautions. All potential breeding sites (compost heaps, dung hills, refuse dumps, etc.) must be destroyed consistently in and around the orchards by deep burying or burning. Alternatively these breeding substrates may be dusted with 5% BHC or carbaryl or sprayed heavily with 0.1% lindane or carbaryl. Antony *et al.* (1958) also suggested using heaps of treated cowdung as traps. Chemical control of this pest, i.e. treating the palm crowns with a mixture of lindane and sawdust (1:9) is not only costly and uneconomical but also not practicable on large scale.

RED PALM WEEVIL

Rhynchophorus ferrugineus (Olivier) (*signaticollis* Chevrolat) is another major pest of date palms confined to old world, including, Pakistan, India, Sri Lanka, Bangladesh, Burma, Thailand, Malaysia, Laos, Cambodia, Vietnam, South China, Taiwan, Indonesia, Philippines, New Guinea (CIE map No. A–258). In India. it is found in Punjab, Uttar Pradesh, Bihar, Bengal, Assam, Karnataka, Kerala, Tamil Nadu and Andaman Islands. Besides date palms and coconut, it is a major pest of toddy palm, areca palm (*Areca catechu*) and number of other ornamental and wild palms. Though the pest is polyphagous, it has preference for coconut and is multivoltine. It is rather difficult to detect the infestation of this pest in the initial stage and when noticed in advanced stage, the loss caused is irreparable. The grubs are voracious feeders; once inside the trunk they seldom come out. They burrow in hard woody portion and tunnel through

Eggs

Grub

Pupa

Adult

Rhynchophorus ferrugineus (Olivier)

Palm trees damaged by the weevil

the tissues in all directions and feed on soft succulent tissues discarding all fibrous material. When mature they make cocoons of of twisted fibres and pupate therein. In case of severe infestation, inside of trunks is completely hollowed and filled with decayed rubbish. As a result of this infestation, in case of young palms, the tops wither, rot and give out offensive smell; while in older trees, the top portions of trunk bend and ultimately breaks at the bend. A thick reddish-brown viscous fluid as also chewed up and discarded fibre pieces are thrown out form the holes.

Adult weevils cause little damage except that they fly and infest other trees, by laying their eggs there. Unlike black palm beetles, these weevils can fly long distances in search of host trees (Leefman, 1920). Like Oryctes grubs, these grubs are also eaten by the people in Burma (Ghosh, 1924).

The various stages of this species have been briefly described by Menon and Pandalai (1958). Eggs are elongated oval, about 2.5 mm long, creamy-white, smooth and shinning. Grubs are fleshy, wrinkled transversely, apodous, pale yellow when freshly formed, later becoming light brown and 50 to 60 mm long when full grown. Pupae are shiny brown and 32 to 35 mm long. Adult weevils are reddish-brown in colour with red spots on thorax, cylindrical, 32 to 36 mm long, flattened with long and slightly curved snout.

The pest breeds throughout the year and all the stages are passed on the host tree. Being a tissue borer, environmental conditions have little influence on pest activity. A female lays 76 to 355 eggs (average 204) during its life span of 2 to 3 months. Eggs are laid singly in soft tissues at the base of leaf sheaths or in the cuts (often made by black palm beetles) or wounds (may be fungal diseases) on the tree trunk. These hatch in 2 to 5 days. Grub and pupal periods are 24 to 61 and 18 to 34 days respectively and the entire life-cycle is completed in 50 to 90 days. After emergence the weevils remain in cocoons for 4 to 17 days till they become sexually mature.

The proverb, "Prevention is better than cure" can be very aptly applied for this pest. Maintenance of orchard sanitation is indispensable. This coupled with prompt destruction of dead and badly infested palms, help in mitigating the pest population. Filling up the leaf axils with 5% BHC or lindane + sawdust or sand (1 : 1) three times a year has also been suggested. This will kill the female weevils that usually rest in leaf axils. Prevent the occurrence of cuts, scars or wounds on tree trunks. If and when these appear, paint the same immediately with lime, coal-tar or lindane. This will prevent the weevils from laying eggs on these trees. If the attack of the pest is detected in the initial stage, remove and kill the grubs mechanically. There is no economical or practical control for the advance stage of infestation, except to cut and destroy the affected palms to prevent the damage from spreading to other trees.

SCALE INSECTS

Atleast four species of armoured scales have been reported damaging date palms in India, *viz.*, *Aspidiotus desrtuctor* Signoret, *Chrysomphalus ficus* Ashmead (*aonidum* Linnaeus); *Parlatoria (Coccus) blanchardi* (Targioni) and *P. zizyphus* (Lucas). Of these, *P. blanchardi* and *A. destructor* are comparatively more common and destructive than the other two species; the former is active during Summer while the latter is found throughout the Winter.

Parlatoria blanchardi, the date palm scale, is indigenous to Iraq (Calcat, 1959) and has by now spread to Iran, Afghanistan, Pakistan, India and even Australia on the eastern side and Turkey, Syria, Israel, Jordan, Arabia, Somalia, Sudan, Egypt, Niger, Tunisia, Algeria, Morocco, Mauritania, USA, Brazil and Argentina towards the West. This is a specific pest of palms and specially that of date palm.

Hot and dry climate is favourable for this pest and it remains dormant during Winter. The pest prefers shade and invariably infests shaded pinnules and the main veins of the fronds. In severe cases even the fruits are attacked. Incidence is higher in trees where fruits have been wrapped with cloth as protection against solar radiation. Similarly, young trees grown in shade or those that are irrigated regularly suffer comparatively more from ravages of this pest. The continuous yellow incrustations of this scale on the leaflets, interfere with photosynthetic, respiration and transpiration activities of the trees which ultimately affect adversely the normal growth of the trees and development of fruits; the fruits become dry deformed and do not mature.

Eggs are rose coloured, 0.04 mm in diameter. Crawlers are also rose coloured and latter become light violet. The scales of males are on an average 1.2 mm long and white in colour and the nymphs below are pale yellow and invisible. The scales of females are brown with white margin. more or less oval, flat and about 1.5 mm long. The mature females are creamy-yellow and when gravid, these gradually become darker and darker till wine-red. An oviparous female may lay 6 to 10 eggs (maximum 15) in her life time. The eggs remain under the female scales and hatch in 8 to 18 days. The nymphs become mature females normally in 30 to 40 days, unless they enter into winter diapause. Male nymphs have to pass through prepupal and pupal stages before becoming winged or micropterous males.   The   life-cycle varies from 50 to 60 days in Summer to 75 to 85 days in Autumn and Spring (Iperti, 1977).

Coconut scale, *Aspidiotus destructor* is a major pest of coconuts, date palms and oil palms. It also attacks avocado, banana, *ber*, citrus, fig, grapevine, guava, *jamun*, mango, papaya, peach, pear, sapota, tamarind, etc. The pest is pantropical in distribution (CIE map No. A–218).

Good crop management restricts the pest development. Growing resistant varieties is also suggested but except in Israel, there has been no attempt to screen the varieties for resistance to this pest. To check, these scale insects, remove and burn all the infested leaflets in the initial stage of attack. This is practicable only when trees are young. In case of older trees, remove all the fronds except those at the very top and pour hot salty water over the crown and remaining leaves (Iperti, 1977).

BLACK HEADED CATERPILLAR

Nephantis serinopa Meyrick -- a major pest of coconut, attacks almost all other palms including date palm. In India, it is commonly found along East-coast and West-coast during April to June and January to May respectively. Eggs are laid in batches on older leaflets. A female lays 100 to 150 eggs. On hatching, the caterpillars feed gregariously on surface tissues of leaflets scraped out from lower surface. They construct silken galleries and feed within these. Attacked leaflets turn brown and dry up. In case of severe attack, the leaflets are reduced to papery tissues and the photosynthetic activity of the fronds is adversely affected. With too many fronds affected, the tree dries up and loses its fruit bearing capacity. Infested trees are easily recognized by dried up patches in the fronds and gradual drying up of these parts. Outbreaks of pest occur during dry months and the pest population declines appreciably during rainy season.

Eggs are scale-like and whitish. Full grown caterpillars are 13 to 18 mm long, slender, elongated, light green with reddish-brown stripes and black head. Adults are ashy-grey moths with uniform pale-whitish forewings; wing expanse is 20 to 25 mm. Egg, larval and pupal periods are on an average 5, 42 and 12 days respectively.

To control this pest, prompt cutting and burning of infested fronds is suggested. Nirula *et al.* (1951) found spraying with 0.2% DDT to be effective. Nair (1975) recommended spraying with an insecticide of short residual action (BHC, trichlorfon) and liberation of parasites, *Trichospilus pupivora* Ferrier, *Bracon brevicornis* Wasmann and *Perisierola nephantidis* Muesbeck.

TERMITES

Odontotermes obesus (Rambur) has been recorded attacking date palms specially the young trees. The termites attack root zone and gradually tunnel their way upwards, making galleries inside the main trunk. If the attack is severe, the entire trunk may be hollowed and ultimately the tree may also die away. Damage to grown up trees is not so serious (Nirula *et al.*, 1953). Soil application with 5% aldrin, chlordane or heptachlor dust before planting the seeds or transplanting the seedlings is quite effective in checking the termites

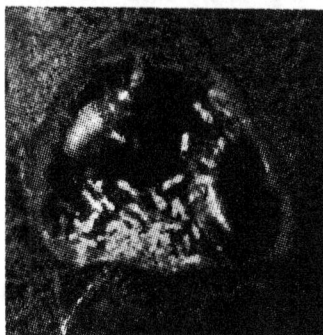

(*Courtesy:* M.G.Jotwani)
Termite colony

3.7

JAMUN

JAMUN, *Syzygium* spp., are native of East India and Malaysia. These are found throughout the length and breadth of Indian subcontinent either growing wild (*S. fruticosum* D.C.) or cultivated (*S. cuminii* Skeels, *S. jambos* Alston) in back-yards, roadsides, borders of fields and wind-breakers around the orchards – hence it is rather difficult to assess the total area under this fruit. *Jamun* is known as poor man's fruit. The fruits contain calcium, iron and proteins in small quantities but no vitamins. Besides eating as fresh fruit, it is also used for making squash and juice. Vinegar is made from juice of unripe fruits. In Goa a wine is made from ripe fruits. It also has some medicinal value. Dried and powdered seeds as also syrup of fruit and decoction of bark are used to cure diabetes, diarrhoea and dysentery. The seed powder also acts as an antidote in poisoning from nux-vomica.

Jamun trees too attract and shelter various insect pests and also serve as alternate host to insect pests of other fruit trees. About 36 insect species have been recorded so far feeding and breeding on *jamun* trees. Of these no single species causes any severe loss but the total damage inflicted by various pests is often substantial and warrants the control measures of these pests. The pests of common occurrence are bark eating caterpillars and mulberry bug. Besides, scale insects, thrips, whiteflies, leaf eating caterpillars, leaf rollers, leaf miners, fruit flies, fruit borers, etc. have also been recorded appearing regularly.

BARK EATING CATERPILLAR

Indarbela tetraonis (Moore) is a polyphagous pest, guava and *jamun* being the preferred hosts. Eggs are laid in clusters in cuts and crevices on the bark during May-June. On hatching, the caterpillars nibble the wood below the bark for couple of days making irregular galleries, then bore inside the stem or trunk. The caterpillars feed therein till December, then overwinter and pupate in late April. Presence of large silken webs comprising of chewed wood particles and excreta of the larvae lying on the trees, specially near the forking of trunk or main branches are the typical symptoms of attack.

Besides, *I.quadrinotata* (Walker) has also been occasionally recorded damaging *jamun* trees causing more or less similar damage.

To check the infestation of these borers, clean the affected portion of the trees by removing all the webby material and insert into the holes, cotton-wool soaked in a good fumigant like carbon bisulphide, chloroform, petrol, etc.

MULBERRY BUG

Halys dentatus (Fabricius) is found all over India. It is a polyphagous pest recorded on number of forest trees as also on *jamun*, mulberry, litchi, mango, etc. Swarms of nymphs and adults may be seen sucking the sap from main stems and trunk of the trees during Summer. Their deceptive colouration, which closely resembles the bark of the trees, makes it difficult to spot the pest.

Halys dentatus (Fabricius)

To control these bugs, spray thoroughly the trunks and branches with 0.05% dichlorvos or trichlorfon.

SCALE INSECTS

A few species of scale insects have also been recorded damaging *jamun* leaves in various parts of India. These include *Aonidiella aurantii* (Maskell), *A. orientalis* (Newstead), *Aspidiotus destructor*

♀   ♂

Aonidiella aurantii Maskell

(Signoret), *Duplaspidiotus tesseratus* (de Charmoy), *Parlatoria pseudopyri* Kuwana, *P. marginalis* Mc Kenzie, *Chloropulvinaria polygonata* (Cockerell), *C. psidii* (Maskell), *Coccus discrepans* (Green) and *Vinsonia stellifera* (Westwood). Besides, if mango trees are nearby, *Drosicha* spp. too may attack *jamun* trees. These are all polyphagous and the loss caused to *jamun* trees is not of any major importance.

PSYLLIDS

Psyllid, *Trioza jambolanae* Crawford forms unilocular galls on dorsal surface of leaves. The galls may be separate or into groups of 2 to 8 galls. The mature galls lose their green colour, harden and split into several lobes. According to Nair (1975) the pest causes malformation on the shoots and foliage. A female lays as many as 500 eggs singly or in clusters scattered on young leaves.

Egg period is 3 to 6 days while nymphal period lasts for 13 to 40 days in Summer and 160 to 170 days during Winter. Overwintering is in nymphal stage inside the galls.

Another psyllid, *Megatrioza vitiensis* (Kirkaldy) has also been reported on *jamun* leaves causing more or less similar damage.

WHITEFLIES

Dialeurodes eugeniae (Maskell) and *Singhiella bicolor* (Singh) are the common whiteflies found on *jamun* leaves, all over India. These are specific pests of *jamun*. Besides these, *Aleurocanthus rugosa* Singh, *Dialeurodes citri* (Ashmead), *D. vulgaris* Singh, *Rhachisphora trilobitoides* (Quaintance and Baker) have also been recorded from various parts of India. They suck the cell sap from leaves and thereby devitalize the trees. A severe infestation may also affect the fruiting capacity of the trees.

Adult females of *D. eugeniae* are 1.8 to 2.0 mm long, having orange coloured abdomen and profuse deposit of white wax on body and wings; hind wings are cinereous and similar coloured bands are on the abdomen. Adult females of *S. bicolor* are slightly smaller in size having dull white coloured abdomen with 5 smoky patches on forewings and two on hind wings. Both species have two generations in a year, summer generation lasts from March to October and winter one from November to February (Singh, 1931).

The whiteflies are normally minor pests of *jamun*, and no control measures are adopted against these. However, in case of young trees, it is desirable to clip off and burn all the infested leaves. In case of severe infestation, spray with 0.03% dimethoate, oxydemeton methyl or phosphamidon.

THRIPS

Leeuwenia ramakrishnae Ananthakrishnan (*karnyi* Ramakrishna), *Mallothrips indica* Ramakrishna, *Rhip iphorothrips cruentatus* Hood, *Teuchothrips eugeniae* Seshadri and Ananth akrishnan and *Thrips florum* Schmutz have been reported feeding on *jamun* trees in India. The first four suck sap from leaves while the last one infests the flowers.

Leeuwenia ramakrishnae – the long tailed thrip, is more common in South India. It is a specific pest of *jamun* and is confined to India. The nymphs are found in large number on ventral surface of leaves feeding motionlessly. Adults move about freely except when tending their eggs. Affected leaves show large yellowish patches which turn reddish-brown and become brittle. Adults are chocolate-brown in colour; wings infumate along the middle with light brown patch; the tube is exceptionally long and clearly pilose. Egg and nymphal periods are 3 to 5 and 7 to 9 days respectively; pre-pupal stage lasts for 1 to 2 days and pupal development takes 3 to 5 days. Total life-cycle is completed in 14 to 21 days (Ananthakrishnan, 1971).

Leeuwenia ramakrishnae Ananthakrishnan

Mallothrips indica is another common species found all over India, inhabiting the dried up psyllid galls on *jamun* leaves. It is found only in well developed galls. The species is characterized by strong enlarged forefemora in both sexes. The adults are dark brown having wings with grey infumation.

Rhipiphororthrips cruentatus is a major pest of grapevine and a minor one of *jamun*. · Its reddish nymphs and dark brown adults are found during April to November on ventral side of leaves, rasping the tissues and imbibing the oozing cell sap, causing brown spots on leaf lamina (Butani, 1976 c).

Teuchothrips eugeniae is a minor pest of *jamun* reported from South India. It is found in leaf galls along with *Leeuwenia ramakrishnae* (Seshadri and Ananthakrishnan, 1954). Adults are dark brown, 2.0 to 2.5 mm long, having wings with slight infumation specially at the base.

Thrips florum is one of the most common blossom thrip found mostly on all ornamental plants. It is highly polyphagous species having a wide range of host plants including *jamun*. It is a minor pest of *jamun*. A large number of individuals may occasionally be seen in flower-buds. In case of severe infestation, which is rare, flower-buds wither and fruit setting capacity is adversely affected.

Spraying the trees with 0.03% dimethoate, oxydemeton methyl or phosphamidon is effective in controlling these thrips.

LEAF ROLLERS

Platypelpa aprobola Meyrick, *Argyroploce mormopa* Meyrick, *Strepsiarates rhothia* (Meyrick), *Polychrosis cellifera* Meyrick, *Homona coffearia* (Nietner) and *Oenospila flavifustata* (Walker) and *Lepidogma* species have been recorded on *jamun* in various parts of India. Though of regular occurrence, these pests seldom cause any substantial loss to *jamun* trees.

Jamun leaf roller, *Polychrosis cellifera* is widely distributed in Indian subcontinent. In India, it has been reported from Punjab, Haryana, Delhi, Uttar Pradesh and Bihar. Eggs are laid in April on tender leaves. On hatching, the caterpillars fold or roll the tender leaves and web the same with silken strands and feed within, eating away patches of leaf lamina upto the midrib. Occasionally the caterpillars may bore into the fruits as well. The summer generation of the pest causes minor damage and usually goes unnoticed. The second generation attacks the fresh flush of leaves during rainy season and causes comparatively more damage.

Homona coffearia – the flush worm is a major pest of tea in certain years and minor pest of *jamun*, specially in nurseries and on young plants. The caterpillars fold the leaves longitudinally and feed within. Pupation also takes place inside these folds. Eggs are scale-like and are laid in overlapping clusters of 100 to 150 on dorsal surface of leaves. Full grown caterpillars are 20 to 25 mm long. Adults are brownish-yellow moths with forewings having oblique band and a few transverse wavy lines; wing expanse is 24 to 28 mm.

Homona coffearia (Nietner)

Egg, larval and pupal periods are 6 to 8, 24 to 28 and 6 to 8 days respectively.

Oenospila flavifustata – the looper caterpillar is more common in South India. It has also been reported from Sri Lanka and Indonesia. Eggs are laid on edges of tender leaf blades. A female lays 20 to 30 eggs in 10 to 12 batches. On hatching, the caterpillars feed on tender foliage and roll up the same. Pupation takes place within rolled up leaves. Eggs are small, circular, disc-like and about 0.25 mm in diameter. Full grown caterpillars are 35 to 40 mm long, greenish in colour and having short scattered bristle like hair arising singly. Pupae are cylindrical, about 5 mm long and green in colour. Moths are medium sized with beautifully patterned wings. Egg, larval and pupal stages last for 2 to 3, 17 to 18 and 7 to 8 days respectively (Venugopal, 1958) and the life-cycle is completed in 26 to 29 days.

Argyroploce mormopa occurs regularly in Kerala as a minor pest of *jamun* trees. The caterpillars web together leaves at shoot tips and feed within. Egg, larval and pupal durations occupy 4 to 5, 18 and 7 days respectively (Aiyer, 1943 b).

Lepidogma species has been reported from South India as a minor pest. The young caterpillars feed gregariously scrapping green leaves and eating away the chlorophyll. Later, the caterpillars web the leaves and feed within. The affected leaves wither and become detached from the stalk but remain sticking there on account of webbing. The eggs are laid in clusters on these webs. Eggs are oval in shape and yellowish-green in colour. Full grown caterpillars are cylindrical, 20 to 25 mm long, brownish in colour with 2 dark coloured longitudinal stripes and inconspicuous setae that arise from black wrats. Pupae are also cylindrical, 10 to 12 mm long and brown in colour. Adults are medium sized moths with dark brownish wings. Egg, larval and pupal periods are 4, 28 and 7 days respectively and total life-cycle occupies about 39 days. (Ananthakrishnan and Venugopal, 1955).

LEAF   MINERS

Acrocercops loxias Meyrick, *A.phaeospora* Meyrick, *A.syngramma* Meyrick and *A. telestis* Meyrick have been reported damaging *jamun* leaves. These are minor pests of sporadic occurrence. Eggs are laid on tender leaves. On hatching, the caterpillars mine the green leaves and cause blister-like swellings on dorsal surface of leaves, each blister being caused by a single larva. These swellings subsequently dry and drop off leaving big holes in the leaves. When full fed, the caterpillars form circular flat cocoons in depressed portions or cavities in the mines, usually besides a raised leaf-vein and pupate therein. The damage is caused during Summer and the pest overwinters in pupal stage. *A. phaeospora*, which is comparatively more common, is specific pest of *jamun* reported from Bihar, Delhi, Haryana and Punjab.

Acrocercops phaeospora Meyrick

To check the infestation of leaf miners, clip off the affected leaves during Winter and burn the same.

LEAF EATING CATERPILLARS

Antispila anna Meyrick, *Idiophantis (Colobodes) acanthopa* (Meyrick), *Bombotelia delatrix* Guenée, *Carea subtilis* Walker, *Chrysocraspeda olearia* Guenée, *Phelgetonia delatrix* Guenée, *Trabala vishnou* (Lefebvre), *Euproctis fraterna* (Moore), *Exelastis atomosa* (Walsingham), *Hyposidra successaria* Walker, *Metanastria hyrtaca* Cramer and *Orthaga* species have been occasionally reported from different parts of India defoliating *jamun* trees. The damage caused is seldom severe.

Carea subtilis Walker

Trabala vishnou has been reported from Karnataka. The egg, larval, pupal and adult stages last for 13, 40 to 50, 13 to 18 and 5 to 6 days respectively and the life-cycle 66 to 73 days (Viswanath and Gowda, 1974). Dusting 10% BHC or spraying with 0.1% BHC + 0.1% DDT is effective in controlling these pests (Butani, 1976 c).

FRUIT   FLIES

Dacus diversus (Coquillett) and *D. correctus* (Bezzi) have been reported damaging *jamun* fruits; the former is found all over India while the latter has been reported from South India. Both are polyphagous, having a wide range of host plants. They breed throughout the year and migrate to *jamun* trees during monsoon. In northern India, during severe cold the pest overwinters in adult stage. Being minor pests of *jamun*, no control measures are usually adopted against these flies.

FRUIT   BORERS

Jamun fruit worm, *Meridarchis reprobata* Meyrick has been reported from Kashmir, Gujarat, Maharashtra and South India. The caterpillars bore into the fruits and feed within. In Kashmir, it bores olive fruits and elsewhere *jamun* is the main host. In South India, it has also been found on *ber* fruits. *Ber* fruit borer, *M. scyrodes* Meyrick has also been occasionally found boring *jamun* fruits. However, both these are of minor importance.

POMEGRANATE

POMEGRANATE, *Punica grantum* Linnaeus is of Iranian origin and is grown extensively in Spain, Morroco and other Mediterranean countries as also in Egypt. Iran, Afghanistan, Arabia and West Pakistan (Singh *et al.*, 1963). It is also grown to some extent in India, Bangladesh, Burma, China, Japan and USA. In India it occupies about 1000 hectares confined mostly to Gujarat and Maharashtra states and a small area in Uttar Pradesh, Andhra Pradesh, Karnataka and Tamil Nadu. It thrives best in areas having hot dry Summer with irrigation facilities. It can also withstand frost to some extent but not humid weather. A wild type is also grown along the foot of Himalayas; the seed of this variety is dried and used as spices (*anar-dana*). The juice of pomegranate is cool and refreshing and has many medicinal properties, specially for patients suffering from leprosy. The rind of the fruit, as also the bark of the trees, are used in dysentery and diarrhoea. The rind is also used as dyeing material for cloth.

Pomegranate trees are attacked by about 45 species of insects in India and unfortunately, in this case, fruit is more vulnerable to the attack of pests than any other part of the tree. The most obnoxious enemy is pomegranate butterfly, next in order being bark eating caterpillars. Besides, termites attack the roots; aphid, whiteflies, scale insects, thrips and leaf defoliators feed on foliage; a few bugs, borers, moths and fruit flies damage the fruits and some scavengers attack even the rotten and over-ripe fruits.

POMEGRANATE BUTTERFLY

Virachola isocrates (Fabricius) is economically the most important of all the butterflies, perhaps the only one that is constantly and regularly injurious. It is a polyphagous pest having a very wide range of host plants, including, *aonla*, apple, *ber*, citrus, guava, litchi, loquat, mulberry, peach, pear, plum, pomegranate, sapota and tamarind. It is widely distributed all over India and is found wherever pomegranates are grown.

The female lays eggs singly on calyx of flowers or small fruits.
On hatching, the caterpillars bore inside the developing fruits and are
usually found feeding on pulp and seeds just below the rind. As
many as eight caterpillars may be found in a single fruit. Subsequently,
the infested fruits are also attacked by bacteria and fungi causing the
fruits to rot. The conspicuous symptoms of damage are offensive
smell and excreta of the caterpillars coming out of entry holes, the
exereta is found stuck around the holes. Sometimes the holes may
also be seen plugged with the anal end of a caterpillar. The affected
fruits ultimately fall down and are of no use; even if the fruits be
picked before falling down; these fruits have no market value.

Eggs are shiny white in colour and oval in shape. Full grown
caterpillars are stout, 17 to 20 mm long, dark brown in colour with
short hair and whitish patches all over the body. Just before
pupation, the caterpillars come out of the fruits and tie the stalk of
fruit with main branch of the tree with fine silken strands to ensure
that the fruit does not fall down, then reenter the fruit and pupate
therein (Narayanan, 1954). Occasionally the caterpillars may pupate
outside also, attaching themselves to the stalk of the fruits. Adult
butterflies are medium sized, glossy-bluish-violet (males) to brownish-
violet (females) in colour; females have conspicuous orange patch on
the forewings. Wing expanse is 40 to 50 mm.

Virachola isocrates (Fabricius)

Incubation, larval and pupal periods range from 7 to 10, 18 to
47 and 7 to 34 days respectively with four overlapping generations
in a year (Lal, 1952). The pest breeds throughout the year.

To reduce the incidence of this pest, remove and destroy all the
affected fruits. The only other effective though expensive method
is bagging of fruits (Pruthi, 1969) and may be practiced, if the

number of fruit trees is limited. Alam (1962) from Bangladesh, suggested spraying with 0.03% phosphamidon just when the fruit formation starts.

Pomegranate fruits are also damaged by pomegranate borer, *Deudorix epijarbas* (Moore), castor capsule borer, *Dichocrocis punctiferalis* Guenée, pyralid borer, *Euzophera punicaella* Moore and fruit sucking moths, *Othreis cajeta* (Cramer) and *O. materna* (Linnaeus). Besides, Ayyar (1924) reported two pentatomid bugs, *Iurtina indica* Dallas and *Halyomorpha picus* (Fabricius) puncturing the tender and ripe fruits. These cause only minor damage to pomegranate fruits. Except *D. epijarbas*, others are polyphagous pests and cause major damage to other economic crops. *D. epijarbas* has been occasionally causing serious damage, specially in Himachal Pradesh and Uttar Pradesh. Dipping of healthy fruits in 2% DDT suspension at an interval of 3 to 4 weeks prevents the attack of this borer (Gupta and Joshi, 1959) but it is a cumbersome method and not practicable on large scale.

BARK EATING CATERPILLARS

Indarbela tetraonis (Moore) and *I. quadrinotata* (Walker) have been recorded boring the bark of pomegranate trees and feeding inside. *I. tetraonis* is comparatively more common and more harmful to pomegranate trees than *I. quadrinotata*. Older trees and the trees in orchards that are not well maintained are more prone to the attack of these pests. Usually there is only one caterpillar in each hole but there may be 10 to 12 holes in a badly infested tree. Such trees will bear no fruits.

Besides *Indarbela* spp., the red borer, *Zeuzera coffeae* Nietner has also been reported boring the stems and trunks of pomegranate trees. This is a polyphagous pest, coffee and tea being its preferred hosts. It is a minor pest of citrus, custard apple, guava, litchi, loquat, pomegranate, etc. A female lays 500 to 1000 eggs in 1 to 2 weeks. Eggs are laid in strings on the bark, branches or stems. On hatching, the larvae bore, usually at the joints between leaf-stalk or twig and main stem, and tunnel straight downwards. The larvae also cut circular holes at various places through which they eject the frass etc. Pupation takes place inside the tunnels. Eggs are oval (1.0×0.6 mm) in shape and reddish-yellow in colour. Caterpillars are pinkish-white with dark brown spots and are about 40 mm long when full grown. Pupae are chestnut-brown in colour and 22 to 28 mm long. Moths are white with pairs of small black dots on thorax; numerous small black spots and streaks on forewings and a few black spots on posterior edges of hind wings. Wing expanse is 35 to 45 mm. Eggs hatch in about 10 days. Larval development takes 60 to 120 days and pupal period lasts for 3 weeks to one month (Beeson. 1941). Total life-cycle lasts for 4 to 5 months in South India and at low elevations, and extends to more than a year at high elevations in North India.

Keeping the orchards clean and avoiding over-crowding of trees helps in minimising the attack by these borers. In case of infestation, clean the affected portions by removing all webs, etc. , and insert in each hole, swab of cotton-wool soaked in carbon bisulphide, petrol or even kerosene and seal the holes with mud.

STEM BORING BEETLES

Celosterna spinator Fabricius and *Olenecamptus bilobus* Fabricius have been reported boring the stems and trunk of pomegranate trees. Both are polyphagous pests causing minor damage. *O. bilobus* is widely distributed in the Oriental region. It is primarily a pest of *Ficus* spp. It prefers breeding in dead wood but also attacks the living branches. The grubs bore inside the trunk and feed on sapwood. Adult beetles are active by day and feed by gnawing the green bark of shoots. The life-cycle is annual with an extended emergence-period from May to November.

C. spinator is a major pest of **babul,** *Acacia arabica*. Adult beetles have pale yellowish-brown body with light grey elytra and are 30 to 35 mm long. These are nocturnal in habit and appear with onset of monsoon; emergence is accelerated by continuous or heavy rains. Egg period is 12 to 15 days, grub 9 to 10 months and pupal 16 to 18 days. There is only one generation in a year and longevity of beetles is 45 to 60 days.

Generally no control measures are required. Nevertheless, the measures suggested against bark eating caterpillars will prove effective in controlling these beetles as well.

SAP SUCKING INSECTS

Drosicha mangiferae (Green), *Hemiaspidoproctus cinerea* (Green),

Icerya purchasi Maskell, *Planococcus lilacinus* (Cockerell), *Andaspis hawaiiensis* (Maskell), *Aonidiella orientalis* (Newstead), *Duplaspidiotus tesseratus* (de Charmoy), *Hemiberlesia punicae* (Signoret), *Lindingaspis greeni* (Brain and Kelly), *L. rossi* (Maskell), *Parlatoria oleae* (Colvée), *Pinnaspis theae* (Maskell) are some of the mealy bugs and scale insects reported from various parts of India on pomegranate leaves. Most of these are polyphagous, having a wide range of host plants. Due to damage by these insects, the trees are devitalised, resulting in shedding of buds and flowers and smaller sized fruits. To prevent the attack from spreading, prune and destroy the affected parts in the initial stage of attack. If the infestation becomes severe, spray with 0.04% diazinon or monocrotophos.

Pomegranate aphid, *Aphis punicae* and pomegranate whiteflies *Siphoninus finitimus* Silvestri and *S. phillyreae* (Haliday) are also minor pests recorded occassionally. They feed by sucking the cell sap from leaves and tender twigs. The affected parts get discoloured and disfigured. These insects also secrete copious amount of honeydew on which sooty mould develops, hindering the photosynthetic activity of the plant.

Thrips, *Retithrips syriacus* (Mayet) and *Rhipiphorothrips creuentatus* Hood have been recorded feeding on leaves, while *Anaphothrips oligochaetus* Karny, *Ramaswamiahiella subnudula* Karny and *Scirtothrips dorsalis* Hood infest the flowers. These are all minor pests of pomegranate. Nymphs and adults lacerate the leaves, flower stalks, petals and sepals and rasp the sap that oozes out of these wounds. As a result, the leaf tips curl and dry away while flowers are shed and ultimately the fruiting capacity of the tree is adversely affected.

To control the aphid, whiteflies and thrips spray with 0.03% dimethoate, oxydemeton methyl or phosphamidon. Two spraying at an interval of 10 to 15 days will be sufficient to keep complete check of these pests.

LEAF EATING CATERPILLARS

Leaves of pomegranate trees are often attacked by one or the other lepidopterous larvae. Those of regular occurrence include, castor semilooper. hairy caterpillars, slug caterpillars and bag worms.

Castor semilooper, *Achaea janata* (Linnaeus) is an important pest of castor. It has also been reported damaging *ber*, citrus, grapevine, guava, pomegranate, etc. The pest has country wide distribution. Eggs are laid singly on tender leaves usually on ventral surface, one to six eggs may be found on a leaf. A female lays on an average 400 eggs. Freshly hatched caterpillars congregate on the leaves of various weeds and a few economic crops, including pomegranate, *ber* and castor, and feed on the chlorophyll. Later, they segregate and feed voraciously devouring the entire leaf lamina. In case of severe infestation, the young trees may be completely defoliated. The moths are also destructive. They suck the juice of various fruits, like citrus, grapes, guava, mango, etc. Eggs are hemispherical, bluish-green and ridged with 40 to 45 striae. Caterpillars are semiloopers about 5 mm when freshly hatched becoming 55 to 65 mm when full grown. The caterpillars show conspicuous colour variations– some are grey with red or brown lateral stripes while others are bluish-grey specked with blue-black and having yellowish lateral stripes. Pupation takes place in the soil 40 to 50 mm deep or even among the fallen leaves or in folds of leaves. Freshly formed pupae are glistering dark green, later becoming brown. Moths are stout, pale reddish-brown with wavy lines on forewings and black hind wings having a medial white band and three large white spots on outer margin. Anal and apical margins of the wings are fringed with hair. Wing expanse is 50 to 65 mm.

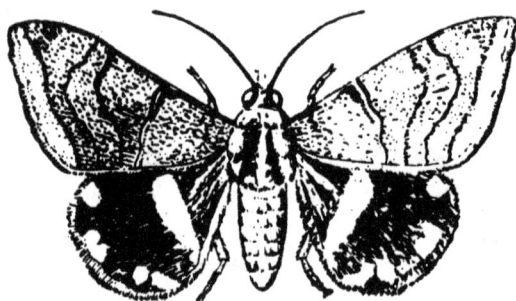

Achaea janata (Linnaeus)

Incubation period is 3 to 5 days while the larval and pupal periods last for 9 to 23 and 7 to 26 days respectively (Pandey *et al.*, 1966). Preoviposition period of moths extends from 6 to 21 days and a life-cycle occupies 3 to 5 weeks during active period of the pest. There are five to six overlapping generations in a year and over-wintering takes place in pupal stage.

To check the damage caused by these semiloopers, collect and destroy the caterpillars mechanically. In case of severe infestation, dust 10% BHC or spray with 0.05% dichlorvos or endosulfan. This will also kill other leaf eating caterpillars, if any.

Ber hairy caterpillar, *Euproctis flava* (Bremer), plum hairy caterpillar, *E. fraterna* (Moore) and castor hairy caterpillar, *E. lunata* Walker are the species reported damaging pomegranate trees. These are highly polyphagous pests, found all over Indian sub-continent. Eggs are laid in clusters on ventral surface of leaves and covered with hair. On hatching, the caterpillars feed gregariously on epidermis of leaves; later they segregate and feed voraciously, defoliating the entire trees. Pupation takes place in hairy cocoons on leaves or on branches. Moths of *E. flava*, have head and thorax bright orange-yellow and abdomen paler. Forewings are orange-yellow in colour with orange spot at end of cell and three subapical black spots; hind wings are paler than forewings. Wing expanse is 30 mm (♂) to 40 mm (♀). Moths of *E. fraterna* are more or less similar to those of *E. flava*, except that these are smaller in size, wing expanse being 24 mm (♂) to 34 mm (♀). *E. lunata* moths are pale ochreous in colour with white ochreous forewings having a large black lunule; wing expanse is 34 mm (♂) to 38 mm (♀).

Biology is these species is more or less same. The eggs hatch in 5 to 10 days, larval period lasts for 4 to 5 weeks and pupal stage occupies 10 to 12 days–thus entire life-cycle is completed in 45 to 57 days.

Porthesia scintillans (Walker) is another hairy caterpillar widely distributed in Indian subcontinent. This is also polyphagous pest, damaging, apple, mango, pomegranate, etc. Caterpillars are dark brown with a series of crimson lateral tubercles on a yellow line bearing tufts of grey hair. Moths are yellowish with reddish line and spots on the edges. Forewings are vinous-brown irrorated with dark scales and hind wings are fuscous-brown with broad yellow margin; wing expanse is 20 to 26 mm and 32 to 38 mm in case of males and females respectively.

Porthesia scintillans (Walker)

Egg, larval and pupal periods last for 6 to 10, 30 to 40 and 8 to 12 days respectively.

Arcyophora dentula Hampson has been reported as a minor pest. The caterpillars gnaw holes in the leaves and ultimately defoliate the tree. Maximum damage is caused during September-October (Mukerjee, 1941).

Caterpillars of *Creatonotus gangis* (Linnaeus) have been reported feeding on leaves during January to March in Andhra Pradesh.

Creatonotus gangis (Linnaeus)

Egg, larval and pupal periods last for 3 to 4, 30 to 33 and 10 to 13 days respectively (Raghunath and Butani, 1977).

Dusting with 10% BHC is effective for checking these hairy caterpillars. It is desirable to carry out the campaign on a cooperative basis and during early stage of infestation when the caterpillars are feeding gregariously; at this stage, even 5% BHC dust will give a good mortality of these caterpillars.

Bagworms are widely distributed all over India. The species recorded feeding on pomegranate leaves, include, *Clania crameri* Westwood and *Acanthopsyche* species. The caterpillars construct a case over their body and live within, nibbling leaf lamina. Females are apterous and devoid of antennae, mouth parts and legs. Males are winged and have bipectinate antennae and short mouth parts.

BEETLES AND WEEVILS

Anomala dimidiata Hope, *Hoplasoma sexmaculata* Hope, *Mimastra cyanura* Hope, *Myllocerus laetivirens* Marshall and *M. undecimpustulatus maculosus* Desbrocher have been occasionally reported feeding on pomegranate foliage. These are polyphagous pests, destructive mostly to temperate fruit trees and of minor importance in case of pomegranate. The grubs live in the soil and feed on the roots and other organic matter. It is only the adults that come out of the soil at night and feed on foliage. Generally no control measures are required against these pests, but if and when necessary, dust with 5 to 10% BHC dust.

FRUIT FLIES

Mango fruit fly, *Dacus dorsalis* (Hendel) and peach fruit fly, *D. zonatus* (Saunders) have been reported attacking pomegranate fruits as well, though the loss caused is of minor importance. These flies are cosmopolitan in distribution and feed on a vast variety of fruits and vegetables. Both the species are found attacking apple, *bael, ber*, citrus, fig, guava, mango, peach, pomegtanate, sapota, etc. In additon, *D. dorsalis* has also been recorded on apricot, banana, jack-fruit, loquat, pear, persimmon, plum, etc., while *D. zonatus* has been recorded damaging custard apple (Kapoor, 1972).

Eggs are laid just below the epidermis of ripening fruits. The affected fruits show dark brown punctures through which juice oozes out. On hatching, the small white grubs feed inside the fruits, as a result of which the fruits rot and fall down.

To prevent population build-up and carry over of these flies, collect and destroy promptly all the fallen and affected fruits. To avoid the infestation, harvest the fruits before they ripen as the flies are not able to puncture the hard rind of unripe fruits.

TERMITES

Odontotermes obesus (Rambur) has been reported attacking pomegranate trees in Rajasthan (Kushwaha, 1960) and Andhra Pradesh (Krishnamoorthy and Ramasubhiah, 1962). This sub-terranean enemy is highly polyphagous and feeds on a large number of economic and wild crops. Its attack is more pronounced in sandy and sandy-loam soils than in clayey or black soils. The termites cannot thrive under conditions of bad aeration and poor drainage.

Soil aplication with 5% aldrin, chlordane or heptachlor dust is quite effective in warding off the attack of this pest and should be regularly practised atleast once in a year in termite infested orchards.

SCAVENGERS

Peach scavenger moth, *Anatrachyntis simplex* Walsingham has been found attacking pomegranate fruits specially over-ripe and malformed ones or those that have been attacked by fungi or bacteria but are still hanging on trees. The caterpillars web together a few seeds with silken strands and feed within.

Scavenger beetles (calvicorn) feed on fermenting or decaying vegetable matter, perticularly those with exuding sap, souring fruits and withering flowers, decomposing bark and sapwood. The species recorded on pomegranate are, *Amphicrosus* species and *Carpophilus dimidiatus* Linnaeus. These polyphagous beetles feed on rotten fruits that have been already damaged by the caterpillars of pomegranate butterfly or fruit flies and have fallen down. *C. dimidiatus* has also been reported feeding on damaged fruits of guava, mango, peach, pear, plum, etc.

To check the population of these scavengers, collect all rotten and infested fruits and destroy the same promptly.

FIG, *Ficus carica* Linnaeus is one of the oldest known fruits. It is said that Adam and Eve used to cover their bodies with fig leaves (may be wild *Ficus* species). A number of ancient songs and stories on fig have been collected by Condit (1947). The original home of this fruit, is beleived to be Asia Minor, from where it spread mostly westward (Mediterranean region), in the form of dried fruit carried by man. Fig seeds in the excreta of birds also account for its wide dissemination. Eastward movement was rather slow because fig trees do not flourish in low wet tropics. These are more adopted to arid conditions and can be grown only at higher elevations and in drier parts of the tropics. At present fig is extensively cultivated in some parts of USA (California State), Algeria, Portugal, Spain, Italy, Greece, Turkey and Afghanistan. In India, commercial cultivation of fig is not yet well developed inspite of the fact that there is a great demand, specially of dried figs. About 200 tonnes of dried figs are imported every year. As it is, the area under this crop in India is decreasing and at present there are hardly about 800 hectares, mostly in Karnataka and a small area around Pune (Maharashtra), Lucknow (Uttar Pradesh) and South Gujarat.

Fig fruit is a wholesome and nutritious food, high in sugar content (45 to 50% in dried figs), usually low in acid content and fairly good source of vitamin A (Hayes, 1966). It also possesses high laxative properties. Talking about insect pests, as many as 50 species have been recorded feeding on fig trees. Those of major importance include, stem boring beetles, chafer beetles, leaf eating caterpillars, coccoids and fruit flies; while the minor ones are, thrips, jassid, psyllids, midges, leaf feeding weevils, leaf rollers, fruit borers, etc. Besides, dried figs are damaged by fig moth, grain moth, Indian meal moth and saw-toothed grain beetle.

STEM BORING BEETLES

Batocera rufomaculata (de Geer) is the most destructive pest of fig trees. It is widely distributed in India, Bangladesh, Sri Lanka, Malaysia and East Africa. In India, it has been recorded on more than 30 host plants, including, apple, fig, guava, jack-fruit, mango, pomegranate and walnut.

Grubs bore into the main stems or branches or even tree trunk and make zigzag tunnels in the wood. The affected stems show holes, from which faecal matter and chewed wood particles keep coming out and may be seen hanging on the affected stems or heaped up on the ground below. Ultimately the infested stems or branches dry up and where trunk is affected the growth and fruit bearing capacity of the tree is adversely affected and old trees may even die away. The various stages of the pest as also its bionomics have been studied by Husain and Khan (1940).

Other stem boring beetles of lesser economic importance include, *Batocera rubus* (Linnaeus), *Aclees cribratus* Gyllenhal, *Apriona cinerea* Chevrolat, *A. rugicollis* Chevrolat (*germari* Hope), *Olenecamptus bilobus* Fabricius and *Rhytidodera* species. An allied species, *Batocera lineolata* Chevrolat, has been recorded as pest of fig trees in Japan (Condit, 1947) but has not yet been reported from India.

Apriona cinerea–the apple stem borer, is an ash-grey beetle, 35 to 50 mm long, with numerous black tubercles at the base of elytra and elytral margins are concolourous. The adults appear around June to August, feed on bark of the trunk or main stems; the latter are often girdled and killed. Eggs are laid on these branches or trunk. On hatching, the grubs bore inside and feed by making tunnels up and down the stems or trunk. From the main tunnel several off shoots are given out which open outside and through which faeces and chewed wood is thrown out. Reddish coloured sap also oozes out through these holes.

Apriona rugicollis has been reported feeding on fig, jack-fruit and mulberry. These beetles resemble *A. cinerea* except that these are slightly bigger in size, being 50 to 60 mm long and grey in colour with elytral sutures and margins bluish-grey These beetles appear from March to October and have similar feeding habits as *A c i ne rea.*

FIG 231

Olenecamptus bilobus is conspicuous due to round white spots on the smooth brown elytra.

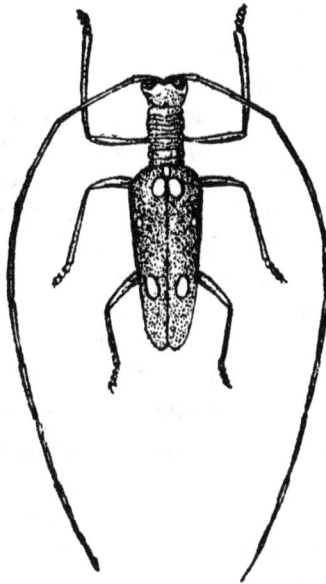

Olenecamptus bilobus (Fabricius)

The beetles are found everywhere in the plains feeding on *gular* and fig trees, though they have also been recorded on jack-fruit and pomegranate trees. The grubs bore into solid wood and the nature of damage caused by these beetles is more or less same as that caused by *Apriona* species.

Rhytidodera species is a medium sized beetle, 20 to 28 mm long with reddish-brown elytra. The grubs bore into the main branches, tunnelling down towards the centre and in the process throw out the frass and sap through the entry holes.

To control these beetles, remove and destroy all the affected branches. In case of main stem or trunk, insert in the holes cotton wool soaked in carbon bisulphide and chloroform (1:1) or insert 2 to 5 gm potassium cyanide and seal the holes with mud (Wadhi and Batra, 1964).

COCKCHAFER BEETLES

Adoretus duvauceli Blanchard (brick red), *A. horticola* Arrow (copper black), *A. lasiopygus* Burmeister (dark brown), *A. versutus* Harold (chestnut red) and *Brahmina coriacea* (Hope) have been reported defoliating fig trees. All these are polyphagous pests having a large number of host plants including, apple, grapevine guava, loquat, peach, pear and plum. *A. versutus* is also found feeding on leaves of apricot, *ber*, walnut, etc. Eggs are laid in soil. On hatching, the grubs feed on rootlets and root hair of various weeds, grasses and economic crops growing around. Pupation takes place deep down (about 300 mm) in the soil. Overwintering also takes place there in pupal stage. On emerging, the adults continue to live in the soil and come out only after the heavy showers of rain and remain active for about two months. These are nocturnal in habit and do all the damage between dusk and dawn, hiding during day, in the upper layer of soil.

Soil treatment with 5% aldrin, chlordane or heptachlor dust is recommended to check the grub population in the soil. Against adults, foliar spraying with 0.2% carbaryl or dusting with 10% BHC is suggested (Butani, 1975 e); repeat the treatment, if necessary, after a week.

COCCOIDS

Drosicha mangiferae(Green), *Icerya purchasi* Maskell, *Perissopneumontam arindus* (Green), *Nipaecoccus viridis* (Newstead), *Planococcus citri* (Risso), *P. lilacinus* (Cockerell), *Aonidiella aurantii* (Maskell), *Aspidiotus destructor* (Signoret), *Hemiberlesia lataniae* (Signoret), *Lepidosaphes conchiformis* Shimer (*ficus* Signoret), *Leucaspis riecae* Targioni, *Parlatoria oleae* (Colvee), *Pinnaspis aspidistrae* (Signoret), *Ceroplastes actiniformis* Green, *C. floridensis* Comstock, *Coccus discrepans* (Green), *Lecanium ramakrishnae* Ramakrishna (nec. Green) and *Saissetia oleae* (Bernard) are the mealy bugs and scale insects recorded feeding on fig trees from various parts of India. One or the other species is always present in every orchard. Most of these species are polyphagous. The nymphs and adult females suck cell sap from leaves, stems and in some cases from fruits as well. As a result of desapping the affected parts dry up and the

FIG   233

trees become weak.

Of the various species recorded, *P. oleae* is comparatively more common. Besides fig, this species has also been reported damaging apple, apricot, *ber*, citrus, grapes, loquat, mango, mulberry, olive, peach, pear, plum, pomegranate, etc. Eggs are broadly oval in shape, 2.5 mm long and 1.8 mm broad. Neonate larvae are oval in shape, 0.2 to 0.4 mm long and yellow in colour with light-reddish tinge. Adult females are broadly oval, 1.0 to 1.4 mm long and wine-red in colour with brownish pygidium. Their scales are circular, distinctly convex and dirty white tinged with brown. Males are winged.

To control these scale insects, remove and destroy the affected parts in the initial stage of attack. If the infestation is severe, spray with 0.04% diazinon or monocrotophos to kill the young nymphs. For full grown nymphs and adult females, use 0.1% diazinon or monocrotophos.

LEAF EATING CATERPILLARS

A large number of lepidopterous larvae have been reported feeding on fig leaves. These include, *Ocinaria varians* Walker, *Diaphania itysalis* (Walker), *D. pyloalis* (Walker), *D. stolalis* (Guence), *Perina nuda* Fabricius, *Aganais ficus* (Fabricius), *Asota alciphron* (Cramer), *Selepa celtis* Moore and *Spodoptera litura* (Fabricius). These are sporadic pests of fig trees and cause severe damage only in certain parts of India.

Ocinaria varians is comparatively more harmful. This is a wild silk moth, more common in Deccan where it feeds on variety of wild figs as well. Full grown caterpillars are 28 to 34 mm long, pale grey in colour with a short horn near the anal end and resemble

Ocinaria varians Walker

fig twigs. Before pupating bright yellow cocoons are spun in rolled leaves within which the larvae pupate. Moths are pale white with wing expanse of 40 to 50 mm.

Diaphania spp. are medium sized moths. D. itysalis moths are comparatively smaller than the other two species. Head is black and white; thorax and abdomen are brownish-grey. Forewings are fulvous and fuscous while hind wings are hyaline white and wing expanse is 24 to 26 mm. D. pyloalis are brown moths, suffused in parts with fuscous and striped with white. Forewings are pale fulvous and hind wings hyaline white; wing expanse being 26 to 30 mm. D. stolalis moths have head, thorax and abdomen whitish with brown stripes. Forewings white and hind wings pearly white. Wing expanse is 32 to 38 mm.

To control these pests, hand-picking and mechanical destruction of caterpillars, in the early stage of attack is suggested. Dusting with 5 to 10% BHC or spraying with 0.1% BHC+0.1% DDT is also effective in checking these caterpillars. This will also control the ak grasshoppers, Poekilocerus pictus (Fabricius) that are occasionally found feeding on fig trees.

FRUIT FLIES

Dacus dorsalis (Hendel), D. correctus (Bezzi) and D. zonatus (Saunders) have been recorded damaging the fig fruits in India. These are polyphagous pests and breed or feed on a large number of host plants including various fruits and vegetables (Narayanan and Batra, 1960). In case of figs, it is only the ripe fruits that are attacked as the latex of unripe fruits is not palatable to the grubs of these flies. But in absence of ripe fruits and any other suitable host around, the flies even attack the unripe fruits by inserting their eggs through calyx, where the amount of latex is comparatively less. On hatching, the maggots bore directly into the pulpy mass of fruits and feed within. From the punctures made by the females on the fruits for oviposition, a dark brown gummy substance keeps oozing out. The affected fruits rot and ultimately fall down. The pest is active throughout the year except during January-February, when it hibernates in pupal stage in the soil. The fig fruits are attacked from second week of June till the end of November.

FIG   235

Mediterranean fruit fly, *Ceratitis capitata* Weidemann, a serious pest of figs in Italy and Spain, has fortunately not yet been reported from any part of India. Strict quarantine measures are being adopted to check its entry in our country (Pruthi, 1969).

To avoid the infestation of fruit flies, pick the fruits while these are still firm (unripe). Remove and destroy promptly all the infested and fallen fruits to prevent the multiplication and carry over of this pest. Plough around the trees during Winter to expose and kill the hibernating pupae. In case of severe infestation, spray a mixture of yeast hydrolysate, sugar and malathion (Gupta, 1958).

SAP SUCKING INSECTS

Fig jassid, *Velu caricae* Ghauri is commonly found in Maharashtra. Nymphs and adults suck cell sap from ventral side of leaves with the result the feeding spots turn yellowish. In case of heavy infestation, the effected leaves curl and dry up, adversely affecting the fruiting capacity of the tree.

Kolla diaphana Distant, *Erythroneura* species and *Nirvana* species are other leaf hoppers reported sucking the cell sap (Nayar *et al.*, 1976). These are also minor pests causing negligible damage.

Thrips, *Gigantothrips elegans* Zimmermann and *Thrips tabaci* Lindemann have been found feeding on fig leaves causing some minor damage. Both nymphs and adults lacerate the leaf lamina and rasp the oozing cell sap. The affected leaves get discoloured and start curling at the tips. Adults of *T. tabaci* are pale yellow to dark brown while those of *G. elegans* are comparatively bigger in size and darker in colour. Life-cycle of these thrips is completed in 10 to 20 days during Summer and it takes a little longer during Winter.

Psyllids, *Dinopsylla grandis* Crawford and *Pauropsylla depressa* Crawford produce galls on leaves.

Spittle bug, *Cosmoscarta niteara* Distant is found occasionally in South India on fig leaves whereas fig bug, *Riptortus linearis* Fabricius has been reported from various fig growing regions, desapping and devitalising the fig trees, but the loss caused by

Dinopsylla grandis Crawford

these pests is of minor importance.

Generally no control measures are adopted against these sucking pests as their infestation is seldom severe. If and when necessary, spray with 0.03% dimethoate, phosphamidon, oxydemeton methyl or quinalphos.

GALL MIDGES

Fig midge, *Anjeerodiplosis peshawarensis* Mani has been reported from Punjab, Delhi and Uttar Pradesh (Batra 1952 b). This species shows remarkable host specificity, being confined to fig only. Adults come out of hibernation with the rise in temperature and the females lay eggs in raw and tender fruits. On hatching, the maggots bore inside the fruits and feed on pulp whithin. As many as 200 to 300 maggots may be seen in a single fruit. When full fed, these maggots come out and pupate in the soil without forming any cocoons. The infested fruits become elongated, soft and wrinkled and ultimately shrivel, wither and drop down prematurely.

Eggs are minute, oval, about 0.2 mm long, hyaline, un-sculptured and pedicellate. Full grown maggots are about 3 to 4 mm long, nearly cylindrical in shape and creamy-white in colour. Pupae are obtect and creamy-white to yellowish in colour. Incubation period is 3 days during Summer and 5 days in Winter.

FIG 237

Maggot duration is about 3 weeks during mid August to October and 4 weeks in January-February. Pupal stage lasts for 10 to 15 days during July to November, extending upto 26 days in January-February (Srivastava and Agarwal, 1966). Adult longevity is 2 to 3 days and there are seven overlapping broods in a year.

Another fig midge, *Udumbarie nainiensis* Grover has been reported form Punjab and Uttar Pradesh. It is active during August to October and again from March to May. A female lays 200 to 250 eggs in cavities on flower-buds. On hatching, the maggots feed on interior contents of infloresceaces and cause discolouration, wrinkling and formation of small protuberances.

To check the carry over of these midges, remove and destroy all the infested fruits and plough around the trees to expose and kill the pupae. In case of heavy infestation, spray with 0.1% endosulfan when fruits are pea-sized; repetition may be necessary after two to three weeks.

LEAF ROLLERS

Phycodes minor Moore and *P. radiata* (Ochsenheimer) have been reported from Pakistan, India, Nepal, Sri Lanka, Indonesia, etc. In India, both the species appear as sporadic pests but *P. radiata* is comparatively more common than *P. minor* specially in South India. Moths of both the species are diurnal and may be seen during March to May sucking nectar from blossoms. Eggs are laid on tender leaves. On hatching, the caterpillars feed on epidermis of fig leaves during Spring (March-April); later they fold the leaves singly or one over another and feed within. These leaves ultimately fall down around August, adversely affecting the plant growth. Hibernation takes place in pupal stage in cracks and crevices of bark.

Phycodes radiata larvae are 18 to 22 mm long, dull-yellowish white, smooth with scattered short hair while those of *P. minor* are 13 to 16 mm long and light green in colour.

Incubation, larval, prepupal and pupal periods in case of *P. minor* last for 5 to 6, 31 to 36, 4 to 5 and 8 to 11 days respectively

Phycodes radiata (Ochsenheimer)

(Simwat and Sidhu, 1974). Adults live for 4 to 11 days and there are four overlapping generations during March to September.

To control these leaf rollers, collect and destroy immediately the rolled leaves with caterpillars within. In case of severe infestation, spray with 0.05% endosulfan, fenitrothion or monocrotophos.

BEETLES AND WEEVILS

Altica caerulescens Baly, *Mimastra cyneura* Hope, *Myllocerus blandus* Boheman and *M. maculosus* Desbrocher have been recorded as minor pests of fig trees. These are more harmful to temperate fruit trees and occur only occasionally on fig trees. If and when necessary, these may be controlled by dusting 5 to 10% BHC.

FRUIT BORERS

Stathmopoda sycastis Meyrick—a most destructive pest of figs in Pakistan has been found occasionally in India, boring ripe fruits of fig and *gular*. Full grown caterpillars are pale yellow, cylindrical, 14 to 16 mm long with small scattered hair. The pupation occurs inside the fruits.

Heliothis armigera (Hubner) is another borer recorded occasionally on apple, citrus and fig. It is a polyphagous pest having wide range of host plants, gram and cotton being the preferred hosts. The pest is widely distributed in tropics, subtropics and warmer temperate regions of old world extending as far north as

FIG 239

Germany and Japan. The species found in new world is *H. zea* (Boddie). Both the species are very much similar in appearance and habits. Moths of *H. armigera* appear in Spring and lay eggs singly on leaves. On hatching, the caterpillars feed on foliage and later bore inside the fruits. The caterpillars are cannibalistic and in absence of suitable host, feed on the younger larvae. Pupation takes place in the soil and overwintering is in pupal stage.

Eggs are spherical, 0.5 mm in diameter and shiny greenish-yellow. Full grown caterpillars are stout, 30 to 40 mm long and variable in colour, often greenish with brown tinge and dark broken stripes. Pupae are shiny brown and 13 to 18 mm long. Adults are stout, light green moths with wing expanse of 30 to 40 mm. Egg, larval and pupal periods last for 2 to 4, 14 to 24 and 10 to 14 days respectively.

To prevent the widespread damage by these borers, collect and destroy the infested fruits in early stage of infestation. Hand-picking of larvae and their destruction is also suggested. If attack is severe, dust 5 to 10% BHC.

No more shall the mite and the gall-making blight
The fruit of fig tree devour,
Of thrushes one troop on their armies shall swoop
And clear them all off in an hour.

ARISTOPHANES (448-385 IC)

FALSA, *Grewia asiatica* Linnaeus is of Indian origin and has been mentioned in the vedic literature as having certain medicinal qualities. There are about 200 hectares under this crop in India, half of which is in Uttar Pradesh. It is grown mostly as a catch crop amidst commercial fruit orchards specially in Haryana, Punjab Rajasthan, Uttar Pradesh and some parts of Maharashtra and Andhra Pradesh (around Hyderabad). A wild species *G.elastica* Royle grows on lower hills all over India. The fruit contains 55 to 65% juice with 2.8 % citric acid, 11.7 % sucrose sugar and traces of vitamin C (Singh, 1967). This is one of the most hardly crops, drought resistant and requires very little post-planting care.

As many as 18 insect species have so far been recorded on *falsa* in India, but except mealy bugs, leaf eating caterpillars, cockchafer and other leaf feeding beetles, loss caused by the remaining is negligible.

MEALY BUGS

Mango mealy bug, *Drosicha mangiferae* (Green) and *ber* mealy bug, *Perissopneumon tamarindus* (Green)–the major pests of mango and *ber* respectively, have also been reported feeding on *falsa*. Of the two, former is more destructive than the latter. A severe attack affects fruit setting capacity of the trees. For control, spray with 0.04% diazinon or monocrotophos as soon as the bugs appear.

LEAF EATING CATERPILLARS

Plum hairy caterpillar, *Euproctis fraterna* (Moore) has been reported from Burma, Bangladesh, India, Pakistan, Sri Lanka, etc. In India, it is a major pest of castor and plum and sporadic pest of apple, apricot, *ber*, citrus, *falsa*, grapevine, mango, mulberry, peach, pear, pomegranate, strawberry, etc., causing occasionally severe damage. Eggs are laid in clusters on ventral surface of leaves and are covered with yellow hair. A single female lays 150 to 300 eggs. On hatching, the caterpillars feed gregariously on leaf lamina, skeletonizing the same completely. Later, the caterpillars segregate and gnaw the leaves. In case of severe infestation, the entire tree may be defoliated.

Eggs are flat, circular in shape and yellow in colour. Early instar caterpillars have whitish hair. Full grown caterpillars are 35 to 40 mm long and have red head and darkish-brown body with white hair on the head and a tuft of long hair at anal end. Pupation takes place in silken hairy cocoons in leaf folds. Adults are yellow moths with pale transverse lines on forewings; wing expanse is 24 to 28 (♂) and 30 to 38 (♀) mm.

Euproctis fraterna (Moore)

Incubation, larval and pupal periods last for 4 to 10, 13 to 29 and 9 to 25 days respectively (Teotia and Chaudhri, 1966). There are three generations in a year. The caterpillars of winter brood feed upto December, hibernate for a couple of months and resume feeding before pupating in February.

Falsa catarpillar, *Giaura sceptica* Swinhoe is a serious pest recorded from India, Bangladesh, Burma, Sri Lanka, etc. Though *falsa* is its main host, the pest has also been found feeding on various ornamental plants and some varieties of beans. The pest is commonly found in Delhi, Bihar, Gujarat and Maharashtra. The caterpillars usually roll the leaves specially the lower ones and feed within. Sometimes, a bunch of leaves may be webbed together and caterpillars found feeding inside the same. Full grown caterpillars are 13 to 18 mm long, yellowish-green in colour and become coppery-brown just before pupating. Pupation takes place in silken cocoons. Pupae are 9 to 10 mm long and brown in colour. Adults are brownish moths of medium size having wing expanse of 20 to 24 mm.

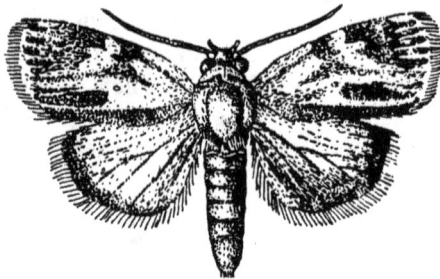

Giaura sceptica Swinhoe

To control these caterpillars, dust 5 to 10% BHC. It will be convenient and economical to kill freshly hatched larvae, as these are gregarious in habit and against these 5% BHC is quite effective (Butani, 1976 d).

LEAF EATING BEETLES

Adoretus spp., *Anomala bengalensis* Blanchard, *Apogonia uniformis* Blanchard, *Holotrichia consanguinea* Blanchard, *H. insularis* Brenske and *Schizonycha* species have been often reported defoliating the *falsa* trees (Varma and Bindra, 1972; Khan and Ghai, 1974). These are polyphagous pests. The grubs feed on roots, while the adults, which are nocturnal and strongly phototropic, feed on foliage during night and hide in the soil or below the big clods during day. Generally the adults appear with first heavy shower of monsoon (May-June) and cause the damage for about two months. Eggs are laid in the soil. Incubation period is about 8 to 22 days and grub duration lasts for 8 to 22 weeks. There is only one generation in a year.

The species commonly found on *falsa* is *Anomala bengalensis*. Besides *falsa*, it has been recorded damaging *ber*, grapevine, guava and *jamun* trees. It has been reported from Gujarat, Rajasthan and Punjab.

Grub   Adult

Anomala bengalensis Blanchard

The most vulnerable stage for control of these beetles is adult stage. Spraying the trees in the epidemic areas with 0.2% BHC or 0.05% endosulfan is suggested. To control the grubs, rake the soil around the infested plants and mix thoroughly, 5% aldrin, chlordane or heptachlor dust, with the soil.

Almond beetle, *Mimestra cyanura* Hope is found all over Indian subcontinent. It is highly polyphagous having a very wide range of host plants—almond being its preferred host. Small (8 to 10 mm long), shinning-yellow beetles with long legs and oblong oval elytra appear in swarms during May-June and defoliate the trees. The peak period of infestation is July-August. Dusting with 5% BHC is quite effective in controlling these beetles.

Falsa beetle, *Oxycetonia versicolor* (Fabricius) has been reported from Andhra Pradesh, feeding on leaf lamina and making round holes. Occasionally it also attacks the flowers.

O. albopunctata (Fabricius) is another cetonine beetle that frequents flowers of *falsa* and citrus trees in South India. The affected flowers wither and die away. Both are minor pests and as such no control measures are adopted against these beetles.

SAP SUCKING PESTS

Falsa bug, *Gargara mixta* Buckton is specific pest of *falsa*. It is a minor pest found all over Indian subcontinent. Eggs are laid on tender shoots during March-April. On hatching, the nymphs move into leaf axils and suck the sap therefrom. The bugs also excrete honeydew, which attracts the black ant, *Camponotus compressus* Fabricius and favours the sooty mould growth.

Besides, a membracid bug, *Leptocentrus taurus* (Fabricius) and a pentatomid bug, *Scutellera nobilia* (Fabricius) have also been reported from Uttar Pradesh. The former is more common on *ber* and citrus whereas the latter is found on *aonla* and grapevine.

Cotton whitefly, *Bemisia tabaci* Gennadius is a major pest of cotton, tobacco and some winter vegetables. In absence of these hosts, the pest migrates to other vegetation growing around and has been recorded feeding on *falsa* and papaya leaves, causing discolouration and curling of leaves.

Generally, no control measures are required against these bugs and whitefly. However, spraying with 0.03% dimethoate, oxydemeton methyl, phosphamidon or quinalphos is effective in controlling these pests.

BARK EATING CATERPILLARS

Indarbela tetraonis (Moore) – a major pest of guava, has also been reported on a number of other fruit trees including *falsa*, specially in Uttar Pradesh. It is a minor pest of *falsa* and is usually found in neglected orchards.

Keeping the orchards clean, is the best way to avoid infestation of this pest.

MELONS

MELONS are grown in the tropical, subtropical and temperate regions of the world. The types commonly grown include, musk melon, *Cucumis melo* Linnaeus and water melon, *Citrulus vulgaris* Schrader – the former is native of Iran (Persia) and North-West Pakistan while the latter is indigenous to tropical Africa (Candolle, 1882). Other relatively less common varieties grown in smaller areas in India are, cucumber, *Cucumis sativus* Linnaeus, *Sarda* melon, *C. melo* Linnaeus, long melon, *C. melo utilissimus* Duthie and Fuller, snap melon, *C. melo momordica* Duthie and Fuller, etc. All these are seasonal creepers that require comparatively less post–planting care and bear fruits during summer months. The fruits of *Cucumis* spp. are wholesome and nutritious but water melons are rather poor in nutritive value though sweet, delicious and juicy and having a cooling effect. The melons are grown mostly in sandy soils or river beds (when there is no water) and require high temperature (25 to 30°C) and dry climate.

All the parts of melon vines including fruits are attacked by a number of pests. In India, about 15 species of insects have been reported damaging these creepers; the loss caused by some of these is substantial and sometimes even render the cultivation of melons unprofitable. The major insect pests include, pumpkin beetles, fruit flies and aphids while the minor ones are, *ak* grasshopper, leaf-footed bug, capsid bug, pumpkin caterpillars, banded blister beetles, water melon weevil, etc.

PUMPKIN BEETLES

Red pumpkin beetle, *Rhaphidopalpa* (*Aulacophora*) *foveicollis* (Lucas), blue pumpkin beetle, *R. intermedia* (Jacoby) (*attripennis*

Fabricius) and grey pumpkin beetle *R. cincta* (Fabricius) (*stevensi* Baly) have been recorded damaging melons in India. The first one is most destructive and is also found in Australia, Burma, Bangladesh, Sri Lanka, Pakistan, Sudan, Mediterranean region, etc. It is widely distributed all over India though it is more common in the North, whereas *R. intermedia* and *R. cincta* are usually found in North and South India respectively and are not as destructive as *R. foveicollis.*

All these beetles are polyphagous and besides melons, they attack various other cucurbitaceous plants, pulses, etc. Nature of damage as also life and seasonal history of the three species are more or less same. The grubs feed on roots and the portion of stems that is below the soil as also bore into the fruits that are touching the soil (Kadam and Patel, 1957). The damaged roots and underground portion of stems start rotting due to secondary infection by saprophytic fungi and unripe fruits of such vines dry up. The fruits infested by the grubs, become unfit for human consumption. Adult beetles feed voraciously on leaf lamina, biting irregular holes. They prefer young seedlings and tender leaves. Often the beetles are also found feeding on flowers. As a result of severe infestation, the growth of creepers is arrested. The pest is active from March to October, though peak period of activity is during Summer (April to June).

Eggs of *R. foveicollis* are elongated, yellowish-brown initially but soon become orange in colour. The grubs are creamy-yellow and

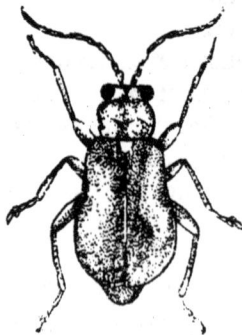

Raphidopalpa foveicollis (**Lucas**)

10 to 14 mm long. Adults are small beetles, 6 to 9 mm long, brilliant red coloured dorsally and black ventrally.

A female lays 150 to 300 eggs singly or in clusters in moist soil around the host plants. The eggs hatch in 5 to 8 days (maximum 15 days in Winter). Grubs take 13 to 25 days to get fully developed and pupate 150 to 250 mm deep in the soil. Prepupal and pupal periods are 2 to 5 and 7 to 17 days respectively (Narayanan, 1953). Adult beetles live for 2 to 4 weeks and the total life-cycle occupies 32 to 65 days. There are five to eight overlapping generations in a year. Overwintering (5-6 months) is in adult stage and is passed hidden under dry and dead leaves or cracks and crevices in the soil. During this period 75% of the adults (mostly males) die away and the remaining become active ($\male : \female :: 1 : 6$) again around February-March.

Clean cultivation, early sowing and use of resistant varieties are the effective cultural practices to ward off the attack by this pest. Deep ploughing after harvest, destroys the grubs in soil. As regards chemical control, Pradhan et al. (1958) as also Bogawat et al. (1969) found gamma BHC and BHC to be effective. But unfortunately, good many organic synthetic insecticides including BHC, endrin and parathion have been reported as phytotoxic to various cucurbits (Banerjee and Chatterjee, 1955; Mookherjee and Wadhi, 1956; Nagaraja Rao, 1959; Kadyan et al., 1971) – as such great care should be taken in selecting the proper insecticide and its concentration. Dusting with 5% endosulfan or oxithion is effective and safe to use (Kadyan et al., 1971). Spraying with 0.04% phosphamidon or dichlorvos (DDVP) has also been suggested by Nayar et al. (1976). This will also kill the banded blister beetles, if these are also present on the crop.

FRUIT FLIES

Melon fruit fly, *Dacus cucurbitae* (Coquillett), Ethiopian melon fly, *D. ciliatus* (Loew) (*brevistylus* Bezzi), Baluchistan melon fly, *Myriopardalis pardalina* (Bigot), guava fruit fly, *Dacus diversus* (Coquillett) and peach fruit fly, *D. zonatus* (Saunders) have been reported attacking melons in India. Of these, the first one is the most destructive.

Dacus cucurbitae is widely distributed in India and has also been reported from East Africa, Mauritius, Pakistan, India, Sri Lanka, Bangladesh, Burma, China, South Japan, Taiwan, Thailand, Malaysia, Indonesia, Philippines, North Australia and Hawaii (CIE map No. A–64). This is a polyphagous pest and has been recorded on more than 70 host plants in India including, citrus, date palm, guava, papaya, peach and mango (Batra, 1953; Gupta, 1960). Outside India, the fruits damaged by this fruit fly include, apple, avocado, Chinese melon, custard apple, fig, mango, oriental pickling melon, pear and strawberry (Narayanan and Batra, 1960). Of course the pest prefers cucurbitaceous fruits specially, musk melon, snap melon, bitter gourd and snake gourd.

Female flies puncture the rind of soft and tender fruits with their ovipositor and lay 4 to 10 eggs inside the fruits. A single female can lay as many as 200 eggs in about 2 months and the same puncture may be used by more than one female for oviposition. On hatching, the maggots feed on pulp of the fruit making the same unfit for human consumption. The full grown maggots come out and pupate in soil. The external symptoms of attack are oozing out of brown resinous juice from the holes made for oviposition. In case of small fruits having too many maggots inside, the fruits get distorted in shape. Late maturing varieties are generally more susceptible than the early maturing ones.

Eggs are white, cylindrical, about one mm long and slightly curved. Full grown maggots are 8 to 10 mm long and pale white in colour. Pupae are barrel-shaped and light brown. Adults are 4 to 5 mm long, reddish-brown flies, with lemon yellow curved vertical markings on thorax. Wings are transparent having brown bands and

Dacus cucurbitae (Coquillett)

grey spots at the apex. Abdomen of males is spherical and that of females is conical (Ranjhen, 1949). Adult flies are strong fliers and can fly 8 to 10 km in a day in search of suitable host.

The pre-oviposition period lasts for 9 to 21 days. Incubation period is only one day in Summer and may extend upto 9 days during Winter. Shortest life-cycle is 10 days in monsoon season and longest is about 3 months during Winter. Lall and Sinha (1959) observed life-cycle to occupy 12.5 to 13.2 days during June-July and 30.3 to 34.1 days during December-January. The pest does not breed during hot months (May, June and September) and overwinters in adult stage.

Ethiopian melon fly, *Dacus ciliatus* is another fruit fly found usually in association with *D. cucurbitae*. It is of African origin and widely distributed in European countries. It is a major pest of melons in Bangladesh where maggots are found boring into the fruits during February to April (Alam *et al.,* 1964). In India its preferred host is *Coccinia indica*; the grubs are often found in the stem galls caused by the midge, *Bimba toombii.* The adult flies are smaller than *D. cucurbitae* and are bright-brown in colour with hyaline wings having no dark markings. Two dark brown, round spots are present on fourth abdominal segment. Other characteristic differences are that the pupation often takes place within the fruits even if these are drying and overwintering is also in pupal stage (Pruthi and Batra, 1960). Egg, maggot and pupal stages last for 1 to 2, 4 to 6 and 8 to 10 days respectively (Cherian and Sundaram, 1939), Total life-cycle occupies 15 to 17 days in October and there are six generations in a year in South India. In North India, where there is a distinct Winter, there are only four to five generations in a year.

Myriopardalis pardalina (Bigot) has been recorded from Punjab and Bihar in India as also from Pakistan, Afghanistan, Israel and East Africa. It attacks all varieties of melons as also other cucurbit fruits. It is a pest of minor importance in India, but major in western countries. Nature and symptoms of damage are same as those caused by *Dacus* species. Egg, maggot and pupal periods are on an average, 3, 15 and 13 days respectively while longevity of adults is 3 to 4 weeks (Coleghorn, 1914). The pest overwinters for about

6 months in pupal stage and has two to three generations during the active season. As many as 130 eggs may be laid by a single fruit fly but of these hardly half a dozen reach maturity.

Dacus diversus and **D.** *zonatus* have also been reported on melons; the former is commonly found feeding and breeding in flower-buds and flowers during Summer. The main hosts of these fruit flies are guava and peach respectively.

To safeguard the melons from ravages of fruit flies, grow resistant varieties. Early maturing varieties also escape the attack by this pest. In endemic areas where the attack is observed every year, change of sowing date is suggested. This coupled with soil application of 5% aldrin or heptachlor dust @ 20-22 kg per hectare will reduce the pest incidence appreciably. As soon as the infestation is detected, remove all the infested fruits and destroy the same promptly either by burning or burying in soil about a metre deep. In addition, rake the soil under the fruits to destroy the pupae lying therein. This will prevent the build-up of pest population. Poison baiting has also been recommended. Mix 0.1% diazinon or malathion in 1% sugar solution or molasses and add to it 1% fermented palm juice or yeast hydrolysate, if available. Keep this solution in shallow dishes at various places in the field to trap and kill the adult flies.

The control measures should always be adopted simultaneously in all the fields around, otherwise, these flies being good fliers will keep on invading inspite of the best individual efforts.

APHIDS

Aphis gossypii Glover (*cucumeris* Forbes *cucurbiti* Buckton), the melon (cotton) aphid, is cosmopolitan in distribution being absent only from colder regions of Asia and Canada (CIE map No. A–18). Besides musk melon and water melon this aphid attacks a large number of cucurbitaceous and malvaceous plants.

Nymphs and adults in small or big colonies are usually found sucking the cell sap from ventral surface of leaves and apical shoots. Younger creepers are more vulnerable to attack of aphids than the older ones. In case of severe infestation, the leaves get curled, cupped or otherwise distorted and gradually dry up. The aphids also secrete honeydew on which sooty mould develops which in turn interferes with photosynthetic activity of the plants. As a result, the plant growth is arrested. The aphids also act as vectors, transmitting more than 50 virus diseases. In case of melons, they transmit the virus causing water melon mosaic (Coudriet, 1962) and the loss caused is irrepairable. The pest is active throughout the year except during severe cold (December-January). On melons the maximum activity has been observed during April to June. Outbreaks are common during spells of dry weather and the pest population dwindles with the onset of rains but again increases during August to October. In Peru (South America), this aphid is regarded as a beneficial insect as there it is the winter host of the egg predator of a more important noctuid pest (Eastop, 1977).

Nymphs are greenish-brown or yellow in colour. Adults are pear-shaped, soft bodied, 1.0 to 1.5 mm long and variable in colour. Those produced on overcrowded leaves at high temperatures are pale yellowish-green in colour and less than one mm in length, whereas winter forms are bigger in size and dark green to almost black in colour. Antennae are only about half the length of body; eyes are red and siphunculi are black. Males have not been recorded and females are mostly wingless; winged forms appear after several generations.

Aphis gossypii (Glover)

Life-cycle is rather complex. Reproduction is parthenogenetic and viviparous. Females reach maturity in 3 to 20 days depending upon the climatic conditions and produce 20 to 140 young ones @ 2 to 9 per day (maximum 22). High humidity and cloudy weather with little rainfall is favourable for rapid multiplication of this pest (Reddy, 1968). In temperate regions overwintering takes place in the egg stage and the adults normally perish during Winter.

Green peach aphid, *Myzus persicae* (Sulzer) has also been reported attacking melons. The preferred host of this polyphagous pest is peach. Besides it is also found on most of the temperate fruits.

To control these aphids, spray with 40% nicotine sulphate (1:600) or spray 0.03% dimethoate, oxydemeton methyl or phosphamidon or 0.05% dichlorvos or fenitrothion. Repeat the spraying, if necessary, after 10 days. Treating the seeds just before sowing, with disyston or phorate has also been found to be effective (Butani 1975 g).

AK   GRASSHOPPER

Poekilocerus pictus (Fabricius) is mainly a pest of *Calotropis* spp. but in absence of the main host it feeds on a large variety of economic crops including melons. Nymphs and adults feed voraciously on leaves and skeletonize the same. Both are conspicuous due to their big size and ornamental body colouration. Nymphs generally appear in melon fields around end of March or early April and become adults in 6 to 10 weeks.

Hand-picking and mechanical destruction of nymphs and adults keeps the pest population under check.

LEAF–FOOTED BUG

Leptoglossus australis (Fabricius) (*membranaceus* Fabricius) is widely distributed in almost all tropical regions of old world (CIE map No. A–243). Though not a serious pest, it is quite common and occurs in large number on various hosts including melons and *Citrus* species. It attacks terminal shoots and developing fruits. The terminal shoots wither and die off beyond point of attack. The fruits show dark spots around feeding punctures.

Adults are big brown bugs, 20 to 25 mm long, with characteristic tibial expansions on hind legs. Other diagnostic character are presence of thin pale orange stripe across the anterior edge of mesonotum and antennae having alternate black and orange segments.

Normally no control measures are required; if and when necessary, hand-picking and mechanical destruction of the bug may be done to check the pest population.

CAPSID BUG

Creontiades pallidifer Walker is another minor pest of melons (Nair, 1975) Which causes small irregular brown spots on growing tips of young leaves. In case of severe attack, which is rare, the leaves droop and dry away.

Eggs are conspicuous having a white ridge or collar and a tag-like prolongation. Adults are about 6 to 9 mm long, ochraceous-green in colour with transparent wings. A female lays 130 to 200 eggs in plant tissues at growing tips, petioles and axils of branches. These hatch in 4 to 7 days. Nymphal period varies between 11 and 18 days. Longevity of males is on an average 5 days and that of females 15 days.

As the loss caused by these bugs is seldom serious, no control measures are generally adopted.

PUMPKIN CATERPILLARS

Diaphania indica (Saunders) was originally described by Saunders (1851) as *Glyphodes indica*. Subsequently the species was placed under genus *Margaronia* and later transferred to *Diaphania* (Wang 1963). The freshly hatched larvae, lacerate the tissues of ventral surface of leaves and feed on chlorophyll; later these larvae fold or bind the leaves together to feed and pupate therein. They also attack ovaries of flowers and bore into young developing fruits which become useless for human consumption (May, 1946).

Caterpillars of *Diaphania indica* are bright green with prominent stripes on the back. Moths are very conspicuous having transparent white wings with dark brown marginal patch.

Diaphania indica (Saunders)

A female may lay from 25 to 365 eggs, singly or in batches on the leaves. The eggs hatch in 5 to 13 days; larval and pupal periods are 11 to 14 and 5 to 13 days respectively (Patel and Kulkarny, 1956).

Diaphania caesalis (Walker) is a major pest of jack fruit. Its caterpillars have also been reported boring into flower-buds and young shoots of melons.

To control these pests hand-picking and mechanical destruction of caterpillars in the early stage of attack is suggested. If attack is severe, dust 5% carbaryl or spray with 0.2% carbaryl or 0.05% endosulfan.

BANDED BLISTER BEETLES

Mylabris spp. have been reported from Africa, India, Bangladesh, Sri Lanka and other countries of South-East Asia. These beetles feed on pollens and petals of flowers and flower-buds of various cucurbitaceous and leguminous crops as also ornamental plants. As a result, the fruit setting capacity of the affected plants is adversely affected.

The species recorded on melons, in India include, *Mylabris* (*Zonabris*) *pustulata* (Thunberg) and *M. phalerata* (Pallas). These are very conspicuous beetles, 22 to 26 mm long and having six alternate black and yellowish-red vertical bands on elytra. *M. pustulata* is on an average slightly bigger than *M.phalerata* and its yellowish-red bands are narrower than the black bands. When touched or disturbed these beetles emit a yellow fluid, containing cantharidine, which is irritant to touch and causes blisters on the skin – hence the name.

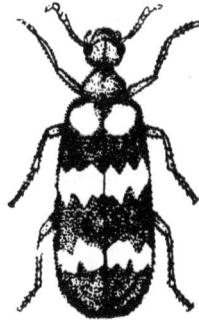

Mylabris pustulata (Thunberg)

Eggs are laid in clusters in soil. Freshly hatched grubs, known as triungulins are very active and feed on eggs of grasshoppers and locusts in the soil. Thus while the adult beetles are harmful, the grubs are beneficial (Singh, 1970). Later instars of the grubs are sluggish with large bodies and reduced legs. Full grown grubs are coarctate and form pseudopupae. This stage is devoid of functional appendages and hibernation also takes place in this stage. The pseudopupae later become pupae. All these stages are passed in the soil and the adult beetles emerge from the pupae and the soil around August and are active till December on various host plants.

To control, hand-collection and mechanical destruction of beetles, in the early stage of attack, is suggested. This may be done during early morning hours when the beetles are comparatively less active.

In case of severe infestation, dust with 5% carbaryl or endosulfan or spray with 0.2% carbaryl or 0.07% endosulfan.

WATER MELON WEEVIL

Acythopius citrulli Marshall is a small dark weevil reported from South India (Ayyar, 1963) boring into ripening water melon fruits. As it is a pest of minor importance, not much work has been done on its life-cycle, bionomics or control, which is seldom needed.

4

PESTS OF TEMPERATE FRUITS

APPLE, *Malus sylvestris* Linnaeus is considered to be the world's oldest fruit crop and its existence can be traced back to the era of Adam and Eve. Its original home is Asia Minor (East of Caspian sea), and it has been cultivated for over 3000 years. Apple is the premier table fruit of the world. It excels the other fruits in having prolonged keeping quality and wide variety of flavour and taste. Moreover, it is not exacting in its climatic requirements – except that it requires minimum chilling (below 5°C) during Winter without which the trees do not flower properly in Spring – hence these are normally grown at an elevation of 1600 to 2700 metres and mostly on hill slopes.

In India, apples are grown along foot-hills of entire Himalayan range; from Shillong (Assam) and Darjeeling (Bengal) in the East through Kumaon hills (Uttar Pradesh), Kangra hills (Punjab), Kulu valley and Simla (Himachal Pradesh) to Jammu and Kashmir in the extreme West. A small quantity is also produced in Nilgiri hills (Tamil Nadu), Panchghani (Maharashtra) and Coorg (Karnataka). It occupies a total area of 75,000 hectares; of which 30,000 and 27,000 hectares are in Himachal Pradesh and Kashmir respectively, 17,000 in Uttar Pradesh and remaining 1,000 spread over Tamil Nadu (200), Punjab (150), Assam (120), Bengal (80), Maharashtra (60), Karnataka (50), Orissa, Haryana and Delhi. This area is increasing day by day.

The apple trees are attacked throughout the year by a number of insect pests. Those of major importance are, San José scale, wooly aphid, root borer, stem borers, tent caterpillar and leopard moth. Thrips, leaf rollers, leaf eating caterpillars as also beetles and weevils occur sporadically as serious pests. Besides, fruit flies, peach borer, shot hole borer, leaf hoppers, aphids and scale insects have been recorded as minor pests while cutworm has been often found causing serious damage in nurseries.

SAN   JOSE   SCALE

Quadraspidiotus perniciosus (Comstock) is widely distributed in all the apple growing countries of the world (CIE map No. A-7). It was introduced in India (Kashmir) from France in 1906 and by now it has been recorded on more than 32 host plants (Rahman and Ansari, 1941), including almond, apple, apricot, cherry. chestnut, citrus, crab apple, grapevine, gooseberry, mulberry, peach, pear, plum, quince, raspberry and strawberry. These tiny insects suck the sap, as a result, the young plants in the nursery become weak and ultimately die away. The leaves, twigs, fruits and sometimes even the entire bark may be seen covered with ashy-grey scales which can be easily scraped off exposing the orange coloured individuals beneath. The fruits present a pink coloured areas around the scales and the market value of such fruits is naturally depreciated.

Quadraspidiotus perniciosus (Comstock)

Climatic conditions have a great effect on incidence of this insect. Sharp fluctuations in temperature as also low temperatures bring about heavy mortality of the pest. In Kashmir, the insects enter into hibernation in all the stages of its life-cycle but only full grown first instar nymphs that have secreted dark grey scales, for their protection, survive the low temperature, the rest perish (Pruthi and Rao 1951). The nymphs hibernate from December to March, resume their activity around end of March and mature in about 4 weeks. The males (winged) fertilize the females (wingless) and die. The females do not lay eggs but produce young ones; 300 to 400 per female, the eggs mature into tiny nymphs in the female ovisac in about a month, during April-May. Nymphal period varies between 40 and 50 days. The number of generations in a year depends mainly on elevation and climatic conditions. Fotidar and Raina (1936) recorded four to five generations in a year while according to Singh (1964) there are six to seven overlapping generations in a year. The pest disseminates by various birds, bats, etc. (Rahman and Kalra, 1940).

To control this pest, Sharma and Bhalla (1962, 1965) suggested spraying the dorment trees in Winter with 3% miscible oil. This is quite effective. Nevertheless, Pradhan (1969) pointed out that due to heavy infestation of the pest on other wild trees in and around the orchards, the good effect of chemical control is soon lost making it obligatory to repeat the spraying. Therefore, it is suggested that in addition to spraying, the parasite, *Prospaltella perniciosi* (Tower) may also be released to check the overwintering population of San José scale on wild host plants growing around. This is integration of chemical and biological control, applied simultaneously in time but separately in space.

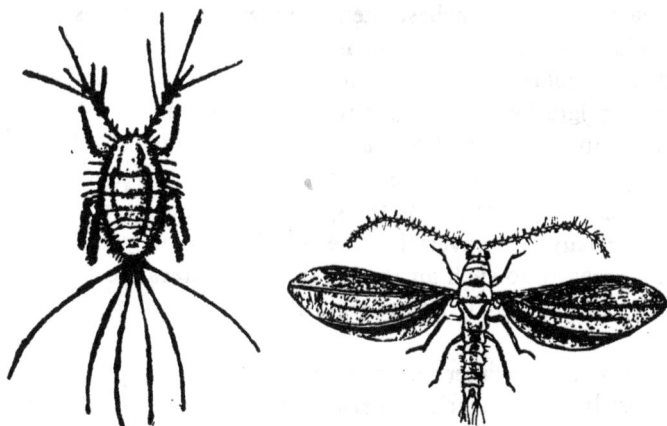

Icerya purchasi Maskell

In addition to San Jose scale, other mealy bugs and scale insects recorded damaging apple trees are, *Drosicha contrahens* (Walker) *D. mangiferae* (Green), *Icerya purchasi* Maskell, *I. seychellarum* (Westwood), *Perissopneumon tamerindus* (Green), *Chrysomphalus dictyospermi* (Morgan), *Dentachionaspis centripetalis* Rao, *Duplaspidiotus tesseratus* (de Charmoy), *Hemiberlesia lateniae* (Signoret), *H. rapax* (Comstock), *Howardia biclavis* (Comstock), *Parlatoria oleae* (Colvée), *P. pseudopyri* Kuwana (*cinerea* Hadden), *Pseudaulacaspis pentagona* (Targioni), *Ceroplastes floridensis* Comstock, *Chloropulvinaria polygonata* (Cockerell), *Eulecanium tiliae* (Linnaeus) and *Parthenolecanium corni* (Bouché) (Rao and Chatterjee, 1948; Nair, 1975); but none of these cause any serious damage to apple trees.

WOOLY APHID

Eriosoma lanigerum (Hausman) is one of the most destructive pest of apple in the world. It is native of America and is cosmopolitan in distribution except for hotter parts of the tropics (CIE map No. A – 17). In India, it was first recorded in 1889 at Conoor (Tamil Nadu) damaging young apple trees (Misra, 1920) and has since then, spread to all the apple growing areas of India. Its alternate hosts in India, include, crab apple, pear and quince; whereas in England and America, this aphid has also been recorded on, almond, elm, hawthorn, mountain ash, pear, etc. The pest is active throughout the year. It attacks primarily the underground roots but winged form also attacks trunk, branches, stems, twigs, leaf petioles and fruit stalks. Upward and downward migrations are accentuated during hottest and coldest seasons respectively (Lal and Singh, 1947). Maximum migration from roots to aerial parts takes place in May and in the opposite direction during December-January. Due to the desapping caused by these pests, the affected trees present a sickly appearance, lose vigour and the growth of these trees as also their fruiting capacity are adversely affected. In case of young trees, the roots disintegrate to such an extent that these trees are easily blown over by even moderately strong winds.

The pest overwinters either as egg or young nymph on the roots of the host tree (In America hibernation is only in egg stage – Pruthi and Batra, 1960). The eggs hatch and the nymphs mature during

Spring. The reproduction during this period is parthenogenetical as well as viviparous. A single female produces 30 to 116 young ones in her

Nymph   Adult (winged)

Eriosoma lanigerum (Hausman)

life time. New nymphs soon settle down in batches and start sucking the plant sap. Within 24 hours these nymphs begin to secrete wooly filaments of wax over their bodies – hence the name, wooly aphid. Nymphal period is about 11 days in June which gradually increases with fall in temperature and becomes 93 days by December (Rahman, 1941). During Summer and early monsoon months there is rapid multiplication both on stems and roots and considerable dispersal of the pest. The winged adults fly away while the wingless forms are blown off by the wind. With the advent of Winter, the sexual forms appear (Kapur and Fotidar, 1943), mate and lay eggs; while the immature nymphs on the trees descend and enter the root zone for hibernation. Reproduction generally is not affected upto 32°C but ceases when the temperature falls below 3°C. Excessive humidity (August-September) retards multiplication of this aphid.

To prevent damage by this aphid, use resistant root-stock like *Golden delicious, Northern spy and Morton stocks* 778, 779 789 or 793 (Javaraya, 1943; Lal and Singh, 1947; Khan, 1955). Attri and Sharma (1971) suggested soil application (80 to 100 mm deep) of dimethoate or thiometon granules @15 gm a.i. per tree during Spring and Summer against the root forms. In addition, foliar spraying with 0.03% dimethoate or phosphamidon or oxydemeton methyl during March-April (Spring) and again in June (Summer) will give complete relief from the aerial forms. The aphid population can also be effectively checked by an exotic parasite, *Aphelinus mali* Hald. This parasite was introduced from American and it is by now well established specially in Himachal Pradesh (Kulu Valley) and Uttar Pradesh (Chaubattia region).

In addition to wooly aphid, other aphids commonly found infesting apple trees are, apple green aphid, *Aphis pomi* de Geer (*mali* Fabricius), melon aphid, *Aphis gossypii* Glover and peach green aphid, *Myzus persicae* (Sulzer). All these are polyphagous pests and as their common names suggest, apple, melon and peach are their respective preferred hosts.

A. pomi has been reported from Uttar Pradesh and Karnataka where besides apple, it also attacks *Citrus* spp. The infestation is more commonly found in the nurseries than in orchards. Nymphs and adults suck the cell sap from leaves, twigs, branches, blossoms, and young fruits. As a result of desapping, the affected leaves curl up, blossoms are shed and the young fruits drop prematurely while the quality of ripening fruits is greatly impaired.

Aphis pomi de Geer

These aphids can be easily controlled by spraing with 0.03% dimethoate, oxydemeton methyl, phosphamidon or quinalphos.

ROOT BORER

Dorysthenes hugeli Redtenbacher is confined to foot hills of Himalayan range and is a serious pest of apples in Kumaon hills.

The main host of this borer is living and dead roots of apple trees, though occasionally roots of apricot, cherry, peach, pear, walnut and a few forest trees are also attacked. Eggs are laid 8 to 12 mm below the soil, sandy soil is preferred. On hatching, the grubs go down into the soil, 100 to 250 mm deep and feed on roots of all kind as well as other organic matter. Later, these grubs bore into main roots or girdle around the roots and feed on the internal tissues. As a result the main roots are severed from the base and the trees, if young, die away while the older ones become weak and fall down with strong winds.

Eggs are avoid, 3×1 mm and yellowish-white in colour. Full grown grubs are creamy-white with black head and mandibles and 75 to 100 mm long. Adult beetles are chestnut-red in colour with head and thorax darker than elytra.

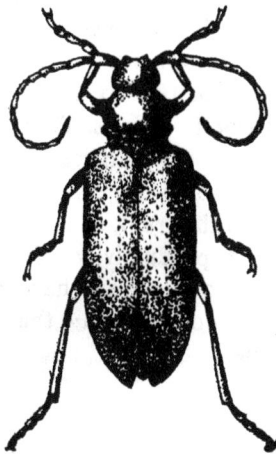

Dorysthenes hugeli Redtenbacher

A female lays on an average 300 eggs. These hatch in 30 to 40 days. Grub duration extends upto $3\frac{1}{2}$ years and the grubs can live even without food for 24 to 90 days. Pupation takes place in earthen cocoons inside the soil and the pupal period lasts for about 3 months. Adult beetles emerge from soil with the advent of monsoon, around end of June and continue doing so till middle of July. Immediately after emergence, they start mating and the males die soon after while the females live for 10 to 12 days. There is always heavy mortality in immature stages and hardly one per cent reach maturity.

To check the incidence of this pest, avoid dry sandy soils for planting apple orchards. Use well rotted Farm Yard Manure and mix thoroughly with soil around the tree 5% chlordane or heptachlor dust @ 200 to 300 gm per tree, soon after the monsoon. This coupled with frequent interculturing will help in killing the grubs and reducing the pest population.

STEM BORERS

Apriona cinera Chevrolat is a destructive stem borer of apple, peach and fig trees and is reported from India, Pakistan and Afghanistan. In India, it is more commonly found in Kashmir, Himachal Pradesh and Uttar Pradesh. The grubs bore through the shoots and make circutous galleries downwards, leading to main stem and trunk of the tree. They also make small circular holes in the outer epidermis at a distance of 100 to 125 mm for throwing out their excreta and chewed wood particles. Each grub makes 8 to 10 holes. Ultimately, all the grubs in different branches come down in the main trunk which is hollowed inside at several places. Six or more grubs may be found in the trunk of a single tree. As a result of this attack the vitality and productivity of these trees is greatly impaired. Adult beetles feed on bark and have an unusual habit of cutting leaves and young shoots as well – thus damaging more than what they actually consume.

Adult beetles are 35 to 50 mm long, ashy-grey with numerous black tubercles at the base of elytra; antennae concolourous and longer than the body. These are found on wings during July-August. Females excavate oval patches on the shoots and lay eggs inside these cavities. Grubs emerge in 7 to 8 days and continue to bore inside. The grubs feed till October, remain quiescent during Winter and resume the feeding activity in March, reaching the tree trunk by Autumn (September-October) again undergo hibernation during Winter and become pupae and adults during the following Summer. Thus the life-cycle is completed in about two years.

To control this borer prune and burn all the attacked shoots and branches during Winter. Inject in the lowest hole about 5 ml of a good fumigant like, carbon bisulphide, petrol, etc., to kill the grubs inside.

Other stem borers recorded damaging apple trees are, *Aeoleathes holosericea* Fabricius, *A. srata* Solsky, *Batocera rufomaculata* (de Geer), *Celosterna scabractor* Fabricius, *C. spinator* Fabricius and *Sphenoptera lifertéi* Thomson. Of these, *S. lafertéi* is a serious pest of peach and *A. holosericea* of cherry. In case of apple trees, the loss caused by these beetles is of minor importance.

Girdler beetles, *Linda nigrosculata* Fairmaire and *Sthenias grisator* Fabricius girdle the twigs of apple trees while their grubs bore into the twigs.

Shot hole borers, *Scolytoplyatypus raja* Blandford and *Pitogenes* species are also occasionally found damaging apple trees. These are sporadic pests. Adults and grubs of *Pitogenes* species mine under the bark and feed on inner bark or wood of the trees while those of *S. raja* tunnel into sapwood and hard wood of branches making pin-holes. Surface of infested branches get perforated followed by yellowing and wilting of leaves. In severe cases the tree may also die. On breaking, infested branches emit foul fermenting odour. These beetles are more common on weak, decaying or dead trees. *Pitogenes* species are small, 4 to 6 mm long, cylindrical, dark brown beetles and *S.raja* are still smaller, 2 to 3 mm long and reddish-black in colour.

Alcidoides mali Marshall is another minor pest. Grubs of this weevil tunnel within the shoots, causing gall-like swellings.

As most of these stem borers are minor pests, normally no control measures are adopted against these. However, removal and prompt destruction of infested shoots and twigs is advisable.

TENT CATERPILLAR

Malacosoma indica (Walker) has been reported from hilly areas of Himachal Pradesh and Uttar Pradesh on a number of fruit and forest trees as also on ornamentals; apple being its preferred host. Besides, almond, apricot, cherry, gooseberry, peach, pear, walnut, etc. are also occasionally attacked. The caterpillars web a tent-like nest at the forking of twigs and hide in this nest during day. At night the caterpillars congregate on leaf lamina and feed voraciously, leaving behind only the midribs and portions of hard veins. Broad-leaved deciduous trees are preferred and in absence of leaves, tender barks too are not spared. The defoliated trees obviously bear little fruit.

Eggs are about one mm long, shinning and elongated thimble-shaped. Full grown caterpillars are 40 to 45 mm long, blackish-brown having a broad greyish band dotted with black on the dorsum and 2 lines of crimson coloured dots on either side Pupae are 18 to 21 mm long, smooth and dark brown in colour. Adults are light brown to light red in colour; forewings traversed by two oblique, broad, whitish stripes enclosing whitish area inbetween. Wing expanse is 29 to 32 mm and 35 to 37 mm in case of males and females respectively. Eggs are laid round the branches of trees as broad bands or rings of 200 to 400 eggs during May to June. These are well protected with a coating of dark brown glue like substance. Incubation period is about 9 to 10 months. The hatching of eggs usually takes place between mid-March and mid-May. Larval period ranges between 39 and 68 days extending from March to May. Pupation takes place in silken cocoons on stems, between the leaves or in dry debris lying on the ground below. Pupal period is 11 to 22 days. Moths emerge from end of May till early June and there is only one generation in a year (Rahman and Kalra, 1943).

To check the infestation of this pest, destroy all the egg bands at the time of pruning in December-January. The nests may also be destroyed mechanically or by spraying heavily with 0.5% DDT. Foliar spraying with 0.05% endosulfan or quinalphos has also proved effective.

LEOPARD MOTHS

Zeuzera spp. are polyphagous pests recorded on a number of host plants including, apple, cherry, citrus, custard apple, guava, litchi, peach, pear, pomegranate, quince, walnut, forest trees and ornamental plants. Eggs are laid during June-July in cracks and crevices of bark or in soil under clods. The newly hatched larvae first feed just below the bark and afterwards bore into the wood.

(*Courtesy: J.N. Sachan*)

An old apple branch damaged by *Zeuzera* species showing lot of excreta pellets

The portion tunnelled is found full of excreta pellets. Incidence is usually confined to lower part of the tree—not more than 3 metres above ground. Pupation takes place within the burrows late in Spring. Moths are white in colour, having beautifully patterned wings They are nocturnal in habit and are seen on wing during end of May to July.

Incubation, larval and pupal periods last for 7 to 8 days, about 20 months and 3 to 7 weeks respectively (Singh, 1964). Adult longevity is about a fortnight.

To combat this pest, kill the caterpillars in their burrows by pushing in a pointed wire, or plug the tunnels with cotton-wool soaked in carbon bisulphide, paradichlorobenzene, petrol, dichlorvos or a mixture of chloroform – creosote (2 : 1) and seal the entry holes with mud. Though effective, this method is not practicable on large scale.

BEETLES AND WEEVILS

Beetles recorded defoliating apple trees in various parts of India include, *Lucanus lunifer* Hope, *Clinteria spilota* Arrow, *Macronota* 4-*lineata* Hope, *Mimela passerinii* Arrow, *M. pectoralis* Blanchard, *Protaetia neglecta* Hope, *Oxycetonia jucunda* (Fald.), *Autoserica* spp., *Brahmina coriacea* (Hope), *Holotrichia longipennis* Blanch., *H.crinicollis* Redtenbacher, *Hilyotrogus holosericeus* Redtenbacher, *Hoplosternus furicaudus* (Ancy), *H. nepalensis* Hope, *Microtrichia* species, *Adoretus bimarginatus* Ohow, *A. duvauceli* Blanchard, *A. horticola* Arrow, *A. lasiopygus* Burmeister, *Anomala dimidiata* Hope, *A. lineatopennis* Blanchard, *A. polita* Blanchard, *A. rufiventris* Redtenbacher, *A. rugosa* Arrow, *Mylabris macilenta* Marseul, *Aplosonyx trifasciatus* Jacoby,

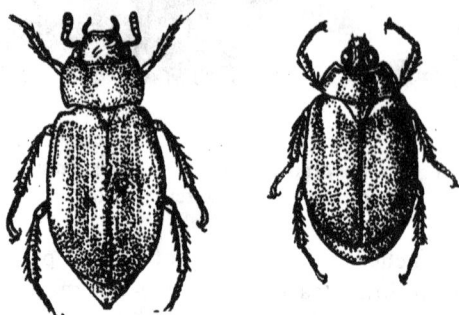

Mimela pectoralis Blanchard   *Hoplosternus nepalensis* Hope

Altica species, *Monolepta erthrycephala* Baly, *Aphthona* species, *Galerucida rutilans* Hope, *Luperus* species, *Mimestra cyanura* Hope, *Popillia cupricollis* Hope and *P. sulcata* Redtenbacher. These are all polyphagous and have a very wide range of host plants. Adult beetles usually appear with the onset of monsoon. These are nocturnal in habit and feed voraciously on leaf lamina during night, hiding during day, in the soil or below the clods. Eggs are laid in the soil near the roots of host trees. On hatching, the grubs feed on roots and other organic matter in the soil. Pupation also takes place in the soil.

Clean cultivation coupled with soil application of 5% aldrin or chlordane dust has been recommended against the grubs. Sharma *et al.* (1971) have suggested spraying with 0.2% carbaryl twice at an interval of 7 days, starting as soon as the adult attack is noticed.

Occasionally *Popillia complanata* Newman and *P. cyanea* Hope have also been reported feeding on flowers of apple, peach and plum trees.

Among the weevils, mention may be made of apple weevil, *Myllocerus discolor* Boheman, lesser apple weevil, *Dereodus pollinosus* Redtenbacher, grey weevil, *Myllocerus undecimpustulatus maculosus* Desbrocher, almond weevil, *M. laetivirens* Marshall, apricot weevil *Emperorrhinus defoliator* Marshall, plum weevil, *Amblyrrhinus*

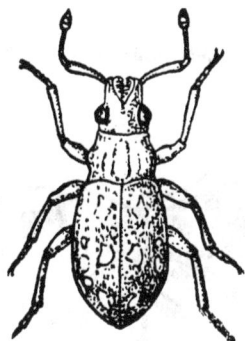

Myllocerus discolor (Boheman)   *Dereodus pollinosus* Redtenbacher

poricollis Boheman and brown weevil, *Tanymecus circumdatus* Wiedemann. Of these, *M. discolor* is comparatively more common and harmful. This weevil is widely distributed in India and Pakistan. It is polyphagous having a wide range of host plants including, apple,

plum, loquat, guava and mango, The grubs feed in the soil on roots
of different plants and the adult weevils feed on foliage.

To control these weevils, dust with 10% BHC. Spraying with
0.04% diazinon or fenitrothion is also effective.

LEAF   ROLLERS

Apple leaf roller, *Archips termias* (Meyrick) (*pomivora* Meyrick,
sarcostega Meyrick) is the most harmful leaf roller. The species was
first described by Meyrick (1924) as *Cacoecia sarcostega* from Punjab
and Uttar Pradesh. Beside apple, the pest has been recorded as a
minor pest of apricot, cherry, mulberry, peach, pear, plum, etc.
(Janjua, 1940). The caterpillars feed on foliage and also bore inside
flower-buds and fruits. Larvae hibernate from end of October till end
of March (86 to 295 days). They crawl out of their hibernacula around
April and feed on developing flowers which are webbed together,
thereby reducing the fruit setting. In case of foliage, the leaves are
cut off from petioles and tied with other leaves or with fruits by
means of silken strands. The caterpillars feed, within this rolled
leafy structure, on soft tissues of the leaves or fruits for 20 to 28 days.
The attacked leaves get partially skeletonized and a gum globule is
seen on the surface of damaged fruit, which seriously affects its market
value. In severe cases 20 to 25 per cent of leaves may be damaged.

Archips termias (Meyrick)

Eggs are oval, about one mm long, flat and pale green. Newly
hatched caterpillars are pale green whereas full grown ones are dark

green in colour and 18 to 24 mm long. Eggs are laid in clusters, 160 to 190 eggs in each cluster on dorsal surface of leaves usually along the midrib. These hatch in 4 to 12 days. Larval period varies between 20 and 37 days (overwintering larvae may take 227 to 258 days); prepupal period is 2 to 10 days while the pupal duration varies from 6 to 15 days. There are three generations in a year (Janjua, 1940).

Gracillaria zachrysa Meyrick has been reported from Assam and Uttar Pradesh as an apple leaf miner and roller. This is a specific pest of apple trees. The young caterpillars make several mines on the ventral surface of leaves and later on they abandon the mines and roll young leaves longitudinally into tubular or cone-shaped pouch of white silk, found on ventral surface of leaves. Maximum damage is caused during Summer (April-May) and Autumn (September-October).

Other leaf rollers recorded attacking apple trees in India include, *Archips micaceana* (Walker) (*Tortrix epicyrta* Meyrick, *T. isocyrta* Meyrick, *T. pensilis* Meyrick) and *A. subsidiaria* (Meyrick). The former is found all over India, while the latter is confined to Jammu-Kashmir. The caterpillars bore into fruits from apical end making a tunnel right through the core. The affected fruits rot and drop down prematurely.

To control these leaf rollers, collect and destroy all the infested and fallen fruits. In case of severe infestation, spray with 0.05% endosulfan or fenitrothion.

LEAF EATING CATERPILLARS

Apple giant caterpillar, apple hawk moth, hairy caterpillars and Indian Gypsy moth have been occasionally reported damaging apple trees in India. None of these cause any severe loss.

Apple giant caterpillar, *Actias selene* Hübner has been reported from entire Himalayan range – Assam to Himachal Pradesh and also from Tamil Nadu, defoliating apple, cherry, pear, walnut, etc. specially shrubs and small trees. Caterpillars are apple-green in colour and cocoons opale brown. Moths have pinkish body and very light–greenish wings that are whitish at the base and pale-yellowish near the margins; hind wings have pinkish tail. Wing expanse is 130 to 160 and 140 to 180 mm in case of males and females respectively.

Actias selene Hübner

As this is not a pest of common occurrence, hardly any work has been done on its biology, bionomics and control.

Apple hawk moth, *Langia zeuzeroides* Moore is found mainly in Assam and Himachal Pradesh defoliating apple, apricot, cherry, pear, etc. during April to August. Eggs are 2.5 × 2.0 mm, full grown caterpillars 125 × 10 mm, pupae 50 × 20 mm and moths have wing expanse of 110 to 150 mm. As this is a sporadic pest and all the stages of the species are conspicuously big in size, merely hand-picking and mechanical destruction of the various stages of the pest is sufficient to check the infestation.

The hairy caterpillars reported defoliating apple trees are apple hairy caterpillar, *Euproctis signata* Blanchard, plum hairy caterpillar, *E. fraterna* (Moore) and *ber* hairy caterpillar, *E. flava* (Bremer) – the names clearly suggest the preferred hosts of these species. The caterpillars are voracious feeders and in severe cases of infestation, the entire trees may be defoliated. *E.signata* is commonly known as gold–tailed hairy caterpillar. The eggs are laid usually in the month of June, in clusters on the dorsal surface of leaves and covered by fine hair. These hatch in 30 to 40 days. Freshly hatched caterpillars feed on chlorophyll, later they fold the leaves and feed gregariously within these folded leaves. After for 80 to 90 days, hibernation takes place sometimes in Autumn. Winter is passed in specially constructed nests called hibernaculae. They are boat-shaped, made of rolled leaves and each hibernacula shelters 100 to 150 caterpillars. The caterpillars come out of the hibernaculae in March and start feeding again for 34 to 55 day. Pupation takes place in cocoons within the rolled leaves. Pre-pupal and pupal stages last for 4 to 7 and 32 to 46 days respectively. Moths emerge in June and their longevity is 5 to 7 days (Janjua, 1947). The hairy caterpillars can be controlled by dusting with 10% BHC if and when they appear in an epidemic form.

Indian Gypsy moth, *Lymantria obfuscata* Walker – a major pest of apricot in Kashmir, has been reported by Rahman (1941) feeding on apple foliage in Himachal Pradesh and Punjab. The caterpillars are gregarious and nocturnal in habit and defoliate the trees completely which results in failure of fruit formation. Eggs are round, shining and light-greyish-brown. Caterpillars are clothed in tufts of hair and are 40 to 50 mm long, when full grown Pupae are reddish-brown and 17 to 20 mm long. Female moths are of apterous type, i e. they have feebly developed wings and relatively bulky body and are, therefore, unable to fly. After being fertilised by winged males, the females lie on the bark of the trees for oviposition and die as soon as their ovaries are empty. To control this pest, collect and destroy sluggish females and the egg-masses that are so conspicuous and easy to locate.

Caterpillars of *Lymantria ampla* Walker, *L. concolor* Walker and *L. dispar* Linnaeus have also been reported feeding on the foliage, but none of these cause any significant damage. *L. ampla* is more common in South India – apple, mango and pomegranate are its preferred hosts; guava and mulberry being alternate hosts. The eggs are spherical, about one mm in diameter and grey in colour. Larvae are 25 to 35 mm (♂) and 50 to 60 mm (♀) long and brownish in colour. Pupae are 15 (♂) to 25 mm (♀) long and dark brown. Adult females are dark-grey with two pairs of rudimentary whitish wings having black spots; mouths parts are atrophied and antennae serrate. Males also have atrophied mouth parts but bipectinate antennae and two pairs of functional wings with small marginal markings. Wing expanse is 33 to 38 mm. Incubation period is 12 days; larval stage lasts for 18 to 22 (♂) and 27 to

32 (♀) days while pupal development takes 11 to 12 (♂) and 7 to
8 days (♀). There are six to seven generations in a year (Veeresh
and Puttarudriah, 1968).

THRIPS

Apple blossoms are usually attacked by the four common
and highly polyphagous species of thrips, namely, *Frankliniella
dampfi* Priesner, *Taeniothrips rhopalantennalis* Shumsher, *Thrips
flavus* Schrank and *T. florum* Shmutz. Eggs are laid in flower-buds
before the buds open. On hatching, the nymphs feed on petals
and vital parts of the flowers, by sucking the sap therefrom,
causing distortion of flowers (Sharma and Bhalla, 1963). As a
result, no fruit setting takes place or weak fruits are formed
which drop prematurely.

To control these thrips, spray with 0.03% phosphamidon,
dimethoate, formothion, thiometon or endosulfan (Ananthakrishnan,
1971).

FRUIT FLIES

Dacus ciliatus (Loew), *D. dorsalis* (Hendel) and *D. zonatus*
(Saunders) have been reported damaging apples in India. The
preferred hosts of the three species are, melons, mango and peach
respectively. In case of apple, *D. ciliatus* is relatively more common
and harmful. Adults of these flies are strong fliers and are active
throughout the year except during severe cold (December–January)
when the pest hibernates in pupal stage.

Dacus ciliatus commonly called Ethiopian melon fly is of
African origin, now widely distributed in European countries, Africa,
Middle East, Pakistan, India, etc. Its main hosts are cucurbits,
including various melons and alternate hosts are, citrus, apple, etc.
Female flies make cavities on fruits with their ovipositor and lay
3 to 8 eggs in each cavity. On hatching, maggots feed on the pulp.
The affected fruits gradually rot and fall down. Eggs are shiny white,
slightly curved and about 2.5 mm long. Maggots are whitish in

colour and 8 mm long. Pupae are cylindrical, brownish to ochraceous in colour and about 5.5 mm long. Adult flies are ferruginous–brown with hyaline wings and have two dark spots on 4th abdominal segment.

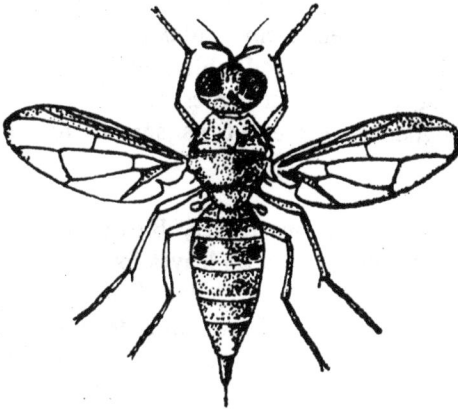

Dacus ciliatus (Loew)

FRUIT BORERS

Bud moth, *Eucosma ocellana* Schiffer is a minor pest of apple, apricot, pear, quince, etc. The moths appear in May-June and live for one to two weeks. The males die soon after mating and females after ovipositing. The eggs hatch in 8 to 11 days. Caterpillars bore into shoots, flower-buds and tender fruits and feed inside the same till September. Overwintering is in caterpillar stage from October to March – thus the larval period lasts for 10 to 11 months. Pupal period is 9 to 13 days and there is only one generation in a year (Singh, 1964).

Caterpillars of *Rapala nissa* Kollar also bore into the flower-buds and tender fruits. This is a minor pest and its presence is easily detected, as the caterpillars are invariably half inside the fruits and half outside.

Codling moth, *Laspercysia pomonella* (Linnaeus) is a notorious pest of temperate fruits, showing marked preference for apples.

It is a polyphagous pest and has also been recorded on citrus, peach, pear, quince, walnut, etc.   The pest is widely distributed throughout Europe, North America and Australia (CIE map No. A–9). Nearer India, it has been found in abundance in Baluchistan (Pakistan) where it causes havoc on apples and other fruits and is stated to constitute a threat to the rest of subcontinent (Pruthi, 1969). In India, its occurrence was reported from Ladak (Kashmir valley) as early as 1900, though Fletcher (1917) doubted this record.   Wadhi (1973) however, confirmed that the pest is present in Ladak region and has suggested strict vigilance to prevent its spread to other parts of India.

Eggs are laid singly on leaves, blossoms and fruits.   The freshly hatched caterpillars feed on leaves for a while then burrow inside the fruits and feed on the pulp.   The entry holes become quite conspicuous as these are filled with dry brown frass and are surrounded by a dark reddish ring.   The infested apples become brighter in colour than those that are not infested and also ripe prematunely.   The fruits that are attacked early in the season, often drop down before the crop is ready for harvest.

Eggs are flattened and white in colour.   Full grown caterpillars are 16 to 22 mm long and pinkish in colour.   Moths are greenish to dark brown with chocolate-brown or copper coloured circular markings near the tip of forewings.   The colour pattern resembles bark of the tree trunk which makes the moths quite inconspicuous. Wing expanse is 18 to 24 mm.

Laspereysia pomonella (Linnaeus)

Egg, larval, and pupal periods are 4 to 12, 28 to 35 and 8 to 14 days respectively.   The caterpillars of third brood overwinter by forming thick silken cocoons in which they pass the Winter under loose scales of the bark of the host trees. When Spring comes, the larvae become pupae inside these cocoons and the moths emerge from the cocoons during March-April.

In view of economic importance of this pest and its restricted distribution, Wadhi and Sethi (1975) have opined that the sterile insect release technique could be successfully applied to eradicate this pest. In this connexion, the necessary basic information regarding suitable and economical laboratory mass rearing techniques is available and it has been established that males can be effectively sterilized with 40 Krad dose.

In addition to these lepidopterous borers, a number of beetles have been observed gnawing the fruits. These include, *Lucanus lunifer* Hope, *Clinteria spilota* Arrow, *Macronota 4-lineata* Hope, *Protaetia neglecta* Hope, *Rhomborrhina glabirhina* Westwood, *Torynorrhina opalina* (Hope), *Anomala lineatopennis* Blanchard, *Hoplosternus (Melolontha) furicaudus* (Ancy), *H indica* (Hope), *Popillia complanata* Newman, *P. cyanea* Hope, *Xylotrupes gideon* Linnaeus, *Autocrates aenleus* Parry, *Catharcius molossus* Linnaeus, *Cerogria nepalensis* Hope and *Mimastra cyanura* Hope. None of these cause any major damage and normally require no control measures. However, dusting with 10% BHC is effective in controlling these beetles.

Dyscerus clathratus (Pascoe), *D. fletcheri* Marshall and *D. malignus* Marshall are the weevils that make small excavations on

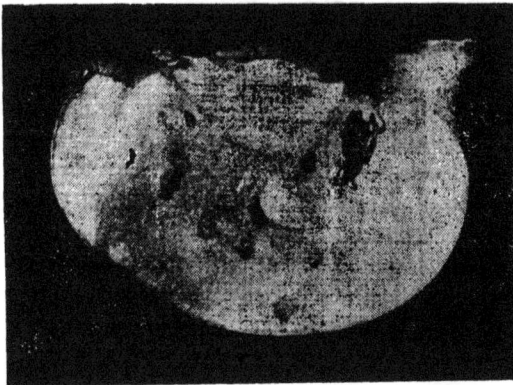

(*Courtesy* : *J.N. Sackan*)
Dyscerus clathratus-grub and adult on apple slice, showing typical damage.

apple fruits for laying their eggs. On hatching, their grubs feed on fruit pulp. All the stages of weevils are seen within the infested fruit and the affected fruits gradually rot and become unfit for human consumption.

Collect and destroy the infested fruits to prevent the carry over of this pest. Application of 0.2% Pyrethrum extract or dusting with 10% BHC is also helpful in checking the weevil infestation. The protective treatment may be applied about ten days before ripening of the fruits.

CUTWORM

Agrotis ipsilon (Hufnagel) (*ypsilon* Rottenburg), a cosmopolitan pest, is found in northern Europe, Canada, Japan, down to Newzealand, South Africa and South America (CIE map No. A–261). It is a winter pest, active from October onwards, peak period of activity being December-January. During Summer, it migrates to hilly regions. Freshly hatched caterpillars feed gregariously on foliage for about a week then become non-gregarious and enter the soil. Carterpillars as well as moths are nocturnal in habit and hide during day in cracks and crevices in soil or below big clods. At night the caterpillars cut seedlings near ground level and eat only the tender portion – thereby the loss caused is much more than what is actually eaten by the caterpillars. Old seedlings are seldom attacked, due to their fibrous and hard stems. The caterpillars are highly cannibalistic and at times bite their own body. They have peculiar habit of coiling up at the slightest disturbance. The pest appears year after year in areas liable to annual inundation.

Eggs are round, dome-shaped, 0.5 mm in diameter, ribbed and greenish-white in colour. Full grown Caterpillars are 25 to 35 mm long, dull brown dorsally with a broad pale grey band along the midline and greyish-green laterally with blackish stripes. Pupae are 18 to 22 mm long and reddish-brown in colour. Adults are dark brown moths with reddish tinge; forewings have dark-brownish-black markings and hind wings are almost white basally with dark terminal fringe. Wing expanse is 45 to 55 mm.

Catterpillar   Moth

Agrotis ipsilon (Hufnagel)

Eggs are laid in clusters on ventral surface of leaves or on moist soil. A single female lays as many as 800 eggs in 16 to 22 clusters of 30 to 40 eggs each. Pupation takes place in soil, 80 to 100 mm deep in earthen cocoons. Eggs hatch is 2 to 13 days, larval period lasts for 4 to 5 weeks and pupal 10 to 30 days. A single life-cycle is completed in 5 to 10 weeks; it takes 32 days at 30°C, 41 days at 26°C and 67 days at 20°C (Hill, 1975).

Hand-picking of larvae has been suggested as a practical control measure (Pradhan, 1969). Nayar *et al.* (1976) suggested soil application of heptachlor dust @ 3.4 kg a.i. per hectare or aldrin or DDT @ 1.7 kg a.i. per hectare before planting to minimize the damage. Breaking of clods and irrigating the infested nurseries and orchards can also help in combating this pest.

4.2

PEAR

PEAR, *Pyrus* species, is an important pome fruit that ranks next only to apple in commercial production. Its origin is shrouded in mystry. The various species are grown extensively in France and all other European countries East of France as also in southern Asia including South China and Japan. In India, the commonly grown varieties are, *Pyrus communis* Linnaeus (Occidental group) and *P. serotina culta* Rehd. (Oriental group); the former is said to be native of Eurasia and the latter originated from eastern Asia. Pear was introduced in India (Simla) by Alexander Coutts during 19th century. At present there are about 1500 hectares under *P. communis* grown mainly in hilly regions at elevation of 1500 to 2500 metres. Heavy soils with considerable humus and good drainage are preferred. *P. serotina* (sand pear) is well adopted in subtropical regions of India, *viz.*, plains of North India and hilly parts of South India. It occupies an area of about 1000 hectares; of this, nearly half is in Nilgiris and Kodaikanal. The fruits have poor keeping quality and are used mainly as dessert.

As many as 70 insect species have been recorded on pear in India. Most of these are pests of peach and plum and appear only occasionally on pear. Fortunately, the pear midge, *Contarinia pyrivora* Riley – a notorious pest of pear in Europe (Pruthi, 1969) has not yet been intercepted from India.

ROOT FEEDING PESTS

Termites, *Odontotermes obesus* (Rambur) and *Microtermes obesi* (Holmgren) have been found damaging pear trees, specially in nurseries and young orchards. They are active throughout the year, but more during November to June.

Apple root borer, *Dorysthenes hugeli* Redtenbacher is another pest found occasionally in pear orchards specially in sandy and moist soils. Eggs are laid in soil. Freshly hatched grubs feed on organic matter available in the soil and the roots of fruit trees. The affected trees become weak and shaky.

Soil application, soon after monsoon, with 5% aldrin, chlordane or heptachlor dust checks the termites as also the grub population of root borer. Irrigating the trees with 0.1% aldrin or heptachlor emulsion is also helpful.

SAP SUCKING PESTS

Aphids, *Aphis fabae* Scopoli, *Aphis pomi* (de Geer) (*mali* Fabricius), *Myzus persicae* (Sulzer) and *Cinara krishni* (George) [*Pyrolachnus pyri* (Buckton)] as also whiteflies, *Aleurocanthus husaini* Corbett and *Siphoninus finitimus* Silvestri have been occasionally found on leaves and tender twigs of pear sucking the sap therefrom.

Pear Psylla, *Psylla pyricola* (Foerster) is a minor pest found all over the hilly regions of India. Nymphs and adults feed inside the central shoot causing twisting of leaves and stunted growth. Eggs are laid on flower-buds during March-April and two generations are completed on fruit trees during March to May; thereafter the adults migrate to other host plants.

Scale insects reported damaging pear trees are, *Drosicha mangiferae* (Green), *Icerya aegyptiaca* (Douglas), *I. seychellarum* (Westwood), *Aonidiella aurantii* (Maskell), *Aspidiotus destructor* Signoret, *Parlatoria boycei* Mc Kenzie, *Pseudaulacaspis pentagona* (Targioni), *Quadraspidiotus perniciousus* (Comstock), *Ceroplastes actiniformis* Green, *C. rubens* Maskell, *Eriochiton theae* Green and *Eulecanium tiliae* (Linnaeus). All these are polyphagous and pear is one of their minor alternate host.

To check the infestation of these sucking pests, clip off and destroy promptly all the infested parts in the initial stage of attack. In case of severse infestation, spray with 0.03% phosphamidon or dimethoate against aphids and whiteflies; for psyllids and scale insects, use 0.04% diazinon or monocrotophos or 0.05% endosulfan.

LEAF EATING CATERPILLARS

Actias selene Hübner, *Antheraea roylei* Moore, *Odina wodier*, *Saturnia simla* Westwood, *Euproctis flava* (Bremer), *E. fraterna* (Moore), *Langia zeuzeroides* Moore, *Malacosoma indica* (Walker), *Spodoptera litura* (Fabricius), *Archips micaceana* (Walker) and *A. termias* (Meyrick) are some of the lepidopterous pests that have been reported from various parts of India, feeding on pear leaves. These are minor pests of pear. If and when any of these appear in large number, spray with 0.05% dichlorvos or endosulfan. For hairy caterpillars apply 10% BHC dust.

BEETLES AND WEEVILS

A number of chafer beetles, namely, *Adoretus duvauceli* Blanchard, *A. horticola* Arrow, *A. lasiopygus* Burmeister, *A. versutus* Harold, *Anomala dimidiata* Hope, *A. rufiventris* Redtenbacher, *A. varicolor* Gyllenhal, *Holotrichia consanguinea* Blanchard, *Brahmina coriacea* (Hope), *Protaetia impavida* Janson and *P. neglecta* Hope have been reported as pests of pear trees.

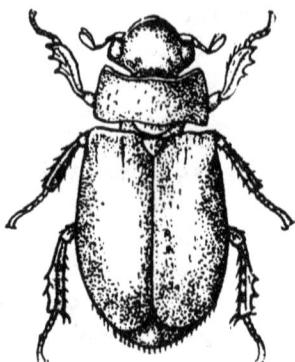

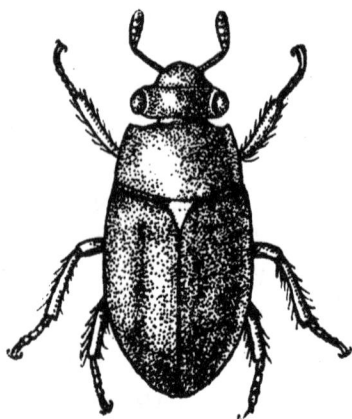

Adoretus duvauceli Blanchard   *Adoretus versutus* Harold

The adult beetles gnaw the leaves and in case of heavy attack the trees may be partially defoliated, arresting its growth and affecting adversely the fruiting capacity. These beetles usually appear around May-June and remain active for 2 to 3 months.

The chrysomelid beetles found on pear trees are, *Altica caerulescens* Baly, *Mimastra cyanura* Hope and *Monolepta erthrycephala* Baly. These are polyphagous, attacking most of the temperate fruit trees. On pear, *A. caerulescens* is comparatively more common. Its grubs and adults feed on underside·of leaves nibbling small holes. The pest though active from February to September, is more so during February to April, when two life-cycles are completed, each occupying five to six weeks.

Number of weevils have also been occasionally observed nibbling leaves of pear trees. These include, *Deiradoleus* species, *Hypera variabilis* Herbest, *Myllocerus laetivirens* Marshall, *M. sabulosus* Marshall, *M. undecimpustulatus maculosus* Desbrocher and *Phytoscaphus triangularis* Olivier. These weevils are polyphagous and active throughout the year except during severe cold when the weevils overwinter in adult stage.

Generally no control measures are adopted against these beetles and weevils on pear. However, dusting with 10% BHC or spraying with 0.05% endosulfan is quite effective in controlling these pests.

FLOWER FEEDING INSECTS

Pear fruits are damaged by Oriental fruit fly, *Dacus dorsalis* (Hendel), hornet wasp, *Vespa orientalis* Linnaeus, fruit sucking moths, *Othreis cajeta* (Cramer) and *O. materna* (Linnaeus), castor capsule borer, *Dichocrocis punctiferalis* Guenée, pomegranate butterfly, *Virachola isocrates* (Fabricius) and the notorious codling moth, *Laspereysia pomonella* (Linnaeus). The last one is restricted to Ladakh region of Jammu and Kashmir. These are polyphagous pests and except *V. orientalis* all others are major pests of various other fruit trees and minor pests of pear, occurring sporadically. Though the damage caused by *V. orientalis* by puncturing a few fruits is negligible, the wasp causes a great deal of annoyance by its very presence

Vinegar fly, *Drosophila* species and scavenger beetle, *Carpophilus dimidiatus* Linnaeus have been reared from fallen and rotting fruits during May-June. Both these are of minor importance.

4.3

PEACH

PEACH, *Amygdalus persica* Linnaeus is of Chinese origin, now grown commercially in temperate and subtropical areas specially South USA, South Europe, Japan and parts of Africa and Australia. Among the various temperate fruits, peach requires relatively warmer climate. In India, there are about 1500 hectares under various varieties of peach, spread along the lower Himalayas and foot-hill areas of Jammu-Kashmir, Himachal Pradesh and Punjab as also submontane region of Uttar Pradesh. A small area is also under cultivation in South India, around Nilgiris. The fruit is mainly used as dessert. It is one of the popular fruit for canning and large quantities are also dried. Inspite of ample potential and possibilities to increase the area under this fruit, there has been hardly any substantial increase in the area due to fear of pests and diseases, that take a heavy toll.

Over 80 insect species have been recorded feeding and breeding on peach trees. The most destructive of these are several species of aphids – those that occur in epidemic form include, peach leaf curl aphid, peach mealy aphid, peach green aphid and peach black aphid (Batra, 1953 b). Besides these aphids, scale insects, peach leaf roller, peach twig borer, peach stem borer and fruit flies also cause substantial loss. Among the minor pests mention may be made of whiteflies, blossom beetles, a few species of bugs, beetles and butterflies that damage the fruits as also termites and scavengers. Besides, grasshoppers and cutworms have been often reported damaging the seedlings.

PEACH LEAF CURL APHID

Brachycaudus helichrysi (Kaltenbach) is one of the most destructive pests of peach trees, reported from almost all peach growing countries of the world. In India, the pest is more serious in Kulu valley, Simla hills and submontane area of Uttar Pradesh. Its preferred host is peach but it is also found on almond, apricot, plum, etc. Local varieties of peach are comparatively more susceptible to the attack of this aphid, than the American varieties. Nymphs and adults suck the cell sap from leaves, petioles, blossoms and fruits. Affected leaves turn pale and curl up; blossoms wither and fruits do not develop into normal size and drop prematurely. The occurrence and breeding of the pest varies substantially with the climatic conditions. Though it is essentially a winter pest, it cannot withstand severe cold which it passes in egg stage. Thus it appears early in February in plains and after middle of March in cooler regions. It usually appears on flower-buds and as soon as the fresh leaves appear, the pest migrates to these tender leaves. The alternate host in cooler region is *Golden rod, Erigeron canadensis* Linnaeus while in plains the aphid breeds on a weed, *Ageratum conyzoides* Linnaeus. Eggs are cylindrical, 0.6 mm long, light green in colour when freshly laid, later turning shiny black. Nymphs are dark green in colour and the adults that feed on leaves are green while those that feed on bark are chocolate coloured. Reproduction is sexual as well as parthenogenetic. Sexual forms appear early in November in cooler regions and lay eggs which hatch in March (Lal and Siddiqui, 1952). At lower altitudes there is no egg-laying and overwintering is in adult stage. With the rise in temperature, there is rapid multiplication (parthenogenetically). A single female gives birth to about 50 young ones in her life time of two weeks and each of these takes about 10 days to mature during March-April and start reproducing. All these young ones are apterous viviparous females. After producing 3 to 4 asexual generations the aphids migrate to pass Summer on its alternate host. The migration takes place during mid May in plains and around July in cooler regions (Butani, 1974 f).

To control this pest, spray with 0.03% dimethoate, oxydemeton methyl, phosphamidon or quinalphos or 0.04% diazinon or dichlorvos just before flowering (pink bud stage) and again after 7 to 10 days. Another one to two sprayings should be given when the fruit is pea-sized. In higher hills only one pre-bloom spraying is sufficient while in mid and lower hills, in addition to pre-bloom spray, a post-bloom spray (8 to 10 days after petal fall) is also necessary (Sharma *et al.*, 1968).

PEACH MEALY APHID

Hyalopterus pruni (Geoffroy) (*arundinis* Fabricius) is another destructive pest reported from India, Pakistan, Italy and USA. In India, it has been recorded from Punjab and Maharashtra on peach, apricot and plum. Nymphs and adults are found congregated on ventral surface of leaves, sucking the cell sap and exuding copious quantity of honeydew, which gives the leaves a glistening appearance. In case of severe infestation, the affected leaves get slightly twisted but not curled as in the case of *Brachycaudus helichrysi*. As a result of its infestation, the fruit size is reduced considerably and half the fruits fall prematurely. The aphid breeds throughout the year but its activity accelerates soon after the disappearance of leaf curl aphid. Reproduction is mainly parthenogenetically though during severe cold season, the sexual forms also appear. Eggs are laid on peach shoots around December-January, a little later than leaf curl aphid. The overwintering is in egg stage and these eggs hatch in Spring, giving rise to apterous viviparous females. Summer is passed on alternate hosts, *i.e.*, *Arundo donax* and *Phragmites karka* (Pruthi and Batra, 1960). The pest is again active during Autumn in peach orchards when both males and females are seen in good number. The activity continues till end of October and may continue beyond, if the climatic conditions are favourable.

To control this aphid, spray with 0.03% oxydemeton methyl or phosphamidon. It is desirable to use systemic insecticides, as contact pesticides may not prove to be satisfactory due to waxy coating on aphid. Moreover, the natural enemies of this aphid, specially the predators are less affected by systemic insecticides.

PEACH BLACK APHID

Pterochlorus persicae Cholodkovsky is widely distributed in India, Pakistan, Afghanistan and entire Mediterranean region. It is a polyphagous pest, found feeding on apricot, cherry, peach, plum, etc. It is larger than other aphids and darker (almost black) in colour, resembling the colour of peach shoots. Innumerable individuals may be seen congregating on stems and tender shoots and sucking the sap therefrom. This continuous drain of cell sap, devitalizes the trees, affecting the fruit setting capacity as also the fruit size. These aphids also produce copious quantity of honeydew which attracts black ants (*Camponotus compressus* Fabricius, *Leptothorax* species and *Phidole* species); muscoids (*Lucilia inducta* Walker and *Chrysomyia aenea*) and wasps (*Polistes herbaeus* Fabricius and *Vespa orientalis* Linnaeus).

♀   ♂
Pterochlorus persicae Cholodkovsky

To control the aphids, spray with any aphicide, like 0.03% dimethoate, oxydemeton methyl, phosphamidon or quinalphos or 0.04% diazinon or dichlorvos.

PEACH GREEN APHID

Myzus (*Neotarosiphon*) *persicae* (Sulzer) is cosmopolitan in distribution (CIE map No. A–45), northwards upto South Scandinavia, North China and Canada and southwards upto Australia, Fiji, New Zealand, Tahiti and South America. It is a polyphagous pest, peach being its primary host. Samuel (1940) recorded this aphid on 21 host plants from Delhi alone. These include, almond, apple, apricot, cherry, citrus, peach, pear and plum. Colonies of the aphid may be seen on leaves, flower-buds and young fruits sucking the sap. The infested leaves become pitted and curled, flower-buds wither and young fruits shrivel and drop prematurely. If the infestation occurs during flowering stage, the fruit setting is adversely affected. Besides doing direct damage by feeding on foliage, the pest causes further and greater loss by transmitting a

number of virus diseases. It would be worthwhile to mention that
M. persicae transmits the maximum number (around 120) of plant
viruses.

These aphids are usually green in colour but may be pale brown
or pinkish, 1.5 to 2.5 mm long with antennae 2/3 to 9/10 as long as
the body and siphunculi fairly long and clavate. Adults appear as
alate and apterous forms at different times of the year depending
upon environmental and climatic conditions. The alatae have a
characteristic dark sclerotic patch on the mid abdominal dorsum.

(*Courtesy*: *Dennis Hill*)

Myzus persicae (Sulzer) – Alate ♀

M. persicae is holocyclic in the temperate regions, having
Prunus spp. as primary hosts. In tropics and subtropics, the species
seems to have lost its sexual cycle and reproduction is parthenogenetic
throughout the year. In temperate regions, gynoparae appear during
September-October. Each gynopara produces 5 to 15 oviparae.
These are fertilized by the males and each ovipara then lays 4 to 14
eggs. The eggs undergo diapause during Winter. Hatching of eggs
generally coincides with vegetational phenology. Newly hatched
young ones (fundatrices) feed on swelling buds, develop rapidly and
start reproducing parthenogenetically (Hill, 1975).

The use of aphicides is effective in checking the pest population
but not the spread of virus disease. Moreover, frequent use of
insecticides results in destruction of the natural enemies – parasites and
predators. Therefore, adoption of integrated control approach is
highly desirable in this case, combining cultural, chemical, biological
and crop management practices.

CHERRY APHID

Myzus serasi (Fabricius) also attacks peach trees but it is not a serious pest of peach and can be easily controlled by application of any aphicide like, 0.03% dimethoate, phosphamidon, oxydemeton methyl or quinalphos

SCALE INSECTS

White peach scale, *Pseudaulacaspis pentagona* (Targioni) is the most destructive scale insect of peach trees. It is cosmopolitan in distribution and has been reported from West Indies, United Kingdom, southern Europe, Indian subcontinent, Japan and Australia. In India, it is commonly found in Assam, Bengal and Uttar Pradesh attacking almond, apple, cherry, peach, plum, etc. Outside India, it has also been reported as a pest of fig, mulberry, papaya and walnut. It sucks cell sap from stems and tender shoots and sometimes from leaves as well. The devitalised trees suffer in growth and fruit setting capacity. A female lays 25 to 55 eggs which hatch in 8 to 12 days. Nymphal period is 8 to 10 days for males and 4 to 5 weeks for females. Life-cycle in case of females is completed in 8 to 9 weeks during Summer and is extended upto 22 weeks in Winter. There are four generations in a year.

Another serious pest of peach trees is plum scale, *Eulecanium tiliae* (Linnaeus). Though peach and plum are the preferred hosts of this pest, it also attacks almond, apricot, apple, cherry, pear, etc.

Eulecanium tiliae (Linnaeus)

A female lays on an average 250 eggs (maximum 400 – Pruthi and Batra, 1938). On hatching, the crawlers move about a bit, settle on a succulent spot preferably on ventral side of leaves or shady side of branches and twigs and start sucking cell sap. They also secrete honeydew which in turn attracts ants, bees and wasps. A single crawler may not cause appreciable damage, but their enormous number drain so much of sap that the vigour of the trees is impaired and in case of severe infestation, the growth of trees is arrested and fruit setting capacity is adversely affected.

Other coccoids recorded on peach trees include, *Drosicha contrahens* (Walker), *D. mangiferae* (Green), *Icerya purchasi* Maskell, *Aonidiella aurantii* (Maskell), *A. orientalis* (Newstaed), *Aspidiotus destructor* Signoret, *Hemiberlesia lataniae* (Signoret), *Howardia biclavis* (Comstock), *Parlatoria oleae* (Colvée), *P. pseudopyri* Kuwana, *Quadraspidiotus perniciosus* (Comstock), *Chloropulvinaria polygonata* (Cockerell) and *Eriochiton theae* Green (Rao and Chaterjee, 1948; Singh, 1964). Of these, *Q. perniciosus*, the cosmopolitan pest, is comparatively more harmful.

To control these scale insects and mealy bugs, prune and destroy the affected plant parts immediately to prevent the population build-up. Spraying with 0.04% diazinon or monocrotophos is also effective against the young nymphs. Against *Q. perniciosus*, spray the orchards with 3% miscible oil during Winter and simultaneously release the parasite, *Prospaltella perniciosi* (Tower) on the wild host plants in the surrounding areas to check the infestation from outside the orchards.

LEAF DEFOLIATORS

A number of chafer beetles have been found feeding on peach foliage. These include, *Adoretus duvauceli* Blanchard, *A. epipleuralis* Arrow, *A. horticola* Arrow, *A. lasiopygus* Burmeister, *A. versutus* Harold, *Anomala dimidiata* Hope, *A. falvipes* Arrow, *A. lineatopennis* Blanchard, *A. rufiventris* Redtenbacher, *Brahmina coriacea* (Hope), *Macronota 4-lineata* Hope, *Hoplosternus furicaudus* (Ancy) and *Microtrichia cotesi* Brenske. Adult beetles appear with break of monsoon (June-July) and feed on leaves during night, hiding during day. They are active for about two months. In case of severe attack even the fruits are scrapped near the apical end. Eggs are laid in the soil; grubs feed on roots and other organic matter available in the soil. Occasionally feeding of the grubs on roots, cause the death of young plants. Overwintering is in grub stage and pupation takes place in the soil.

Other beetles recorded feeding on peach leaves are, blue beetle, *Altica* species; walnut blue beetle, *Monolepta erthrycephala* Baly; almond chrysomelid, *Mimastra cyanura* Hope and peach chrysomelid, *Hoplasoma sexmaculata* Hope. Loss caused by these is usually negligible.

Among the weevils mention may be made of almond weevil, *Myllocerus laetivirens* Marshall; *ber* weevil, *M. sabulosus* Marshall; grey weevil, *M. undecimpustulatus maculosus* Desbrocher; plum weevil, *Amblyrrhinus poricollis* Boheman; apricot weevil, *Emperorrhinus defoliator* Marshall, etc. Grubs of these weevils also live underground and only the adults feed on leaves, nibbling from margins, forming zigzag marginal patterns. The damage caused in case of peach trees is of minor importance.

Lepidopterous defoliators include, jute hairy caterpillar, *Dasychira mendosa* Hübner; plum hairy caterpillar, *Euproctis fraterna* (Moore) and tent caterpillar, *Malacosoma indica* Walker, which occasionally lay eggs on peach trees and on hatching, the caterpillars feed voraciously on leaves. Eggs of *D. mendosa* are spherical (0 8 mm diameter) in shape and creamy-pale-yellow in colour. Caterpillars are 30 to 38 mm long, greyish-yellow with red stripes on prothorax and paired lateral tufts of greyish white plumose hair on each segment. Pupae are 20 to 25 mm long and creamy-white. Moths are pale yellow with irregularly patterned brownish forewings. Wing expanse is 26 to 30 mm ($\male$) and 36 to 40 mm ($\female$). Egg, larval, pre-pupal and pupal periods last for 8 to 12, 20 to 38, 2 to 3 and 5 to 7 days respectively. A single life-cycle occupies 40 to 58 days with five to six generations in a year.

If and when these defoliators appear, dust with 5 to 10% BHC. For chafer beetles, deep ploughing around the trees to expose and kill the pupae and hibernating grubs coupled with soil application of 5% aldrin or chlordane dust is suggested. Sharma *et al.* (1971)

recommended two sprayings at 7 days' interval with 0.2% carbaryl wettable powder against adult beetles.

PEACH LEAF ROLLER

Archips micaceana (Walker) (*Tortrix epicyrta* Meyrick) is found all over India, Pakistan, Sri Lanka and Indonesia attacking, apple, apricot, citrus, guava, peach, pear, plum, etc. Overwintering caterpillars, become active in March-April, web together a few leaves and flower-buds and feed within. Caterpillars of subsequent generations roll the leaves breadthwise and feed from inside on leaf margins. Occasionally, the caterpillars fasten the leaves with fruits and feed on fruits as well. As a result of infestation, the fruit setting capacity of the trees is reduced and the fruits attacked lose their market value. Pupation takes place in silky cocoons on the leaves. Moths appear around April, June and early September. Egg, larval and pupal periods last on an average for, 8, 30 and 10 days respectively with three generations in a year.

Spraying in April with 0.05% endosulfan or fenitrothion has been found effective in controlling this pest.

PEACH TWIG BORER

Anarsia lineatella Zeller was first recorded in India from Kashmir (Fletcher, 1932) and later from Pakistan (Janjua and Samuel, 1941). It is also found in South China, Asia Minor, Iraq, central and southern Europe, North Africa and North America. Besides peach, it damages apricot and plum. The young caterpillars bore into twigs and later into developing fruits. The affected shoots dry up while the fruits rot. Secondary attack by bacterial and fungal organisms also takes place. There is also gummy exudation at the entrance hole in the fruits.

Eggs are oval, 0.4 to 0.5 mm long and white in colour. Full grown larvae are 12 to 18 mm long and reddish-pink in colour. Pupae are 6 to 7 mm long and brown in colour. Moths are steel-grey. Moths usually fly during dusk and as such often go unnoticed.

Eggs are laid on shoots, twigs or fruits either singly or in batches of 2 to 3 eggs during May to September. A female lays on an average 170 eggs. Pupation takes place under flimsy webs in cracks in the bark or in curled leaves. Egg, larval, prepupal and pupal durations last for 4 to 11, 18 to 35, 3 to 9 and 6 to 14 days respectively (Janjua, 1942); a single life-cycle occupies 40 to 69 days. There are three generation in a year. The third instar caterpillars of 3rd generation hibernate from end of September and their activity and feeding restarts from end of March.

To control this pest, remove and destroy all the affected shoots and fruits and spray 0.05% endosulfan or fenitrothion. This will also check the population of leaf rollers.

STEM BORERS

Sphenoptera lafertéi Thomson is a major pest of peach widely distributed in Afghanistan, Pakistan and India. Besides peach, the grubs also bore into the stems and main branches of almond, apricot, cherry, loquat, pear and plum trees. Occasionally grubs of *S. dadkhani* Obenberg have also been recorded boring peach stems but the loss caused by this species is negligible. Adult beetles appear around middle of March and feed on foliage, Eggs are laid singly scattered all over the tree trunk and main branches. On hatching, young grubs feed below the bark making minute irregular galleries. A number of grubs feed near each other, as a result of which the bark gets loosened and splits. Full grown grubs bore into the wood, make a small chamber (4–6 × 2–4 mm), about 10 mm deep in the wood and pupate therein. Outside on the bark, gum globules ooze out from the entrance holes. Gradually the leaves turn pale and the growth of the tree is arrested. Young as well as full grown trees are attacked alike but the incidence is higher in water-logged areas or in orchards with poor drainage. In case of severe infestation the attacked branches dry up resulting in failure of fruit formation on that branch and in extreme cases even the entire tree may die. The pest is more active during warmer part of the year.

Eggs are small, spherical in shape and white in colour. Full grown grubs are 18 to 24 mm long and club-shaped. Adults are small (10 to 13 mm long), blackish-bronze beetles.

Sphenoptera lafertéi Thomson

Incubation period is about 20 days. Grub stage occupies 2 months (Summer) to 6 months (Winter). Prepupal and pupal periods last for 1 to 2 and 8 to 12 days in Summer; overwintering is in grub stage and there are three generations in a year (Batra and Renjhen, 1950).

The dead or heavily gummed branches should be cut and destroyed. Even if the main trunk is showing innumerable gum globules indicating severe attack of this pest, the entire tree may be topped or beheaded and cut portion burnt. Swabbing the infested trunks and branches with high concentrations of aldrin or dieldrin has also been suggested (Sharma *et al.*, 1963).

Cherry stem borer, *Aeolesthes holoserica* Fabricius and Quetta borer, *A. sarta* Solsky also sometimes bore the stems and branches of peach trees. Chewed wood particles mixed with excreta of the grubs are found coming out of freshly made holes. These are minor pests. Nevertheless, to avoid serious damage, clean the affected spots and insert into the holes cotton-wool soaked in a good fumigant like, carbon bisulphide, chloroform, petrol, etc. This will kill the grubs inside and prevent the damage from spreading.

Another cerambycid, *Macrotoma crenata* Fabricius has been reported from Pakistan, India, Burma and Sri Lanka on almond; fig and peach trees. The grubs bore into main stems and sometimes trunks as well, making zigzag tunnels deep upto core; these tunnels are packed with fairly compact mass of fibres and excreta. There are no external symptoms of damage.

Apriona cinerea Chevrolat is another minor pest widely distributed along the foot-hills of western Himalayas damaging apple, fig, mulberry and peach trees; apple being its preferred host. Eggs are laid under the bark of trees. On hatching, the grubs bore into the stems and kill the same outright. To control this pest, prune and burn all the infested stems, shoots and branches during Winter.

FRUIT FLIES

Number of fruit flies have been recorded damaging peach fruits in India. These include, *Dacus ciliatus* (Loew), *D. correctus* (Bezzi), *D. dorsalis* (Hendel), *D. cucurbitae* (Coquillett), *D. duplicatus* (Bezzi), *D. maculipennis* (Doleschall) and *D. zonatus* (Saunders) (Singh, 1964; Kapoor, 1970; Sharma *et al.*, 1973). Of these, *D. zonatus* is comparatively more common and more harmful. All these flies are cosmopolitan in distribution. In India, they remain active all through the year except for extreme cold months (December to February) when these overwinter in pupal stage. Eggs are laid just below the epidermis of fruits in clusters of 2 to 10 eggs. On hatching, the grubs feed on pulp of the fruits. Affected fruits show dark coloured punctures through which the juices ooze out while fungi and bacteria get in. As a result, the affected fruits start rotting, get wrinkled, present brown patches and ultimately fall down. The grubs leave the fallen fruits, to pupate in the soil, 30 to 75 mm deep. Even if the infested fruits do not fall down, the grubs, when full fed, come out, drop down and pupate in the soil. Eggs hatch in 3 to 4 days ; grub and pupal stages last for about a week each during Summer and a little longer during Winter. Adult flies survive for one to four months.

To control these flies, Gupta (1958) found bait–spray of yeast hydrolysate, sugar and malathion to be effective. Sharma *et al.*

(1973) recommended two sprayings, four and two weeks before fruit harvest, with fenthion or fenitrothion 2.5 ml a. i. per tree. Harvesting the fruits while these are still hard (not fully ripe); collecting and promptly destroying all the infested and fallen fruits and ploughing around the trees during Winter are also helpful in checking the fruit fly infestation. Deol *et al.* (1977) have suggested growing early maturing varieties, such as, *16-33* and *Flordasum*. These varieties escape the attack of fruit flies under field conditions.

WHITEFLIES

Citrus blackfly, *Aleurocanthus husaini* Corbett and pomegranate whitefly, *Siphoninus finitimus* Silvestri have been reported as sporadic pests sucking the cell sap from leaves and tender shoots of peach trees. They also exude honeydew. The presence of these whiteflies is noticed only after the sooty mould has developed on leaves and twigs.

Spraying 0.03% dimethoate, monocrotophos or phosphamidon is effective in controlling the incidence of this pest.

BLOSSOM   BEETLES

Cetonine beetles, *Clinteria spilota* Arrow, *Epicometis squalida* Linnaeus, *Protaetia impavida* Janson and *P. neglecta* Hope have been occasionally found feeding on peach blossoms in northern India. The affected flowers shrivel and dry away. The damage caused is usually light and the pest is thus of minor importance.

FRUIT   DAMAGING   PESTS

Antestia cruciata (Fabricius), *Spilostethus* (*Lygaeus*) *pandurus* (Scopoli) and *S. hospes* (Fabricius) have been reported occasionally sucking the sap from ripening fruits. Besides, the juice also oozes out from the tiny punctures made by these bugs. These droplets of juice start fermenting, as a result the fruits start rotting and ultimately fall down. These are minor pests albeit *S. pandurus* is relatively more common and causes comparatively more damage.

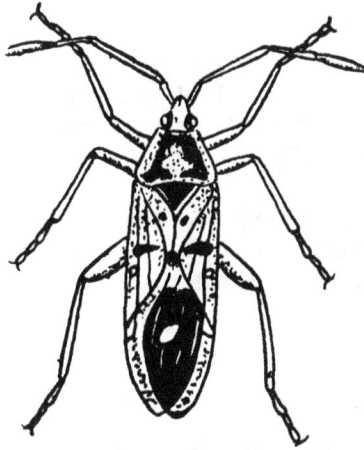

Spilostethus pandurus (Scopoli)

To prevent the damage, pluck the fruits before ripening and destroy promptly all the infested and fallen fruits.

A few species of beetles, namely, *Cerogria nepalensis* Hope, *Catharcius molossus* Linnaeus, *Xylotrupes gideon* Linnaeus, *Popillia complanata* Newman and *Lucanus lunifer* Hope have also been found

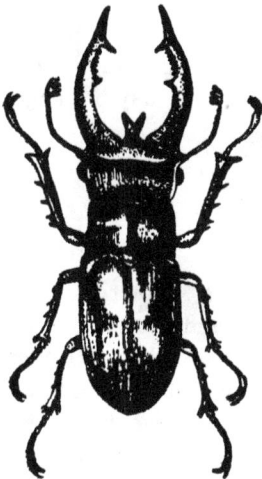

♂

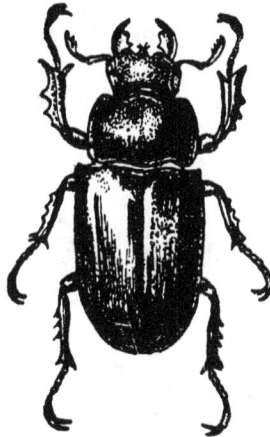

♀

Lucanus lunifer (Hope)

boring and/or gnawing peach fruits. Singh (1964) has suggested spraying with 0.2% pyrethrum extract to control these beetles.

Castor capsule borer, *Dichocrocis punctiferalis* Guenée – a polyphagous pest has been reported boring into terminal shoots, flower-buds and developing fruits of peach. To control this pest, remove and destroy immediately the affected parts and spray with 0.1% BHC + 0.1% DDT.

Codling moth, *Laspeyresia pomonella* (Linnaeus) is a world wide notorious pest of temperate fruits – apple being its preferred host. Its presence in India is confined to extreme north–western region (Ladakh). Strict quarantine measures and vigilance need be kept to prevent spread of this obnoxious pest in India (Butani,1974 f).

Pomegranate butterfly, *Virachola isocrates* (Fabricius) lays eggs on developing fruits. On hatching, the caterpillars bore inside the fruits and feed within. This is also a polyphagous pest having a very wide range of host plants; pomegranate is its preferred host.

Butterfly, *Kallima inchus* Boisduval mimics a dry leaf and sucks juice from ripening fruits. Citrus fruit sucking moths, *Othreis cajeta* (Cramer) and *O. materna* (Linnaeus) also attack peach fruits. The moths puncture the ripe fruits and suck the juice therefrom ; the affected fruits shrivel, wrinkle and ultimately fall down.

Ripe fruits are also attacked by hornet wasp, *Vespa orientalis* (Linnaeus). The damage caused by these wasps may be negligible but their nuisance value is rather great.

TERMITE

Microtermes mycophagus Desneux is a subterranean pest. The attack is more pronounced in sandy and loamy soils and the nurseries often suffer a severe setback when infested by termites. It is a polyphagous pest having a wide range of host plants. It is usually present throughout the year but is more active during August to December. The termites attack the root zone and tunnel upwards through the main stem or trunk. Leaves of affected trees, turn pale-yellow and gradually dry up. In the nurseries it is easy to pull out the affected plants while in case of young trees, these fall down when the strong winds blow. The pest also feeds on bark of the trees. The infestation is indicated by prominent mud galleries on the tree trunks in which the workers move about avoiding light.

Apple root borer, *Dorysthenes hugeli* Redtenbacher – a major pest of apple, is occasionally found damaging the peach roots as well. Generally this is a minor pest of peach trees. The grubs bore into the main roots ; as a result the trees become weak and shaky and the young trees may even fall down with strong winds.

Soil treatment with 5% aldrin, chlordane or heptachlor dust @ 22 to 25 kg per hectare keeps these pests at bay. The treatment need be repeated once every year. Addition of aldrin emulsion @ 2 to 2.5 litres a.i. per hectare in the irrigational water also drives away the termites.

SCAVENGERS

Peach scavenger moth, *Anatrachyntis simplex* Walsingham is found attacking deformed, disfigured and rotten fruits hanging on the trees.

Scavenger beetles, *Carpophilus dimidiatus* Linnaeus and *Amphicrosus* species are common in Himachal Pradesh, Jammu and Kashmir feeding on rotten fruits of guava, peach, pear, plum, pomegranate, etc. Normally only those fruits are attacked that have already been damaged by fruit flies or fungus etc. and have fallen down. But the real loss is caused when the fruits kept for drying are attacked by these beetles and spoiled.

Collect and destroy all the affected and fallen fruits to kill and prevent the spread of these scavengers.

NURSERY PESTS

In nurseries, where peach seedlings are raised, damage is often caused by tobacco grasshopper, *Atractomorpha crenulata* Fabricius and surface grasshoppers, *Chrotogonus* species. The former is relatively more destructive because of its climbing habit whereas damage by *Chrotogonus* spp. is confined only to lower leaves. Generally the damage caused is of minor importance and warrants no control measures. However, if and when necessary, dust with 10% BHC.

Cutworms, *Agrotis ipsilon* (Hufnagel) and *A. segetum* (Schiffer–Müller) may also cause serious damage in the nurseries. Both are polyphagous and attack peach seedlings soon after germination. The seedlings are cut at ground level, which means total loss of the plants. Once the seedlings become more than 35 to 40 mm tall, these are seldom attacked as by then their stems become fibrous and hard. The caterpillars live 50 to 70 mm deep in the soil and are nocturnal in habit. *A.segetum* moths are pale-whitish-brown with forewings ochreous-brown having double waved sub-basal ante-and post medial lines and marginal series of specks; hind wings are iridescent white with dark marginal line. Wing expanse is 40 to 48 mm.

Agrotis segetum (Schiffer-Muller)

To check the loss caused by cutworms, rake the soil around the seedlings and mix thoroughly with the soil 10% BHC or 5% aldrin, chlordane or heptachlor dust. Besides, dust the seedlings and the soil around with 5 to 10% BHC.

PLUM

PLUM, *Prunus* species, is commercially grown in eastern Europe and South-West Asia. It was introduced in India (Simla) by Alexander Coutts in 1887. Due to its immense adaptability to climatic conditions, plum is grown throughout temperate zone as well as in subtropical regions where about 200 chilling hours are completed. Two types of plums, namely, European plum, *Prunus domestica* Linnaeus and Japanese plum, *P. salicina* Lindley are commonly grown in India. The former originated from Caucasus and trans-Caucasus region where as the latter is of Chinese origin. The total area under plums in India is about 1200 hectares; of which over 500 hectares are in Jammu-Kashmir. *P. domestica* is generally cultivated at higher elevation and *P. salicina* at lower elevation. Sandy-loam, well drained soil is most suitable for its cultivation.

Fruits have a wide range of colour, flavour, size and texture. They also attract a great deal of attention on account of their beauty in bloom, the aroma of the blossoms and the beauty of the foliage as also highly coloured fruits. The fruits are usually eaten fresh, however, these are also used for making jams, jellies as well as for canning. Besides the fruit juice is used for making brandy (not in India).

Over 60 species of insect pests have been recorded damaging various parts of plum trees – roots, stems, leaves, flowers and fruits. Most of these are polyphagous pests, causing severe damage to other fruit trees and cause only minor damage to plum trees. The seedlings in nurseries suffer relatively more damage.

NURSERY PESTS

Grasshoppers and cutworm often cause severe damage in the nurseries. Although grasshoppers appear more regularly than cutworm, the damage caused by the latter, if and when it appears, is more devastating.

Atractomorpha crenulata Fabricius is a polyphagous pest enjoying countrywide distribution. It is a serious pest in nurseries and also attacks a wide variety of other crops, just after germination. Nymphs and adults gnaw the leaves and the young seedlings are destroyed completely. Adults have long and slender body with smooth pinkish abdomen and short stout antennae. Tegmen are pointed, green in colour and longer than hind wings. Hind wings are also pointed, about twice as long as broad and hyaline with base and nervures pinkish. Length is 18 to 26 mm and tegmina 15 to 20 mm.

Atractomorpha crenulata Fabricius

The pest breeds profusely and is most active during July to September. Eggs are laid in the upper layer of soil, about 10 to 15 mm deep. These hatch in 10 to 15 days during April to August and take over two months in Winter. Nymphal period is 30 to 40 days during Summer and extends upto 70 days during November to January. Longevity of adult males and females is 25 to 35 and 40 to 70 days respectively.

Surface grasshoppers, *Chrotogonus* species are commonly found in plains of North India and Peninsular region. These are polyphagous pests and though of regular occurrence, the damage caused is usually of minor importance. They prefer low vegetation and unlike *A. crenulata*, these grasshoppers have no climbing habit. Life-history is more or less same as that of *A. crenulata*.

Agrotis ipsilon (Hufnagel) has occasionally been reported cutting young seedlings at ground level and eating away the tender leaves. The caterpillars are nocturnal in habit and hide in the soil during day. The pest appears soon after the germination of seedlings and if not checked in time may cause severe loss.

To control these pests, dust 5 to 10% BHC. For cutworms, the soil around the plants need also be dusted with 10% BHC.

SAP SUCKING PESTS

Aphids, *Brachycaudus helichrysi* (Kaltenbach), *Hyalopterus pruni* (Geoffroy), *Myzus persicae* (Sulzer) and *Pterochlorus persicae* Cholodkovsky have been occasionally recorded sucking the sap from tender shoots and leaves. The preferred host of these aphids is peach. The affected leaves get curled and crumpled.

Whitefly, *Aleurocanthus husaini* Corbett – a polyphagous pest, having citrus as its preferred host, has been occasionally recorded on plum leaves as a minor pest.

A few species of scale insects, namely, *Drosicha contrahens* (Walker), *D. mangiferae* (Green), *Hemiberlesia lataniae* (Signoret),

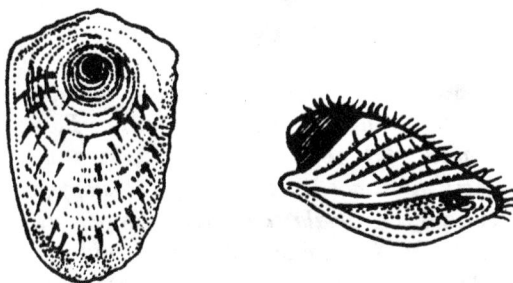

Hemiberlesia rapax (Comstock)

H. rapax (Comstock), *Parlatoria pseudopyri* Kuwana, *Pseudaulacaspis pentagona* (Targioni), *Quadraspidiotus perniciousus* (Comstock), *Eriochiton theae* Green, *Eulecanium tiliae* Linnaeus, *Parthenolecanium corni* (Bouché) have also been recorded from time to time causing minor damage to plum trees.

Plum fruits are damaged by various bugs, including, *Spilostethus pandurus* (Scopoli), *S. hospes* (Fabricius) and *Antestia cruciata* (Fabricius). Nymphs and adults suck the juice from developing fruits and the minute punctures made by these bugs for their feeding serve as entry to various bacterial and fungal organisms. As a result of this damage the affected fruits become shrivelled and wrinkled while those infested by fungi or bacteria start rotting and drop prematurely.

Fruit sucking moths, *Othreis cajeta* (Cramer), *O. materna* (Linnaeus) have also been observed sucking the juice from plums. These are major pests of *Citrus* spp. and only of minor importance in case of plum.

LEAF EATING CATERPILLARS

Hairy caterpillars, *Euproctis fraterna* (Moore), *E. flava* (Bremer) and *E. lunata* (Walker) as also leaf rollers, *Archips micaceana* (Walker) and *A. termias* (Meyrick) have been reported as sporadic pests, defoliating and damaging plum trees. These are polyphagous pests having a wide range of host plants. To control these pests dust with 10% BHC or spray 0.05% endosulfan or fenitrothion.

BEETLES AND WEEVILS

Adult beetles of *Mimastra cyanura* Hope, *Monolepta erthrycephala* Baly, *Protaetia impavida* Janson, *P. neglecta* Hope, *Brahmina coriacea* (Hope), *Adoretus duvauceli* Blanchard, *A. epipleuralis* Arrow, *A. horticola* Arrow, *A. lasiopygus* Burmiester, *A. versutus* Harold, *Anomala dimidiata* Hope, *A. lineatopennis* Blanchard, *A. varicolor* Gyllenhal and *Altica caerulescens* Baly have been recorded defoliating the plum trees. These are polyphagous pests and except *A. lineatopennis*, others cause only minor damage to plum trees. The eggs are laid in soil; grubs feed on roots and other organic matter available in the soil whereas the adults are leaf feeders. The adults are active during April to August and the peak period of activity is June-July. The incidence is relatively higher in sandy and sandy-loam soils than in clayey soils.

The weevils found on plum trees include, *Amblyrrhinus poricollis* Boheman, *Myllocerus laetivirens* Marshall, *M. sabulosus* Marshall and *M. undecimpustulatus maculosus* Desbrochers. Of these *M. laetivirens* is comparatively more common and sporadically severe. Adults are abundant in August, feeding voraciously during night. They live for 2 to 3 months. Eggs hatch in 4 to 5 days; grub stage lasts for about 10 months and pupal 5 to 6 days. There is only one generation in a year.

To control the grub stage of these pests, mix thoroughly with the soil, 5% aldrin, chlordane or heptachlor dust. Against adults, use 10% BHC dust or spray with 0.05% endosulfan or 0.075% dichlorvos.

FLOWER FEEDING BEETLES

Blister beetles, *Mylabris macilenta* Marseul, and *M. phalerata* (Pallas) have been found infesting the flowers of pear and plum while *Popillia cyanea* Hope have been reported feeding on flowers of apple and plum, occasionally boring the small developing fruits as well. Eggs of *M. phalerata* are laid in the soil. On hatching, the grubs feed on eggs of grasshoppers, etc. available in the soil. Adult beetles appear in large number and feed on the pollens and petals of flowers and flower-buds. As a result, the fruit setting is adversely affected.

Collect and destroy mechanically the adult beetles. In case of severe infestation, dust with 10% BHC.

FRUIT BORERS

Anarsia lineatella Zeller, *Eucosma ocellana* Schiffer-Müller, *Dichocrocis punctiferalis* Guenée, *Virachola isocrates* Fabricius, *Stathmopoda* species and *Zeuzera* species have been recorded boring the plum fruits. First two are serious pests of plums in Pakistan whereas *D. punctiferalis* and *V. isocrates* are major pests of guava and pomegranate respectively and *A. lineatella* is a serious pest of peach.

Eucosma ocellana is commonly known as eye-spotted bud moth. Its caterpillars bore into shoots, flower-buds and pea-sized fruits of plum. The caterpillars feed till September-October, then overwinter till March-April and pupate. Moths emerge from early May onwards and start laying eggs. Eggs hatch in 8 to 11 days; larval period lasts for 10 to 11 months and pupal stage occupies 9 to 13 days – thus there is only one generation in a year.

To control these pests, remove and destroy immediately all the affected shoots and fruits. In case of severe infestation, spray 0.05% endosulfan or fenitrothion.

Besides the above mentioned borers, Oriental fruit fly, *Dacus dorsalis* (Hendel) and hornet wasp, *Vespa orientalis* Linnaeus have also been found hovering in plum orchards. The preferred host of *D. dorsalis* is mango, while *V. orientalis* is attracted to sweetness of fruits. The damage caused by the wasps is negligible, but their very presence is a nuisance.

APRICOT

APRICOT, *Armenica vulgaris* Lamarck is native of western China where it still grows wild over wide areas. It was being cultivated in China as early as 2000 B.C., from where it soon spread to India, Persia, Armeniya (USSR), Egypt, etc. It is susceptible to frost and is therefore grown commercially only in warm temperate regions of China, Japan and North Africa. In India, there are only about 700 hectares under this fruits; of this over 500 hectares are in Jammu and Kashmir. The fruits are eaten fresh in localities where these are grown but elsewhere these fruits are available in form of dried, frozen, canned or candied fruits. Oil extracted from apricot seed is a good substitute for almond oil.

Over 30 insect species have been recorded on apricot trees. The major pests include, stem borer, leaf defoliating beetles, weevils and caterpillars, peach twig borer and fruit boring chalcid. Those of lesser importance include, aphids, scale insects, leaf rollers, bud moth, apple root borer and mango fruit fly. Cutworms often cause substantial loss in nurseries. In addition, a few store grain bettles, namely, *Rhizopertha dominica* Fabricius, *Tribolium castaneum* (Herbest), *Oryzaephilus mercator* Fauvel, *O. surinamensis* (Linnaeus) and *Laemophelus* species as also moths of *Cadra cautella* (Walker), *Plodia interpunctella* (Hübner) and *Sitotroga cerealella* Olivier have been recorded as pests of dried apricots.

STEM BORERS

Peach stem borer, *Sphenoptera lafertêi* Thomson is one of the most destructive pest of apricot trees. It is a polyphagous pest; peach and plum being its preferred hosts. The grubs bore into trunk and main branches and in case of severe attack, the entire tree dries up suddenly. The affected branches should be cut and destroyed promptly. In case of damaged trunk, the entire tree may be topped and the cut portion should be chopped into small pieces and burnt.

Quetta borer, *Aeolesthes sarta* Solsky is another stem borer reported from South-West China, Afghanistan, Pakistan, western Tibet and India. It is polyphagous pest and besides apricot, it has been recorded on apple, almond, peach, quince, walnut, etc. Eggs are laid in clusters inside the wounds or cuts on bark of the trees. On hatching, the grubs bore inside the trunk or main branches, making horizontal tunnels. Pupation takes place within these tunnels, generally in the centre of the trunk or main branch. Eggs hatch in 10 to 14 days while grub and pupal durations last for 12 to 14 and 4 to 5 months respectively.

BEETLES AND WEEVILS

Cetonid beetle, *Mimela fulgidivirdata* Blanchard; melolonthids, *Brahmina coriacea* (Hope), *Holotrichia crinicollis* Redtenbacher, *H. longipennis* (Blanchard), *Hoplosternus furicaudus* (Ancy) and *Serica* species as also rutelids, *Adoretus duvauceli* Blanchard, *A. versutus* Harold, *Anomala dimidiata* Hope, *A. lineatopennis* Blanchard, *A. polita* Blanchard and *A. rugosa* Arrow are polyphagous pests that attack apricot also. The grubs of these species feed on roots of host plants and other grasses growing around whereas the adult beetles feed, during night, on foliage. The beetles usually appear

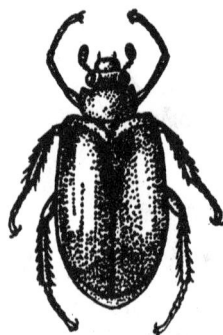

Anomala polita Blanchard   *Holotrichia longipennis* (Blanchard)

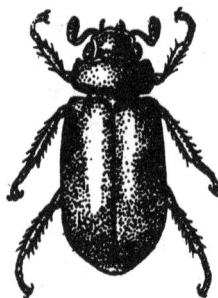

with the first heavy shower of monsoon and are active for about two months; maximum damage to foliage is caused during June-July.

Almond beetle, *Mimastra cyanura* Hope is a minor pest of apricot trees. These beetles appear in large number early in May and keep on nibbling the leaves till August. The grubs as usual live in soil feeding on roots of host plant and also of other weeds and grasses growing around the host trees.

Apricot weevil, *Emperorrhinus defoliator* Marshall is widely distributed along the foot of Himalayan range from Khasi hills (Assam) to Kangra and Kulu valley (Punjab and Himachal Pradesh). It is a polyphagous pest having wide range of host plants including apricot, apple, cherry, peach and pear – alder (*Alnus nitida*) being the preferred host. The pest is active in Summer, when swarms of these weevils may appear and defoliate the trees.

Other weevils found occasionally defoliating the apricot trees are *Amblyrrhinus poricollis* Boheman, *Myllocerus discolor* Boheman *M. laetivirens* Marshall, *M. maculosus* Desbrochers. These are all polyphagous and of minor importance in case of apricot.

To control these beetles and weevils, dust with 10% BHC. Keeping the orchards clean, ploughing around the trees and mixing 5% aldrin or chlordane dust with the soil, help in mitigating the grub population.

LEAF EATING CATERPILLARS

Lepidopterous defoliators include, Indian Gypsy moth, plum hairy caterpillar, tent caterpillar, apple hawk moth and apricot leaf caterpillar. Of these, first two are comparatively more common and harmful.

Indian Gypsy moth, *Lymentria obfuscata* Walker is one of the most destructiue pest found in all apricot growing regions. Hampson (1892) was first to record this pest from North-West Himalayas. The caterpillars are gregarious and nocturnal in habit.

They feed only during night and hide during day. In severe cases
of infestation the trees are completely defoliated and bear no fruit.
A female lays 200 to 400 eggs in batches during June-July, under
the bark of the trees and cover them with yellowish-brown hair;
initially each batch is covered separately and finally all the batches
are covered together. These eggs hatch during the following
March-April i.e. after 8 to 9 months. Larval period is 66 to 100
days while the pupal period varies between 9 and 21 days during
May to mid July (Rahman and Kalra, 1943). Thus there is only one
generation in a year ; longevity of male and female moths is 4 to 7
and 11 to 31 days respectively. Males are active fliers but females
are incapable of sustained flight owing to weak wings.

To control this pest, collect and destroy the egg-masses during
August to March. The egg-masses being conspicuous are
easy to locate. Spraying with 0.1% BHC+0.1% DDT or dusting
with 10% BHC dust is effective in controlling the larval population
of this pest as well as that of leaf rollers and to some extent of peach
twig borer. It is desirable to carry out organised campaign during
early stage of attack (March-April), when the caterpillars are feeding
gregariously.

Plum hairy caterpillar, *Euprocits fraterna* (Moore), as the name
suggests, is a major pest of plum but it also attacks apple, apricot,
quince, etc. Apricot trees are damaged during May-June and
again from August to November. The caterpillars feed voraciously
on the leaves and also attack the raw fruits. As a results, the trees
are defoliated and fruits become corky. Dusting with 10% BHC is
quite effective for keeping this pest under check.

Tent caterpillar, *Malacosoma indica* Walker is a major pest
of apple trees. It is commonly found in Kumaon hills (Uttar
Pradesh) and Simla hills (Himachal Pradesh). The eggs are laid in
broad bands around the tender twigs. The caterpillars web silken
tent like nests at the base of small twigs and feed on tender leaves.
Look out for egg-bands and destroy the same specially at the time
of pruning (December-January). If and when the nests appear,
these should be removed and destroyed mechanically.

Apple hawk moth, *Langia zeuzeroides* Moore has been reported from Assam, Sikkim, Himachal Pradesh and Punjab. Its main host is apple. All the immature stages of this moth are conspicuously big and can be easily hand-picked and destroyed.

Apricot leaf caterpillar, *Latoia repanda* (Walker) is a minor pest of apricot reported from Simla hills. The freshly hatched caterpillars feed gregariously on leaves and later segreggate. In case of severe infestation. which is rare, the entire tree may be defoliated. To control, clip off and destroy such leaves as are having the caterpillars feeding gregariously.

PEACH TWIG BORER

Anarsia lineatella Zeller is another major pest of apricot which also attacks peach and plum (Fletcher, 1932). The caterpillars bore into tender twigs and feed inside the pith and inner bark. The affected twigs dry up, droop and break off. The pest has two summer broods of 6 to 10 weeks each and a winter brood that lasts for 8 to 9 months.

To control this borer, remove and burn the infested twigs. In case of severe infestation, spray with 0.05% endosulfan or fenitrothion.

APRICOT CHALCID

Eurytoma samsonovi Vasiljev is widely distributed in Central Asia, Afghanistan, Pakistan and India. Though apricot is its main host, peach, plum and prunes etc. are also attacked but to a much lesser extent. Eggs are laid inside the tender fruits. On hatching the grubs bore into the kernel and feed in the inner contents, leaving the papery coat intact. As a result, fruit development is arrested and the fruits fall prematurely with grubs still feeding within the fruits.

Grubs are translucent white in colour having soft and fleshy body and strong mouth parts. After feeding during April-May they remain quiescent from June to February and then pupate inside the fruits. Thus grub period is generally about 10 months but in some cases it may extend to 2 or even 3 years (Pruthi and Batra, 1939). Pupal period lasts for 2 to 3 weeks. Adults emerge from dry fruits from end of February to middle of March. Both pupae and adults are dimorphic. Females are bigger in size and much brighter in colour than males.

To control this chalcid, collect and burn all infested and fallen fruits.

APHIDS

The common aphids attacking temperate fruits have also been found attacking apricot. These include, *Brachycaudus helichrysi* (Kaltenbach), *Hyalopterus pruni* (Geoffroy), *Myzus persicae* (Sulzer), *Pterochlorus persicae* Cholodkovsky. These species are widely distributed in subtropical and temperate regions of the world causing severe damage to several crops. However, in case of apricot, the aphids are of minor importance. They usually attack fresh tender leaves and blossoms. Infested leaves become pale and curled while the blossoms wither.

Spraying with 0.03% dimethoate, oxydemeton methyl, phosphamidon or quinalphos is quite effective in controlling the aphids.

COCCOIDS

Drosicha contrahens (Walker), *D. mangiferae* (Green), *Icerya purchasi* Maskell, *Parlatoria oleae* (Colvée), *P. pseudopyri* Kuwana (*cinerea* Hadden), *Pseudaulacaspis pentagona* (Targioni), *Quadraspidiotus perniciosus* (Comstock) and *Eulecanium tiliae* (Linnaeus) have been reported sucking sap from leaves and tender shoots of apricot trees. Of these, *Q. perniciosus* is relatively more destructive and is found on almost all temperate fruit trees, apple being its preferred host.

Spraying 3% miscible oil on dormant trees is quite effective against *Q. perniciosus* though it is little expensive. It is suggested to prune and destroy the infested leaves and twigs in the initial stage of attack. If the infestation is severe, spray 0.04% diazinon or monocrotophos.

LEAF ROLLERS

Archips micaceana (Walker) and *A. termias* (Meyrick) have been reported damaging the apricot trees. Both are distributed widely all over India but *A. micaceana* is relatively more common and harmful to apricot while *A. termias* (Meyrick) is a major pest of apple and minor one of apricot, cherry, mulberry, reach, pear, plum, etc. (Janjua, 1940). The caterpillars cut leaves and petioles and tie the same with other leaves or fruits and feed inside on soft tissues. The attacked leaves get partially skeletonized while the ripening fruits are punctured and the caterpillars feed on inner ripe pulp. A gum globule is often formed over the puncture on the fruit and the affected fruits become unfit for human consumption.

Collect and destroy the rolled leaves and webbed flowers. Spraying with 0.05% endosulfan or fenitrothion is effective in controlling these leaf rollers.

4.6
CHERRY

CHERRY is a delicious, delicate and highly attractive fruit. It originally came from the region between Caspian and Black seas and being a favourite food of many birds, the seeds were scattered over various parts of the continents of Eastern hemisphere, possibly before the arrival of man (Marshall, 1954). At present this fruit is grown in limited areas at high altitude preferably on slopes. In India its cultivation is confined mostly to Kashmir, Kulu valley and Kumaon hills with a small area spread along the foot of entire Himalayan range. There are two types of cherries – sweet cherry, *Prunus avium* Linnaeus and sour cherry *P. cerasus* Linnaeus; the former is used as dessert and the latter is mostly used for canning or cooking.

There are some 40 species of insects recorded from India which may cause minor or major damage, by attacking foliage, fruit or woody portions. These include, stem borers, aphids. coccoids, fruit fly, root borer, leaf rollers, leaf defoliators and cherry blossom beetles. Of these, only 2 or 3 cause severe loss to cherry trees and all others are of minor importance.

STEM BORERS

Aeolesthes holoserica Fabricius is the most destructive pest of cherry reported from India, Sri Lanka, Bangladesh, Burma, Malaysia, Thailand, etc. It is a polyphagous pest and besides its main hosts – apple, cherry and guava, it has also been reported damaging apricot, crab-apple, mulberry, peach, pear, plum, walnut, etc.

A female lays upto 100 eggs, inserted in the cuts, cracks and crevices in the bark of the trees and preferably in dead bark or fallen trees. On hatching, the grubs feed on inner layer of bark and make zigzag galleries; later these bore inside and feed on outer layer of sapwood. Fibrous and faecal matter is thrown out by these grubs which is a clear indication of their presence in those galleries. Full grown grubs bore further inside the wood of tree trunks for pupation. As a result of their feeding, vitality and productivity of the trees is greatly impaired. Adult beetles are nocturnal in habit and do little damage (Butani, 1974 a).

Eggs are elliptical in shape about 2.5 mm long and white in colour. Full grown grubs are 70 to 80 mm long, yellowish in colour and clothed with fine bristles. Adults are dark brown beetles, 38 to 45 mm long having a short mottled yellowish or silvery pubescence on the elytra. Antennae of males are 1½ times their body length while in females these are about the length of their body.

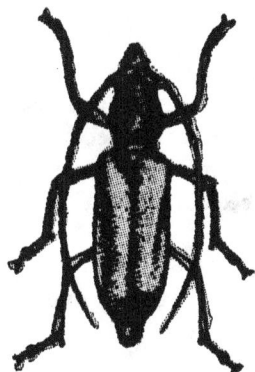

Aeolesthes holoserica Fabricius   *Aeolesthes sarta* Solsky

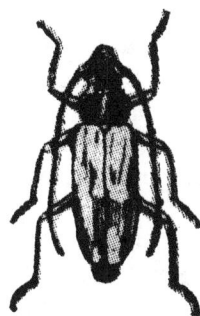

Adult beetles appear from end of April to mid July and remain active till the end of October. Eggs are laid during May to October. Incubation period is 7 to 12 days. The development of grubs is completed in 27 to 32 months. Pupation takes place either in October-November or in March-April. Prepupal period lasts for 3-150 days. Pupal period varies between 40 and 100 days. Adult beetles emerging from pupae formed in October remain within the stems throughout Winter and Spring; while those beetles that emerge from pupae formed in April rest for only 6 weeks – thus a single life-cycle is completed in 31½ to 36 months (Rahman and Khan, 1942).

Quetta borer, *A. sarta* Solsky; peach stem borer, *Sphenoptera lafertéi* Thomson and leopard moth, *Zeuzera* species have also been occasionally found boring the stems and trunks of cherry trees. However, these are all minor pests.

To control these stem borers, collect and destroy the grubs and beetles. Fumigate the holes with paradichlorobenzene (Alam, 1962) or insert in the holes cotton wick, about 150 mm long, soaked in a good fumigant like carbon bisulphide, chloroform, etc. Sharma and Attri (1969) have also suggested use of 0.1% dichlorvos or 0.025% phosphamidon, oxydemeton methyl or thiometon. The treated holes should be sealed with mud.

APHIDS

Cherry aphid, *Myzus cerasi* (Fabricius); peach black aphid, *Pterochlorus persicae* Cholodkovsky and peach green aphid, *Myzus persicae* (Sulzer) have been reported damaging cherry in India. Of these *M. cerasi* is more common.   The affected leaves become dwarf, curl badly and are covered with honeydew secreted by these aphids. It also affects the size and flavour of fruits besides making the fruits sticky and unattractive.

To control these aphids, spray with 0.03% dimethoate, oxydemeton methyl, phosphamidon or quinalphos.

COCCOIDS

Mango mealy bug, *Drosicha mangiferae* (Green); white scale, *Pseudaulacaspis pentagona* (Targioni); San José scale, *Quadraspidiotus perniciosus* (Comstock) and a soft scale *Eulecanium tiliae* (Linnaeus) have been recorded infesting cherry trees.   All these are polyphagous and of minor importance in case of cherry.   These insects suck the cell sap from leaves and tender shoots.   In case of severe infestation, due to loss of vitality, the growth and fruiting capacity of the affected trees is impaired.

Spray with 0.05% dichlorvos, fenitrothion, carbophenothion or 0.1% malathion to suppress the population of these coccoids.

FRUIT FLY

Oriental fruit fly, *Dacus dorsalis* (Hendel), a major pest of mango, has occasionally been recorded attacking cherry fruits as well. The affected fruits rot and fall down. It is a minor pest and normally no control measures are required except to collect and destroy the affected fruits.

ROOT BORER

Dorysthenes hugeli Redtenbechar has been occasionally found feeding on roots of cherry trees. This is a major pest of apple trees and a minor pest of various other fruit trees. The affected trees become weak and as a result of severe infestation the entire tree may get uprooted and fall down.

Frequent interculturing or raking the soil and mixing 5% aldrin or chlordane dust in the upper layer of soil keeps the pest population under check.

LEAF ROLLER

Archips termias (Meyrick) is a major pest of apple and minor pest of cherry and various other fruits. Caterpillars roll the leaves and feed within; the damaged leaves get skeletonized. In case of severe infestation, leaves and fruits are webbed together and larvae feed inside on the soft tissues and ripening fruits.

Collect and destroy rolled leaves and webbed bunches of flowers and fruits to prevent the infestation from spreading. When the infestation is severe spray with 0.03% phosphamidon or 0.05% endosulfan or fenitrothion.

LEAF DEFOLIATORS

Apple giant caterpillar, *Actias selene* Hübner and tent caterpillar, *Malacosoma indica* (Walker) are major pests of apple that are occasionally found feeding on foliage of cherry trees. Besides these, a number of beetles and weevils have also been recorded feeding and defoliating the leaves of cherry trees. These include, cherry black beetle, *Stalagosoma albella* Pallas; cherry green beetle, *Mimela passerini* Arrow; cherry gold lustre beetle, *Anomala flavipes* Arrow; chafer beetles, *Adoretus bimarginatus* Ohaus, *Holotrichia longipennis* Blanchard, *Hoplosternus furicaudus* (Ancy), *Serica clypeata* Brenske, *S. maculosa* Brenske, *S. marginella* Hope; flea beetles, *Altica caerulescens* Baly and *Pachnephorus impressus* Rosenbrug. Among the weevils, mention may be made of apricot weevil, *Emperorrhinus defoliator* Marshall and cherry weevil, *Myllocerus lefroyi* Marshall. None of these is a major pest and no control measure is generally warranted, however, in case of severe attack, spray 0.1% carbaryl or 0.05% endosulfan.

BLOSSOM BEETLES

Protaetia impavida Janson has been observed feeding on cherry blossoms as well as fruits. *P. neglecta* Hope feeds on leaves, flowers and fruits. Besides cherry these beetles also feed on apple, peach, pear, plum, etc. In India their occurrence has been reported from the foot of entire Himalayan range from Kashmir to Assam including Nepal. Beetles of *P. impavida* are stout, rather convex, shining blueblack having elytra closely set with large crescentic punctures and minute white markings. Beetles of *P. neglecta* are slightly bigger, 20 to 22 mm long, stout, convex, bronze coloured with elytra strongly and rugosely punctured and having intricate pattern of grey lines. Both are minor pests and no control measures are necessary for these beetles.

ALMOND

ALMOND, *Amygdalis communis* Linnaeus is the most favourite nut that fetches exorbitant price. It is native of central Asia (Afghanistan to Iran). In India, there is limited area under almond, confined mostly to Kashmir valley with a small area in Tamil Nadu. About 5000 tonnes of almond nuts are imported every year for local consumption. Attempts are underway to grow almonds in Himachal Pradesh along the Tibetian border, but the main limiting factor in expanding its cultivation in India, is the specific climatic requirements. It requires fairly warm and dry weather during ripening of fruits. Almond is usually taken as dry fruit. It is the most nutritious fruit containing over 20% proteins and a rich source of calcium (0.25%), phosphorus (0.5%), iron (3.5%) and vitamin B_1. Like any other crop, these trees are also vulnerable to attack of various insect pests. About two dozen insect species have been recorded so far on almond trees. Of these, a few species of beetles and weevils and occasionally some aphids cause severe damage whereas scale insects, stem borers, leaf eating caterpillars and fruit borers are of minor importance. Cutworms sometimes cause severe damage to seedlings. Besides, dried nuts are attacked and spoiled by *Cadra cautella* (Walker), *Plodia interpunctella* (Hübner) and *Sitotroga cerealella* (Olivier) as also *Laemophloeus* species, *Oryzaephilus mercator* Fauvel, *O. surinamensis* (Linnaeus) and *Tribolium castaneum* Herbest (Butani, 1974 e).

BEETLES AND WEEVILS

Almond beetle, *Mimastra cyanura* Hope, *ber* beetle, *Adoretus nitidus* Arrow and a chafer beetle, *Holotrichia consanguinea* Blanchard have been recorded nibbling the leaves of almond trees. Of these, first one is comparatively more common and has been reported from Jammu-Kashmir, Himachal Pradesh and Uttar Pradesh on almond, apple, apricot, chestnut, *falsa*, fig, grapevine,

mulberry, peach, pear, plum, pomegranate, etc. The grubs live in
soil and swarms of adult beetles may be seen defoliating the trees.
The beetles also emit an acrid yellow fluid from their head. The pest
appears around May but maximum damage to almond trees is caused
during July-August. Dusting 5 to 10% BHC is quite effective in
checking these beetles.

Among the weevils, almond weevil, *Myllocerus laetivirens*
Marshal is a major pest whereas, apricot weevil, *Emperorrhinus
defoliator* Marshall; apple weevil, *Myllocerus discolor* Boheman;
grey weevil, *M. maculosus* Desbrocher and plum weevil, *Amblyrrhinus
poricollis* Boheman are of minor importance.

Myllocerus laetivirens is polyphagous and beside almond, it
has been recorded on apple, apricot, *ber*, citrus, *falsa*, loquat,
mango, peach, pear, plum, pomegranate, etc. Almond and pear are
its preferred hosts. Adult weevils congregate on ventral surface of
leaves, nibble irregular holes and gradually eat away the entire leaf
laminae leaving only midribs. Tender leaves are first attacked and
when these are consumed, the weevils move on to the older leaves,
which are skeletonized.

Eggs are broadly oval, about 0.6 mm long, creamy-yellow,
smooth, transparent and shiny. These are fixed to substratum by
means of sticky substance. Freshly hatched grubs are creamy-white,
stout and finely setose; full grown ones are about 4 mm long, stout,
with large conspicuous spiracles and having no legs but short erect
setae which help them in locomotion. Pupae are 2.5 to 3.5 mm long,
dirty white when freshly formed, later becoming dark brown, sparsely
setose and anal armature well developed. Adult weevils are small,
3 to 4 mm long and pale-metallic-green in colour. Adult weevils
are poor fliers but very active feeders both during day and night. They
appear in May – the population increases with on-set of rains and
becomes maximum around August after which it declines and by
October the weevils are rare. Egg-laying starts from end of July and
continues till beginning of September. Eggs are laid in batches of
40 to 50 in the soil. Due to the small size and cryptic colouration
these eggs are difficult to locate in the soil. Incubation period is 4
to 5 days. On hatching, the grubs burrow deeper in the soil, 200 to
300 mm deep and feed on roots. When full grown they come up to

pupate in the upper 25 mm of the soil. The grub duration lasts for about 10 months and pupal period is 4 to 5 days. Overwintering is in pupal stage in the soil.

To check these weevils, rake the soil around trees and mix thoroughly with soil 5% aldrin or chlordane dust. In case of severe infestation, dust the trees with 10% BHC dust.

APHIDS

Black aphid, *Pterochlorus persicae* Cholodkovsky, leaf curl aphid, *Brachycaudus helichrysi* (Kaltenbach) and green aphid, *Myzus persicae* (Sulzer) have been recorded damaging almond trees; the first one being comparatively more common and harmful. These are polyphagous pests with wide range of host plants. Nymphs and adults suck sap from leaves, petioles, blossoms and fruits. Infestation is higher during Spring (February-April) when there is fresh flush of leaves. The infested leaves become curled, pale and fall down; development of fruits is arrested and these drop down prematurely. Besides, there is exudation of honeydew on which sooty mould develops, which further hinders the growth and fruiting capacity of the trees.

To control the aphids, spray with 0.03% diazinon, dimethoate, oxydemeton methyl, phosphamidon or thiometon.

SCALE INSECTS

Icerya purchasi Maskell, *Chrysomphalus ficus* Ashmead, *Pseudaulacaspis pentagona* (Targioni), *Quadraspidiotus perniciosus* (Comstock) and *Eulecanium tiliae* (Linnaeus) have been found sucking the sap from leaves and tender shoots; the mature ones migrate to stems and branches and occasionally attack fruits as well. These are all polyphagous and destructive to one or the other fruit crop. In case of almond, only *Q. perniciosus* – San José scale, is of major importance, as it often causes serious damage specially in nurseries and to young trees.

Prune and destroy the infested parts in the initial stage of damage and spray 0.04% diazinon or monocrotophos.

STEM BORERS

Peach borer, *Sphenoptera lefertéi* Thomson and Quetta borer, *Aeolesthes sarta* Solsky have been recorded damaging almond trees; the preferred hosts of these pests are peach and apricot respectively.

LEAF EATING CATERPILLARS

Malacosoma indica Walker – a major pest of apple, also attacks number of other temperate fruit trees including almond. The caterpillars prefer the newly opened leaves – thus hindering the growth and development of the tree. Collect and destroy the egg-masses.

Trabala vishnou (Lefebvre) – a sporadic pest of castor, has also been found feeding on foliage of almond, guava, *jamun* and pomegranate. The caterpillars feed at night and hide during day under the leaves or in cracks and crevices in the stems. Full grown caterpillars are 60 to 68 mm long, pinkish-brown in colour with tufts of dark hair arising from blue spots. Cocoons are tent-shaped, tough and papery interwoven with larval hair. Moths are pale green with faint dark markings and a large reddish-brown patch and a dark spot on forewings. Wing expanse of male moths is 48 to 55 mm. There are four generations in a year – two during hot weather, one in monsoon (August to October) and one during Winter (Beeson, 1941).

FRUIT BORER

Heliothis armigera Hübner, a serious pest of gram and pulses, has also been observed on almond, boring the tender fruits. Bunch varieties like, *Mercede* and *IXL* are more prone to attack of this borer. It is a polyphagous pest and appears on almond trees during April. Spraying with 0.05% dichlorvos or endosulfan is effective in controlling this borer.

WALNUT, *Juglans regia* Linnaeus originated from South-East Europe and is one of the most important temperate fruits. It is grown extensively in France, Italy, China and USA (California). In India, it is grown mostly in Kashmir valley. There are very few regular plantations, most of the trees are scattered and estimated area under walnut is about 5000 hectares; of which over 3000 hectares are in Jammu and Kashmir and the rest is spread over the entire foot-hills of Himalayas, extending from Simla to Darjeeling. Concerted efforts are underway to increase the area under walnut in India specially in Himachal Pradesh and Kashmir. The nutritive value of walnut is next only to almond and cashewnut, which are quite costly fruits as compared to walnut. Walnut contains about 16% protein as against 21% in almond and cashewnut, 3% in dates and less in other fruits. It is a rich source of vitamin B_1 and nicotinic acid. Besides, the wood of the tree (timber) is of excellent quality and is used for carvings, making butts of guns and furniture of superior quality.

Walnut trees are attacked by over 30 species of insect pests in India. Of these hardly four – aphid, bark boring beetle, leaf eating beetle and walnut weevil cause severe damage; the others, namely, San Jose scale, leopard moth, leaf eating caterpillars, chafer beetles and apple root borer are of minor importance. Indian meal moth and saw-toothed grain beetle cause substantial loss in storage.

APHIDS

Aphis pomi (de Geer), *Chromaphis juglandicola* Kaltenbach and *Panaphis juglandes* (George) have been reported as pests of walnut. Of these, *C. juglandicola* – walnut green aphid, is comparatively more common and sporadically severe. The aphids are usually active during April to September. Nymphs and adults suck the sap confining themselves on the ventral surface of leaves and tender shoots. The affected parts become pale, curl and disfigured. In case of severe infestation, even the growth of the tree is affected and the plant is stunted. The aphids also secrete honeydew, on which sooty ˙mould develops causing hinderance in photosynthetic activity of the plant and further retarding its growth.

Disulfoton 5% granules mixed thoroughly with the soil around the trees is effective in checking the aphid infestation. Quantity of granules will vary with the age of the tree. For young trees 200 gm per tree is sufficient whereas the older ones may require upto 500 gm per tree. Spraying with 0.03% dimethoate, oxydemeton methyl, phosphamidon or quinalphos is also effective in controlling the aphids.

BARK BORING BEETLES

Long horned beetle, *Batocera horsfieldi* Hope is a serious pest in Darjeeling, Kumaon hills, Kulu valley and Simla hills. Eggs are laid singly on the bark on the lower one meter portion of the bole. On hatching, the grubs feed on inner side of the bark making zigzag tunnels. Later, the grubs bore down to the surface of sapwood and go upto the centre of heartwood. Usually the infestation is limited to one or two tunnels near the base of the tree but as many as 16 tunnels have been observed upto 6 metres length of tree trunk (Beeson, 1941). Adult beetles feed on bark of young twigs but the damage caused is negligible. Attacked trees are not killed nor is the vigour of such trees affected but the timber is so damaged that it is useless for all purposes except for firewood which hardly fetches the cost of cutting the timber.

Eggs are oval in shape and brown in colour. Full grown grubs are 90 to 150 mm long, pale yellow in colour. Pupae are 40 to 60 mm long and creamy-white. Adult beetles are 45 to 65 mm long, black with fine ashy or yellowish-grey pubescence. Pronotum has two elongate yellowish-white spots. Elytra have numerous shining black tubercles at base and several rounded or broken white marks extending upto apex. A female lays 55 to 60 eggs. These hatch in 8 to 15 days. Grub, prepupal and pupal periods last for 20 to 25 months, 50 to 182 days and 40 to 90 days respectively (Rahman and Khan, 1942b). Adult longevity is 4 to 5 months (early June to early November) and a single life-cycle occupies 23 to 32 months. Full grown grubs overwinter from October to March but the grubs that are less than a year old continue feeding during Winter as well.

Quetta borer, *Aeolesthes sarta* Solsky, is a major pest of walnut in Baluchistan (Pakistan) and often causes severe damage in Kashmir valley, to various temperate fruits. The eggs are laid in the cuts or wounds on the bark of the trees. On hatching, the grubs bore in the bast and sapwood and tunnel downwards. The tunnelling destroys the quality of the timber and the bark over the attacked areas often falls down. These beetles appear generally in May and lay eggs during May–June. Eggs hatch in 10 to 14 days. Grubs are active during entire Autumn, become dormant during Winter and again active from next Summer till Winter. Pupation takes place just before Winter and lasts for 6 to 7 months, till the temperature is high enough for emergence. A single life-cycle takes two years.

Cherry stem borer, *Aeolesthes holosericea* Fabricius has also been occasionally noted boring in walnut trees in Himachal Pradesh and Jammu-Kashmir. This, however, is a minor pest of walnut.

Hand–picking and mechanical destruction of grubs and adult beetles as also treating the holes in trunk with carbon bisulphide, chloroform–creosote mixture can easily control these borers.

LEAF EATING BEETLES

Walnut blue beetle, *Monolepta erythrocephala* (Baly) has been reported from Himachal Pradesh as a major pest of walnut. It is also found in Punjab and Uttar Pradesh. Besides walnut, the beetle has been recorded feeding on leaves of apple, grapevine, pear, etc. The beetles feed on epidermis of leaves, leaving behind a net work of veins. These are active from April to September though severe damage is caused only during June to August. Adult beetles are 4 to 5 mm long having reddish-brown head, black body and bluish elytra which are broader at the base than at prothorax.

Other chrysomelid beetles found feeding on walnut leaves are *Altica caerulescens* Baly, *Cerogria nepalensis* Hope, *Dactylispa lohita* Maulik, *Hispa daina* Chapman, *Galerucida rutilans* Hope, *Platypria hystrix* Fabricius, *Phyllotreta chotanica* Duvivier and *Triclina* species. All these are of minor importance, occurring sporadically and occasionally causing economic losses.

To control these beetles, if and when necessary, the trees may be sprayed with 0.5% dichlorvos or endosulfan.

WALNUT WEEVIL

Alcidodes porrectirostris Marshall is the specific pest of walnut in India and has been recorded from all the walnut growing areas. Eggs are laid in the pits excavated by the female weevils on the fruits at the rate of 1 to 4 (maximum 15) eggs per fruit. On hatching, the grubs bore deeper and feed on kernels, reducing the same into useless black mass. The grubs also cause extensive premature dropping of the fruits, which starts from middle of April and continues till August, maximum being during June-July. The affected fruits also show dark brown spots which are in fact, dried resinuous excretions. Pupation takes place inside the fruits and adult weevils emerge from these fruits by bitting circular holes. Adult weevils feed on petioles, female floral buds, tender shoots and even on young fruits. Grubs are more harmful than adults.

Eggs are 1.0 to 1.5 mm long and transluscent. Full grown grubs are 12 to 16 mm long and creamy-white in colour. Pupae are also creamy-white and 9 to 12 mm long. Adult weevils are 9 to 12 mm long, jet-black when young turning dark brown with age, with a prominent snout bending downwards. Eggs hatch in 3 to 5 days; grub and pupal durations last for 13 to 22 and 9 to 17 days respectively (Husain and Khan, 1949). Adult longevity is 4 to 6 weeks whereas hibernating adults live for 6 to 7 months. Two generations are completed between the end of April and September. Thereafter, the weevils continue to feed as long as twigs are green but with shedding of leaves early in November, the weevils enter into hibernation (November to March) hiding under the bark of trees, debris, clods or even in cracks and crevices in the soil.

Myllocerus viridanus Fabricius has also been occasionally reported nibbling the walnut leaves – its alternate host is guava. On walnut, this pest is of minor importance.

Collect and destroy the fallen fruits in May and June to prevent the population build-up of the pest. To control these weevils, dust with 5 to 10% BHC early in April so that the weevils feeding on young shoots die before ovipositing. Spraying with 0.04% quinalphos or 0.05% dichlorvos is also effective.

SAN JOSE SCALE

Quadraspidiotus perniciosus (Comstock) is becoming an increasing menace in the orchards along the entire Himalayan range. Though apple is its preferred host, the pest attacks almost all the temperate fruit trees. It passes several generations, each of 5 to 6 weeks' duration, during its active period and overwinters in nymphal stage from mid-October to mid-March.

LEOPARD MOTH

Zeuzera species has been reported attacking shoots and stems of walnut trees. It is a polyphagous pest. Eggs are laid singly in the cuts and wounds on the stems or in main branches. On hatching, young caterpillars feed just below the bark and later bore into tender shoots and feed within for 8 to 9 months when they come out and bore the thick stems or main branches, wherein they feed for another 10 to 11 months and pupate within. As a result, young trees may succumb while in case of big or older trees, leaves of the attacked shoots wither and eventually the branches die – thereby affecting the growth of the tree. A single life-cycle lasts for about two years or longer depending upon the climatic conditions.

Eggs are oval, about one mm long and dark yellow in colour. Full grown caterpillars are 50 to 70 mm long and pinkish-red in colour. The moths are white with beautifully patterned wings having numerous dark blue spots on forewings and marginal dots on hind wings. Wing expanse is 30 to 40 mm.

To control this pest, insert in the hole, cotton-wool soaked in carbon bisulphide, paradichlorobenzene or ethyl acetate or chloroform–creosote mixture and seal the hole with mud.

LEAF EATING CATERPILLARS

Actias selene (Hübner), *Belippa laleana* Moore, *Oxyambulyx sericeipennis* Butler have been reported from Assam (Fletcher, 1921) defoliating walnut trees. Besides, caterpillars of *Malacosoma indica* Walker and some wild silkworm moths, namely, *Antheracea roylei* Moore, *Odina wodier* and *Saturnia simla* Westwood have also been recorded from Himalayan range, feeding on walnut foliage. These are all minor pests. Dusting with 10% BHC or spraying with 0.05% endosulfan is effective in controlling these pests.

CHAFER BEETLES

Adoretus versutus Harold, *Anomala rugosa* Arrow, *Brahmina coriacea* (Hope), *Holotrichia longipennis* Blanchard and *Mimela pusilla* Hope have been recorded gnawing leaves of walnut trees. The beetles are nocturnal in habit and hide during the day, usually in the soil. They generally appear during June-July and remain active for about two months. To control these beetles, dusting with 10% BHC is recommended.

ROOT BORER

Dorysthenes hugeli Redtenbacher – a major pest of apple, has been reported occasionally damaging the roots of walnut trees. Eggs are laid in the soil; dry and sandy soils being preferred. After hatching, grubs feed for a few weeks on organic matter available in upper 100 mm of soil. Later, the grubs move down in the soil and attack rootlets and roots upto a depth of 200 to 250 mm and even gnaw the underground portion of main stem. The grubs usually live for 3 to 4 years. The affected trees become weak and may even fall down in due course, if the grubs are not killed by frequent interculturing coupled with soil application with 5% chlordane or heptachlor dust @ 200 to 300 gm per tree.

5
EPILOGUE

EPILOGUE

INSECTS AND FRUITS are ubiquitous. Wherever there is vegetation there are fruits as well as insects. Go to any nook and corner of the world – tropical, subtropical or temperate, there will always be one or the other fruit tree growing in that area. Similarly, the insects will also be present feeding on different parts of those trees, causing slight to severe damage. The losses resulting from such damage may be both qualitative as well as quantitative. A good percentage of this loss can be easily avoided by adopting proper and timely pest management methods.

Prevention is better than cure. This adage fits well for pest control also. It is far more desirable, easier and economical to prevent the occurrence of insect pests, than to control the pest once the population has built-up. Destroying sources of infestation and carry over, protecting the susceptible parts by suitable treatments, avoiding conditions that are likely to contribute the rapid multiplication of insects, are the possible methods which can mitigate the incidence and spread of the pest.

Orchard sanitation is the first and foremost method which can prevent the flare up of pests. It is the rundown or neglected orchards that are most susceptible to attack of insect pests. Similarly, orchards that are over-crowded with trees or are water-logged, harbour more insects. Besides keeping the orchards clean, by removing weeds and bushes, removal of loose bark from the trees and other trash from the soil is a good cultural practice to keep the insects pests at bay. This wise counsel has a sound scientific basis. The studies on bionomics of various insect species have clearly shown the important role played by alternate and collateral hosts as well as shelter afforded by the trash to adult and quiscent stages of insects. Growers should avoid sandy or sandy-loam soils for raising their orchards and should prefer as far as possible insect resistant varieties and root-stocks before planting the orchard.

Mechanical methods when adopted, at the initial stage of attack by insect pests, are the most economical and effective. Removing and destroying promptly all the infested and affected plant parts (leaves, twigs and fruits); hand-collection and mechanical destruction of eggs, larvae, pupae and adults are some of the methods which check the population build-up at hardly any cost. These methods are specially useful in case of lemon butterflies, apple hawk moth, tent caterpillar, etc. Bagging of fruits can be useful to check damage by pomegranate butterfly and fruit sucking moths. Ploughing around the trees, spiking the bark and trunk borers and use of slippery or sticky bands are age old but still effective methods for exposing the immature stages of pests in the soil, killing the borers inside the tree trunks and preventing the ascend of pests like mango mealy bugs. Recently, hot water treatment – a physical method, has been suggested for mango stone weevil. These methods no doubt are cumbersome and time consuming but there is ample scope to modify these methods to fit in better with the recommended crop husbandry.

Biological method of pest control has not yet received adequate attention in India. There are a few reports of successful biological control against some fruit pests, like, citrus mealy bug, fluted scale, San José scale, wooly aphid, etc , but by and large this aspect has remained neglected.

Chemical control measures are often unavoidable, specially when there is severe infestation. The method should be adopted if and when necessary. The pest control with some of the modern insecticides is so spectacular that, in general, use of pesticides is considered by various orchardists as the only feasible method of pest control. This has often led to injudicious use of pesticides. Some of the orchardists indiscriminately treat their trees, ignoring the recommendations and warnings, specially when the trees are in full bloom or the fruits are at the ripening stage and ready to be harvested.

A word about insecticides. Selection of proper insecticide is of paramount importance. The insecticide selected should not be phytotoxic. It should be as far as possible, fresh, easily available and reasonably priced; above all it should have low mammalian toxicity and non-persistent when it is to be applied at flowering and fruiting stages. It is not advisable to use the same insecticide repeatedly for a long time as it may lead to development of resistance in the insect species against which the insecticide is used. Insecticides of different groups may be used alternatively. Unfortunately, the storage facilities at most of the places are not suitable, as a result, with passage of time even the best pesticides gradually lose their efficacy. Therefore, use of old insecticides should be avoided. If at all it is necessary to use insecticides stored for some time, the dosage may be suitably increased after determining their bio-efficacy.

Orchardists may sometimes think of mixing together two or more pesticides (insecticides, fungicides, miticides) and even a fertilizer (urea, etc.) or hormone (NNA etc.), to save on time and cost of application. There is possibility that in such admixtures chemical or physical properties of compounds are altered which may result in loss of suspensibility, emulsifiability or even the biological efficacy of the insecticide and may have antagonistic effects or even prove to be phytotoxic when applied on foliage. Hence, it is essential to know about the compatibility or otherwise of various chemicals. Organophosphates, carbamates and some chlorinated hydrocarbons like BHC, DDT and lindane are highly susceptible to alkaline reactions. Mixing of these insecticides with other chemicals having high alkalinity like, Bordeaux mixture or lime sulphur should be avoided.

It is being increasingly realised that the popular single method of control approach of using insecticides has to go. Orchardists in advanced countries have tested the consequences of using enormous quantities of insecticides. Toxic residues on harvested fruits, destruction of pollinators and natural enemies, resurgence of minor pests, killing of birds and last but not the least, development of resistant strains of insect species, are some of the serious side effects of frequent use of insecticides. Integrated control for Pest Management is the strategy which has been evolved to overcome these problems. This is a system which utilises various suitable and compatible techniques of pest control, with a view to reducing the pest population and maintain the same below the economic injury level. Environments are so manipulated that the pest population is prevented from causing any felt losses. Such an integrated

approach is more efficient and economical. The rational and
judicious use of pesticides in conjunction with cultural, mechanical
and biological methods avoids upsetting of balance in nature,
minimizes residue hazards of pesticides, delays the development of
pesticide resistant strains of pests.

The rapid strides of horticulture all over the country are
bound to bring about a corresponding increase in pest complex.
Keeping in view the fast pace of development of horticulture and
rapid growth of orchards, it will be wise policy to fully visualize
the magnitude of pest problems and strengthen the pest control
research to match the expected pest problems and not to allow
them to become the limiting factor in our production programme.

Fore-warned is fore-armed, is an apt adage for our struggle
against insect pests. Surveillance is an important component of
the strategy to forecast impending pest outbreaks and undertake
suitable methods for pest management. It is also necessary to
identify the key factors favouring population build-up of a particluar
pest. Fortunately, an average orchardist is relatively more
enlightened than an average farmer growing conventional crops. He
can easily grasp the significance and importance of Pest Management
and implement the programmes based on the integrated methods.

Lab to Land – it is for the research workers to develop and
pin-point the specific methods of control for major pests and for the
extention worker to properly explain and convincingly demonstrate
the efficacy of these methods. Only by adopting this approach the
real break-through in the control of pests of fruit trees can be
achieved. This is no longer a distant dream, it is a reality which has
been successfully demonstrated by the orchardists in some of the
advanced countries.

6

REFERENCES

REFERENCES

ABRAHAM, E.V., 1953 : A pest of new importance of pineapple in Wynas area of Madras State - the mealy bug *Pseudococcus brevipes* Cockerell. *S. India Hort., 1* : 99-104, Coimbatore (83)

AIYER, PADMANABHA, K.S., 1943 a: On three caterpillars destructive to mango flowers. *Indian J. Ent., 5* (1-2) : 53-57, New Delhi (118)

AIYER, PADMANABHA, K.S., 1943 b : Notes on two major caterpillar pests of *Eugenia jambos* (rose apple). *J. Bombay nat. Hist. Soc., 48 :* 673-675, Bombay (216)

ALAM, M. ZAHURUL, 1962 : *Insect and Mite Pests of Fruits and Fruit trees in East Pakistan and Their Control,* 115 pp. Department of Agriculture, East Pakistan, Dacca (68,71,96,221,318)

ALAM, M. ZAHURUL, AHMAD ALAUDDIN, ALAM SHAMSUL AND ISLAM AMEERUL MD., 1964 : *A Review of Research, Division of Entomology (1947-1964),* 272 pp. Department of Agriculture, East Pakistan, Dacca (96,109,189,249)

ALAM, M. ZAHURUL AND M.A. WADUD, 1964 : On the biology of litehi mite, *Aceria litchi* Keifer (Eriophyidae : Acarina) in East Pakistan. In : *A Review of Research, Division of Entomology (1947-1964)* : 157-166, Department of Agriculture, East Pakistan, Dacca (109,189)

ALI, S. MOHAMMAD, 1961 : A new record of *Icerya pulcher* (Leonardi) in India and its identity (Coccidae). *Indian J. Ent., 23* (2) : 151-152, New Delhi (19)

ALI, S. MOHAMMAD, 1962 : Coccids affecting sugarcane in Bihar. *Indian Jour. Sug. Cane Res., 6*(2) : 72-75, New Delhi (101)

ALI, S. MOHAMMAD, 1964 : Some studies on *Pulvinaria cellulosa* Green a mealy scale of mango in Bihar, India. *Indian J. Ent., 26* (3) : 361-362, New Delhi (19)

ALI, S. MOHAMMAD, 1958 : Coccids (Cocccidae : Hemiptera : Insecta) affecting fruit plants in Bihar (India). *J. Bombay nat. Hist. Soc., 65* (1) : 120-137, Bombay (19,101)

ANANTHAKRISHNAN, T.N., 1955 : Notes on *Pseudodendrothrips dwivarna* (R. & M.) from S. India. *Indian J. Ent., 17*(2) : 213-216, New Delhi (105)

ANANTHAKRISHNAN, T. N., 1971 : Thrips (Thysanoptera) in agriculture, horticulture & forestry - diagnosis, bionomics and control. *J. Sci. industr. Res., 30* (3) : 113-146, New Delhi (51,84,105,143,164,213,276)

ANANTHANARAYANAN, K. P. AND E.V. ABRAHAM, 1955 : The slug caterpillar, *Parasa lepida* Cram., and its control. *J. Bombay nat. Hist. Soc., 53* (2) : 205-209, Bombay (34)

Figures in parentheses at the end of each citation refer to page numbers of this book.

ANANTHANARAYANAN, K.P. AND S. VENUGOPAL, 1955 : Notes on a new pest, *Lepidogma* sp. (Pyralidae) on *Eugenia jambolana* in Coimbatore. *Indian J. Ent.*, *17* (2) : 183-185, New Delhi (216)

ANANTHASUBRAMANIAN, K.S. AND T.N. ANANTHAKRISHNAN, 1975 : Taxonomc,i biological and ecological studies of some Indian Membracids (Insecta : Homoptera), Part II. *Rec. Zool. Serv. India*, *68* : 305-340, Calcutta (74)

ANONYMOUS, 1903 : Notes on insect pests from the Entomological section, Indian Museum : II - Insect pests of fruit trees. *Indian Mus. Notes*, ·5 (3, : 117-127, Calcutta (37)

NSARI, A.R., 1942 : Occurrence of *Bourbon aspidiotus (Aspidiotus destructor* Sign.) in the Punjab. *Indian J. Ent.*, *4* (2) : 233, New Delhi (19)

ANSARI, M. ATIQUR RAHMAN, 1945 : *Dichocrocis punctiferalis* Gn. as a pest of guava. *Indian J. Ent.*, *7*(1 & 2) : 241, New Delhi (59)

ANTONY, J., CHANDY KURIAN AND K.P.V. MENON, 1958 : Trap the rhinocercs. *Cocon. Bull.*, *12* (9) : 331-335, Ernakulam (204)

ANUFRIEV, G.A., 1970 : Description of a new genus *Amritodus* for *Idiocerus atkinsoni* Leth. from India (Hemiptera : Cicadellidae). *J. Nat. Hist.*, *4* (3) : 375-376, London (10)

ATTRI, B.S. AND P.L. SHARMA. 1971 : Granular systemic insecticides for control cf wooly aphid, *Eriosoma lanigerum* (Hausm.) on apple (*Malus pumilla* Mill.). *Indian J. agric. Sci.*, *41* (7) : 627-631, New Delhi (263)

ATWAL, A.S., 1961 : Some ecological problems relating to insect pests of mango. *Gardening*, *3* (10) : 75-76, Lucknow (14)

ATWAL, A.S., 1963 : Insect pests of mango and their control. *Punjab Hort. J.*, *3* (2-4) : 235-258, Patiala (21,27)

ATWAL, A.S., 1976 : *Agricurtural Pests of India and South-East Asia*, 502 pp. Kalyani Publishers, Ludhiana (166,169)

ATWAL, A.S. AND G.S. JOSAN, 1962 : Citrus pests round the year. *Punjab Hort. J.*, *2* (1) : 8-16, Patiala (129)

ATWAL, A.S. AND G.C. VARMA, 1968 : Studies on the control of citrus psylla, *Diaphorina citri* Kuwayama (Hemiptera : Psyllidae) by foliage spray and soil application. *J. Res. PAU.*, *5* (2) : 240-243, Ludhiana (131)

AYYAR, RAMAKRISHNA T.V., 1922 : Weevil fauna of South India with special reference to economic species. *Imperial Agricultural Research Institute, Pusa, Bull No. 125*, 21 pp., Calcutta (33)

AYYAR, RAMAKRISHNA T.V., 1924 a : A further contribution to our knowledge of South Indian Coccidae. *Rept. Proc. 6th Ent. Mtg.*, *Pusa (Bihar), February 1923* : 339-344, Calcutta (19)

AYYAR, RAMAKRISHNA T.V., 1924 b : Short notes on some South Indian insects. *Rept. Proc. 6th Ent. Mtg., Pusa (Bihar) : February 1923* : 263-269, Calcutta (221)

AYYAR, RAMAKRISHNA T.V., 1928 : A contribution to our knowledge of the Thysanoptera of India. *Mem. Dept. Agric. India, Ent. Ser.*, *10* (7) : 217-316, Calcutta (116)

AYYAR, RAMAKRISHNA T.V., 1930 : Contribution to our knowledge of South Indian Coccidae (scales and mealy bugs). *Imperial Council of Agricultural Research, India, Misc. Bull. No. 197*, 73 pp., Calcutta (89, 136)

AYYAR, RAMAKRISHNA T.V., 1932 : An annotated list of the insects affecting the important cultivated plants in South India. *Madras Dept. Agric. Bull. No. 27,* 95 pp., Madras (83)

AYYAR, RAMAKRISHNA T.V., 1938 : An annotated conspectus of the insects affecting fruit crops in S. India. *Madras agric. J.*, *26* (9) : 341-351, Madras (69)

AYYAR, RAMAKRISHNA T.V., 1942 : A nettle grub pest of the banana plant in South India (*Miresa deceder* Wlk.). *Indian J. Ent.*, *4* (2) : 171 - 172, New Delhi (46)

AYYAR, RAMAKRISHNA T.V., 1963 : *Handbook of Economic Entomology for South India*, 516 pp. Government of Madras, Madras (21,60,108,256)

AYYAR, RAMAKRISHNA T.V. AND V. MARGABANDHU, 1931 : Notes on Indian Thysanoptera with brief descriptions of new species. *J. Bombay nat. Hist. Soc.*, *34* (4); 1029-1040, Bombay (51,105)

BAKER, C.F., 1915 : Studies of Philippine Jassoidea - IV : The Idiocerini of the Philippines. *Philippines J. Sci.*, *10* (6) : 317-342, Manila (10)

BANERJEE, S.N. AND B.K. CHATTERJEE, 1955 : Studies on the phytotoxic action of BHC and DDT on the common cucurbitaceous plants in West Bengal. *Proc. Indian Acad. Sci.*, *41-B* (5) : 227-239, Bangalore (247)

BASHA, JANAB MOHAMED GHOUSE, 1952 : Experiments on the control of fruit borers of jujube (*Zizyphus* spp.), — *Carpomyia vesuviana* Costa and *Meridarchis scyrodes* Meyr. in South India. *Indian J. Ent.*, *14* (3): 228-238, New Delhi (186)

BATRA, H N., 1952 a : Occurrence of three banana pests at Delhi. *Indian J. Ent.*, *14* (1) : 60, New Delhi (43)

BATRA, H.N., 1952 b : Record of the fig midge at Delhi. *Indian J. Ent.*, *14* (1) 60, New Delhi (236)

BATRA, H.N., 1953 a : Biology and control of *Dacus diversus* Coquillett and *Carpomyia vesuviana* Costa and important notes on other fruit flies in India. *Indian J. agric. Sci.*, *23* (2) : 87-122, New Delhi (56,174, 248)

BATRA, H.N., 1953 b : Aphids infesting peach and their control. *Indian J. Ent.*, *15* (1) : 45-51, New Delhi (286)

BATRA, H.N., 1955 : Ak grasshopper, *Poecilocerus pictus* (Fabr.) (Acrididae) as a pest of papaya and some other plants in Delhi. *Indian J. Ent.*, *17* (1) : 132, New Delhi (76)

BATRA, H.N., 1961 : First record of *Haltica cyanea* Weber and *Bagous* species on singhara crop. *Indian J. Ent.*, *23* (1) : 66-68, New Delhi (106, 108)

342

FRUIT PESTS

BATRA, H.N. 1968 : Biology and bionomics of *Dacus* (*Zeugodacus*) *hageni* de Maijere (*D. caudatus* Fabr.). *Indian J. agric. Sci*, *38* (6) : 1015-1020, New Delhi (90)

BATRA, H.N. AND P.L. RENJHEN, 1950 : Preliminary observations on the biology of the buprestid borer (*Sphenoptera lafertei* Thomson) of certain fruit trees in North-west India. *Indian J. Ent.,12* (2) : 159-165, New Delhi (296)

BATRA, R.C., O.S. BINDRA AND B.S. SOHI, 1973 : New record of some chafer beetles as pests of grapevines. *Indian J. Ent.*, *35* (2) : 177, New Delhi (162)

BEESON, C.F.C., 1941 : *The Ecology and Control of Forest Insects*, 1007 pp. C.F.C. Beeson, Forest Research Institute, Dehra Dun (37,38,93,97,98,99, 117,120,160,222,321,324,326)

BERLESE, A. AND G. LEONARDI, 1896 : Le cocciniglie Italiane viventi sugli agrame. *Rev. Patol. veg.*, *3* : 49-171; *4* : 74-179,195-292, Paris (138)

BESS, H.A. AND F.H. HARAMOTO, 1977 : *Dacus dorsalis* Hend. In : *Diseases Pests and Weeds in Tropical Crops* : 526-528, Verlag Paul Parey, Berlin (20)

BHADURI, AJOY SHANKAR, 1958 : A list of butterflies found on date palms tapped for toddy. *J Bombay nat. Hist. Soc.*, *55* (2) : 375-376, Bombay (202)

BHAT, S.S. AND H.G. PATIL, 1933 : The gholwad chiku : its cultivation, varieties, economics, etc. *Poona Agric. Coll. Mag.*, *24* (4) : 32-38, Poona (91)

BHATIA, D.R. AND H.S. SIKKA, 1952 : The singhara beetle (*Galerucella birmanica* Jacoby Chrysomelidae) in Delhi area and its control. *Indian J. Ent.*, *14* (3) : 239-242, New Delhi (108)

BHATIA, L.J., 1967 : Chemicals that take care of mango hoppers. *Indian Fmg.*, *16* (10) : 16, New Delhi (12)

BHATTACHARJEE, N.S., 1959 : Studies on *Lygaeus pandurus* Scopoli (Heteroptera : Lygaedidae) — 1. Bionomics, description of the various stages, biology and cantrol. *Indian J. Ent.*, *21* (4) : 259-272, New Delhi (191)

BINDRA, O.S., 1959 ; Mode of carry over in singhara beetle (*Galerucella birmanica* Jacoby). *Indian J. Hort.,16*, (2) : 113-116, Bangalore (106)

BINDRA, O.S., 1966 : Role of insects and other animals in relation to citrus decline. *Punjab Hort. J.,6* : 108-116, Patiala (130)

BINDRA, O.S. 1967 : Fighting pests of commercial fruits *Indian Hort.*, *11* (4) : 71-74, New Delhi (137)

BINDRA, O.S., B.S. SOHI AND R.C. BATRA, 1973 : Chemical control of chafer-beetles on grapevines. *Indian J. Ent.*, *35* (3) : 277-279, New Delhi (162, 163)

BINDRA, O.S., G.C. VARMA, G.S. SANDHA AND H.K. CHHABRA, 1970: Studies on the control of citrus pests by soil application of systemic insecticides. *J.Res. PAU.*, 7 (2) : 197-202, Ludhiana (131)

BINGHAM, C.T., 1907 : *The Fauna of British India including Ceylon and Burma. Butterflies*, 2 : 417-419, Taylor & Francis, London (182)

BIRAT, R.B.S., 1963 : Comparative efficacies of organophosphorous insecticides for the control of *Drosicha mangiferae* Green (Homoptera : Coccidae). *Indian J. Ent.*, 25 (4) : 377-379, New Delhi (14)

BODENHEIMER, F.S., 1951 : *Citrus Entomology in the Middle East*, 663 pp. Dr. W. Junk, Publishers, The Hague (The Netherlands) (135,136,138)

BOGAWAT, JATAN KUMARI, C.L. KAUL AND B.K. SRIVASTAVA, 1969 : Susceptibility of two species of pumpkin beetles to some insecticides. *Indian J. Ent.*, 31 (2) : 170-172, New Delhi (247)

BOURNIER, ALEXANDRE, 1977 : Grape insects. *Ann. Rev. Entomol.*, 22 : 355-376, California (159)

BRAHMACHARI, K., 1938 : On the bionomics of a bagworm (*Kophene cuprea* M.) on banana. *J. Bombay nat. Hist. Soc.*, 40 (1) :56-61, Bombay (45)

BRANYOVITS, F., 1953: Some aspects of the Biology of armoured scale insects. *Endeavour*, 12 :202-209, London (137)

BUTANI, DHAMO K., 1962 : Save your mango crop from insect pests. *Indian Hort.*, 6 (2) : 6-7 & 19, New Delhi (29)

BUTANI, DHAMO K., 1964 : Pests that pester your mango crop. *Indian Hort.*, 8 (1) : 28, New Delhi

BUTANI, DHAMO K., 1973 a: Insect pests of fruit crops and their control-1 : Ber. *Pesticides*, 7 (7) : 33-35, Bombay (158,177)

BUTANI, DHAMO K., 1978 b : Insect pests of fruit crops and their control-2 : Banana. *Pesticides*, 7 (8) : 29-31, Bombay (42)

BUTANI, DHAMO K., 1973 c : Insect pests of fruit crops and their control-3 : Apple . *Pesticides*, 7 (10) : 13-17, Bombay

BUTANI, DHAMO K., 1973 d : Insect pests of fruit crops - 4 : Citrus. *Pesticides, 7* (12) : 23-27, Bombay (128,131,157)

BUTANI, DHAMO K., 1973 e : Les ravageurs et les maladies des Citrus en Inde. *Fruits, 28* (12) : 851-856, Paris (157)

BUTANI, DHAMO K., 1974 a : Insect pests of fruit crops and their control-5 : Cherry. *Pesticides*, 8 2) : 33, Bombay (317)

BUTANI, DHAMO K., 1974 b : Insect pests of fruit crops and their control-6 : Water nut (Singhara). *Pesticides*, 8 (2) : 34, Bombay

BUTANI, DHAMO K., 1974 c : Les ravageurs de la vigne en Inde. *Fruits, 29* (2) : 149-152, Paris

BUTANI, DHAMO K., 1974 d : Insect pests of fruit crops and their control-7 : Mango. *Pesticides*, 8 (3) : 37-41, Bombay (12,14)

BUTANI· DHAMO K,, 1974 e : Insect pests of fruit crops and their control-8 : Almond & Apricot. *Pesticides, 8* (8) : 28-30, Bombay (321)

BUTANI, DHAMO K,, 1974 f : Insect pests of fruit crops and their control-9 : Peach· *Pesticides, 8* (9) : 25-31, Bombay (287.300)

BUTANI, DHAMO K., 1974 g : Insect pests of fruit crops and their control-10 : Grapes. *Pesticides, 8* (10) : 25-29, Bombay (161)

BUTANI, DHAMO K., 1974 h : Les insectes parasites du palmier-dattier en Inde et leur controle. *Fruits, 29* (10) : 689-691, Paris (91)

BUTANI, DHAMO K·, 1974 i : Insect pests of fruit crops and their control-11 : Guava. *Pesticides, 8* (11) : 26-30, Bombay

BUTANI, DHAMO K., 1974 j : Pests of fruit crops in India and their control-12: Loquat· *Pesticides. 8* (12) : 17-18, Bombay (199)

BUTANI, DHAMO K., 1974 k : Pests damaging roses in India and their control. *Pesticides, 8* (12) : 40-42, Bombay (200)

BUTANI DHAMO K., 1975 a : Insect pests of fruit crops and their control-13 : Pineapple . *Pesticides, 9* (1) : 21-22, Bombay

BUTANI, DHAMO K., 1975 b : Insct pests of fruit crops and their control-14 : Papaya. *Pesticides, 9* (2) : 23-24, Bombay

BUTANI, DHAMO K., 1975 c : Parasites et maladies du manguier en Inde. *Fruits, 30* (2) : 91-101, Paris (12,33)

BUTANI. DHAMO K·, 1975 d : Insect pests of fruit crops and their cnotrol-15 : Date palm· *Pesticides, 9* (3) : 40-42, Bombay

BUTANI, DHAMO K·, 1975 e : Insect pests of fruit crops and their control - 16 : Fig· *Pesticides, 9* (11) : 32-36, Bombay (232)

BUTANI, DHAMO K., 1975 f : Insect pests of fruit crops and their control - 17 : Sapota. *Pesticides, 9* (11) : 37-39, Bombay (89)

BUTANI, DHAMO K., 1975 g : Insect pests of fruit crops and their control - 18 : Melons. *Pesticides, 9* (12) : 39-43, Bombay (252)

BUTANI, DHAMO K., 1976 a : Insect pests of fruit crops and their control - 20 : Custard apple. *Pesticides : 10* (5) : 27-29, Bombay (72)

BUTANI, DHAMO K·, 1976 b : Insect pests of fruit crops and their control - 21 : Pomegranate. *Pesticides, 10* (6) : 23-26, Bombay

BUTANI, DHAMO K·, 1976 c : Insect pests of fruit crops and their control - 22 : *Jamun* (Black plum)· *Pesticides, 10* (9) : 33-34, Bombay (214,218)

BUTANI, DHAMO K·, 1976 d : Insect pests of fruit crops and their control - 23 : *Falsa. Pesticides, 10* (10) : 25-26, Bombay (242)

BUTANI DHAMO K. 1977 a : *Indarbela quadrinotata* (Wlk.). In : *Diseases, Pests and Weeds in Tropical Crops* : 438-439, Verlag, Paul Parey, Berlin (151,181)

BUTANI, DHAMO K·, 1977 b : Insect pests of guava in India and their control. *Fruits, 32* (1) : 61-66, Paris

BUTANI, DHAMO K·, 1977 c : Pests of litchi in India and their control· *Fruits, 32* (4) : 269-273, Paris (191)

BUTANI, DHAMO K., 1977 d : Insect pests of water nut in India and their control. *Fruits, 32* (9*)* : 569-571, Paris

BUTANI, DHAMO K , 1977 e : Insect pests of fruits crops and their control-24 : Litchi. *Pestisides, 11* (2) : 43-48, Bombay

BUTANI, DHAMO K., 1978 a : Insect pests of *falsa* (*Grewia asitatica* Mast.) in India and their control. *Fruits, 33,* (2) : 121-124, Paris

BUTANI DHAMO, K:. 1978 b : Pests and diseases of jackfruit in India and their control. *Fruits, 33* (5) : 351-367, Paris (96)

BUTANI, DHAMO K., 1978 c : Insect pests of fruit crops and their control-25 : Mulberry. *Pesticides, 12* (8) : 53-59, Bombay

BUTANI, DHAMO K., 1978 d : Insect pests of tamarind and their control. *Pesticides, 12* (11) : 34-41, Bombay, 41 (111)

BUTANI, DHAMO K. AND PREM NIDI BAJPAI, 1965 : The destructive borers of fruit crops. *Indian Hort., 9* (8) : 31, New Delhi

BUTANI, DHAMO K. AND M. G. JOTWANI, 1975 : Trends in the control of insect pests of fruit crops in India. *Pesticides Annual 1975* : 139-149, Bombay

BUTANI, DHAMO K., AND B.D. TAHILIANI, 1974 : Red ants on mango in South Gujarat. *Entomologists' Newsletter, 4* (7) : 37-38, New Dehli (28, 105)

CALCAT, A., 1959 : Diseases and pests of datepalm in Sahara and North Africa. *FAO Pl. Prot. Bull. 8* : 5-10, Rome (207)

CAPOOR, S P., D G. RAO, AND S. M. WISWANATH, 1967 : *Diaphorina citri* Kuwayama a vector of the greening disease of citrus in India. *Indian J. agric. Sci., 37* (6) :572-576, New Delhi (130)

CAPOOR, S.P. ND P.M. VARMA, 1948 : A mosaic disease of *Carica papaya* L. in the Bombay Province. *Curr. Sci. 17* (9) : 265-266, Bangalore (78)

CHARI, M.S. S.N. SERHADRI AND H.K. PATEL, 1969 : Control of mango hoppers. *Pesticides, 3* (12) : 33-35, Bombay (12)

CHEEMA G.S., S.S. BHAT AND K.C. NAIK, 1954: *Commercial Fruits of India,* 422 pp. Macmillan & Co. Ltd., Calcutta (11,22)

CHERIAN, M.C. AND P. ISRAEL, 1939 : Notes on *Perina nuda* Fabr. (Lyman) and its natural enemies. *Madras agric. J., 26* (6) : 203-207, Madras (102)

CHERIAN, M.C. AND C.V. SUNDARAM, 1939 : Notes on the life-history and habits of *Dacus brevistylus* Bezzi (Family trypetidae) a pest of *Coccinia indica* fruits. *Indian J. agric. Sci.,9* (1) : 127-131, New Delhi (249)

CHOWDHURY'S. AND S. MAJID, 1954 : *Handbook of Plant Protection,* 117 pp. Department of Agriculture, Assam, Shillong (66,95,146,151,183)

CLEGHORN, JAMES, 1914 : Melon culture in Peshin, Baluchistan and some account of the melon-fly pest. *Agric. J. India, 9* (2) : 124-140, Pusa (249)

COLLINS, J.L., 1960 : *The Pineapple,* 294 pp. Leonard Hill (Books) Ltd., New York (82)

346 FRUIT PESTS

CONDIT, IRA J., 1947 : *The Fig*, 222 pp. Chronica Botanica Co., Waltham, Mass, USA (229, 230)

CORBETT, G.H., 1939 : A new species of Aleurodidae from India *(Aleurocanthus husaini* sp. n., on citrus in Punjab). *Indian J. Ent.*, *1* (3) : 69-70, New Delhi (133)

COUDRIET, D.L., 1962 : Efficiency of various insects as vectors of cucumber mosaic and water melon mosaic viruses in Cantaloupe. *J. econ. Ent.*, *55* (4) : 519-520, Wiscounsin (251)

CUILLE, J., 1950 : Recherches sur le charancon du bananier. *Inst. Fr. Agr. Col.*, *Ser. tech.*, *4*, 225 pp (42)

DAVID, B.VASANTHARAJ., 1961 : Some new hosts of *Sylepta lunalis* Guen. in South India. *Indian J. Ent.*, *23* (4) : 301, New Delhi (167)

DAVID, B. VASANTHARAJ AND P.C. SUNDRA BABU, 1964 : Preliminary investigations on the varietal incidence and control of the mango nut weevil, *Cryptorrhynchus mangiferae* (Fb.). *Madras agric. J.*, *51* (2):91, Madras (22)

DAVID, H. AND A.N. KALRA, 1966 : Some observations on *Holotrichia serrata* F., a beetle pest of sugarcane in Hospet area of Mysore State. *Indian Sug.*, 16 (2) : 195-198, Calcutta (65)

DAVID, LEELA A., 1961 : Notes on the biology and habits of the red tree ant *Oecophylla smaragdina* (Fabricius) *Madras agric. J.*, *48* (2) : 54-67 Madras (27, 28).

DAVID, S. KANAKARAJ, 1956 : Additional notes on some aphids in the Madras State. *Madras agric. J.*, *43* (3) : 103-107, Madras (104)

DEOL, I.S., G.S. SANDHU AND GURCHARAN SINGH, 1977 : Field screening of peach varieties against fruit fly. *Punjab Hort. J.*, *17 (3-4)* : 152-154, Patiala (298)

DEV, H.W., 1964 : Preliminary studies on the biology of the Assam thrips, *Scirtothrips dorsalis* Hood on tea. *Indian J. Ent.*, *26* (2) : 184-194, New Delhi (115,164)

DIETZ, H. F. AND J. ZETEK, 1920 : The blackfly of citrus and other subtropical plants. *U.S.A.D. Bull. No. 885*, 55 pp. Washington, D.C. (132)

DUTT, N. AND B.B. MAITI, 1972 a : Bionomics of banana pseudostem weevil *Odoiporus longicollis* Oliv. (Coleoptera : Curculionidae). *Indian J. Ent.*, 34 (1) : 20-30, New Delhi (43)

DUTT, N. AND B.B. MAITI, 1972 b : Studies on the control of banana pseudostem weevil, *Odoiporus longicollis* Oliv. *Indian J. Ent.*, *34 (4)* : 272-289, New Delhi (43)

DUTTA, S, 1966 : *Horticulture in the Eastern Region of India*, 264 pp. Directorate of Extention, Ministry of Food & Agriculture, New Delhi (80,144,146)

EASTOP, V.F., 1977 : *Aphis gossypii* Glov. In : *Diseases, Pests and Weeds in Tropical Crops* : 328-329, Verlag Paul Parey, Berlin (251)

EBELING, WALTER, 1959 : *Sub-tropical Fruit Pests,* University of California Press, Los Angeles (123,138, 147)

ESSIG, E.C., 1949 : Aphids in relation to quick decline and Tristeza of citrus. *Pan-Pacif. Ent., 65* (1) : 13-22, San Francisco (114)

ESTALILLA, H., 1921 : The atis moth borer, *Heterographis bengalensis* Rag. *Philipp. Agric., 10* (4) : 169, Manila (71)

F.A.O., 1970 : Grape culture. *Report to the Government of India. Report No. T.A. 2825,* 99 pp. Rome (159)

FEAKIN, SUSAN D., 1977 : *Pest Control in Bananas.* PANS Manual No. 1, 126 pp. Centre for Overseas Pest Research, London (46,52)

FLETCHER, BAINBRIGGE T., 1914 : *Some South Indian Insects,* 565 pp. Superintendent Government Press, Madras (31,34,95)

FLETCHER, BAINBRIGGE T., 1917 : Fruit-trees. *Rept. Proc. 2nd Ent. Mtg. Pusa (Bihar), February 1917* : 209-257, Calcutta (9,43,94,278)

FLETCHER, BAINBRIGGE T., 1920 a : Annotated list of Indian crop pests. *Rept. Proc. 3rd Ent. Mtg., Pusa (Bihar), February 1919, 1* : 313-314, Calcutta (67,108)

FLETCHER, BAINBRIGGE T., 1920 b : Life-histories of Indian insects. Micro-lepidoptera : VI — Gracillaridae. *Mem. Dept. Agric. India, 6* (6) : 137-167, Calcutta (92)

FLETCHER, BAINBRIGGE T., 1921 : Annotated list of Indian crop pests. *Agricultural Research Institute, Pusa, Bull. No. 100,* 246 pp. Calcutta (38,81, 123,330)

FLETCHER, BAINBRIGGE T., 1932 : Life-histories of Indian microlepidoptera (II series) - Alucitidae (Pterophoridae), Tortricina and Gelechidae. *Imperial Council of Agricultural Research Sci. Monog. No. 2,* 68, pp., New Delhi (294,318)

FOTIDAR, M.R. AND A P. KAPUR, 1941 : *Aspidiotus destructor* a pest of mango at Jammu (Kashmir). *Indian J. Ent., 3* (1) : 142, New Delhi (19)

FOTIDAR, M R. AND J.L. RANIA, 1936 : *San Jose Scale in Kashmir and its Control,* 8 pp Karachi (261)

FROHLICH, G. AND W. RODEWALD, 1970 : *Pests and Diseases of Tropical Crops and Their Central,* 371 pp. Pergamon Press, Oxford (42)

GANDHI, S.R., 1955 : *The Mango in India.* Farm Bull. No. 6, Indian Council of Agricultural Research, New Delhi (21)

GANGOLLY, S.R., RANJIT SINGH, S.L. KATYAL AND DALJIT SINGH, 1957 : *The Mango,* 530 pp. Indian Council of Agricultural Research, New Delhi (11,23)

GARCIA, C.E., 1935 : A field study on the citrus green bug *Phynchocoris serratus* Donovan. *Philipp. J. Agric., 6* (3) : 311-325, Manila (27)

GARCIA, C.E., 1939 : The citrus rind borer and its control. *Philipp. J. Agric., 10* (1) : 89-92, Manila (147)

GHAI, SWARAJ, 1976 : Phytophagous mites and their control in India. *Proc. Nat. Acad. Sci*, *India, 46-B* (1-2) : 49-59, Allahabad (189)

GHOSH, C.C., 1914 : The lemon caterpillars *Papilio demoleus* Linn. and *Papilio pammon* Linn. Life-histories of Indian insects V — Lepidoptera (Butterflies). *Mem. Dept. Agric. India, Ent.*, *5* (1) : 33-52, Calcutta (125)

GHOSH, C.C., 1924 : A few insects used as food in Burma *Proc. Rept. 5th Ent. Mtg., Pusa (Bihar), February 1923* : 403-405, Calcutta (203,206)

GOLDING, F.D., 1945 : Fruit piercing Lepidoptera in Nigeria. *Bull. ent. Res.*, *36* (2) : 181-184, London (66)

GREEN, ERNEST E., 1903 : Remarks on Indian scale insects (Coccidae) with descriptions of new species. *Indian Mus. Notes, 5* (3) : 93-103, Calcutta (13,14)

GROVER, P., 1965 : Studies on Indian gall midges (Diptera : Cecidomyiidae) : XII - The mango blossom midge, *Dasyneura amaramanjirae* n. sp. *Ann. ent. Soc. Amer., 58* (2) : 202-206, Baltimore (24)

GUPTA, K.M. AND J.C. JOSHI, 1959 : Prevention of damage to pomegranate fruits by borer *(Deudorix epijarbas* Moore) in the hills of Uttar Pardesh, *Proc. 46th Indian Sci. Cong., Delhi,* III (10) : 502, Calcutta (221)

GUPTA, R.L., 1958 : Preliminary trials of bait spray for the control of fruit flies in India. *Indian J. Ent., 20* (4) : 304-306, New Delhi (197,248,,235,297)

GUPTA, R.L. AND G.A. GANGRADE, 1955 : The life cycle and seasoanl history of chikoo-moth (*Nephopteryx eugraphella* Rag.), *Indian J. Ent., 17* (3) : 326-336, New Delhi (88)

HAMPSON, G.F., 1892 : *The Fauna of British India including Ceylon and Burma. Moths, 1* : 405-406,460-461, Taylor & Francis, London (183,311)

HAMPSON, G.F., 1894 : *The Fauna of British India including Ceylon and Burma. Moths, 2* : 376,562, Taylor & Francis, London (37,152)

HAMPSON, G.F., 1896 : *The Fauna of British India inclcuding Ceylon and Burma. Moths, 4* : 70,77,356, Taylor & Francis, London (71,87,96)

HAMPSON, G.F., 1912 : Phalaenae in the British Muleum. *Catalogue of the Lepidoptera 11,* : 68, London (37)

HAYES, W.B., 1966 : *Fruit Growing in India,* 512 pp. Kitabistan, Allahabad (69,81,95,146,229)

HILL, DENNIS S., 1975 : *Agricultural Insect Pests of the Tropics and Their Control,* 516 pp. Cambridge University Press, Cambridge (54,61,63,77, 80,83,103,140;143, 202,281,290)

HOOD, J.D., 1919 : On some new Thysanoptera from southern India. *Insecutor Inscit. Menstr., 7* (4-6) : 90-103, Washington, D.C. (115)

HUKAM SINGH, 1974 : *Aproderus blandus* Faust· (Coleoptera : Curculionidae), a new pest of litchi (*Litchi chinensis* Sonn·). *Indian J. Ent., 36* (3) : 239-240, New Delhi (194)

HUKAM SINGH, 1975 : *Acrocercops cramerella* Snell· (Gracilleriidae : Lepidoptera) as a pest of litchi in Uttar Pardesh and its control· *Indian J. Hort., 32* (2 - 4) : 152-153, Bangalore (192)

HUKAM SINGH, 1978 : Pest complex of litchi in Dehradun and Saharanpur districts of Uttar Pardesh· *Indian J. Ent., 40* (4) : 464, New Delhi (191)

HUSAIN, M· AFZAL AND M· ABDUL WAHID KHAN, 1940 : Bionomics and control of fig tree borer, *Batocera refomaculata* De Geer (Coleoptera : Lamiidae). *Indian J. agric. Sci., 10* (6) : 945-959, New Delhi (230)

HUSAIN, M· AFZAL AND ABDUL WAHID KHAN, 1945 : The citrus Aleurodidae in the Punjab and their control· *Mem. Ent. Soc. India, 1* : 44, New Delhi (134)

HUSAIN, M· AFZAL AND M· ABDUL WAHID KHAN, 1949 : Bionomics and control of the walnut weevil (*Alcides porrectirostris* Mshll·) (Alcidinae : Coleoptera). *Indian J. Ent., 11* (1) : 77-82, New Delhi (328)

HUSAIN, M· AFZAL AND DINA NATH, 1927 : The citrus psylla *Diaphorina citri* Kuw. (Psyllidae : Homoptera). *Mem. Dept. Agric. India, 10* (2) : 5-27, New Delhi (131)

HUSAIN, M· AFZAL AND HEM SINGH PRUTHI, 1921 : Preliminary note on winter spraying against mango hopper (*Idiocerus* spp·), vernacular name *tela. Rept. Proc. 4th Ent. Mtg., Pusa (Bihar), February 1921* : 148-152, Calcutta (11)

HUSAIN, MASOON, 1974 : Some observations on the biology and control of (*Phyllophaga consanguinea* Blanch·), a potent pest of groundnut in Andhra Pradesh· *Indian J. Pl. Prot., 2* (1 & 2) : 107-110, Hyderabad (94)

HUTSON, J.C· AND M·P·D· PINTO, 1934 : Two caterpillar pests of citrus. *Trop.Agric., 83* (3) : 180-183, Paradeniya (128)

IPERTI, G·, 1977 : *Parlatoria blachardii* Targ. In : *Diseases, Pests and Weeds in Tropical Crops* : 359-361, Verlag Paul Parey, Berlin (208)

JACOBY, M·, 1889 : List of the Phytophagus Coleoptera by Sig· L· Fea at Burmah and Tenasserin with description of new species· *Ann. Mus. Stov. nat., 27* : 147-273, Genova (106)

JANJUA, NAZEER AHMED, 1940 : On the biology of *Cacoecia sarcostega* Meyr· in Baluchistan· *Indian J. Ent., 2* (2) : 145-154, New Delhi (272,273,315)

JANJUA, NAZEER AHMED, 1942 : On the biology of *Anarsia lineatella* Zeller in Baluchistan· *Indian J. Ent., 4* (2) : 137-144, New Delhi (295)

JANJUA, NAZEER AHMAD, 1947 : The biology of hairy caterpillar (*Euproctis signata* Blanchard) in Baluchistan. *Indian J. Ent., 9* (1) : 159-166, New Delhi (275)

JANJUA, NAZEER AHMED, 1948 : The biology of *Dacus* (*Strumeta*) *ferrugineus* (Fab·) (Trypetidae : Diptera) in Baluchistan. *Indian J. Ent., 10* (1) : 55-61, New Delhi (20)

JANJUA, NAZEER AHMED AND C.K. SAMUEL, 1941 : *Fruit Pests of Baluchistan.* *Misc. Bull. No. 42*, 41 pp. Imperial Council of Agricultural Research, New Delhi (294)

JARAMILLO, ROBERTO, 1946 : La Pina. *Universidad de Antiquia,19* : 441-447 (82)

JARVIS, E., 1914 : A new fruit-boring caterpillar of banana occurring at Tweed Heads, *Heteromicta latro. Queensland Agric. J., n. s., 1* : 280-284, Brisbane (54)

JAVARAYA, H.C., 1943 : Bi-annual cropping of apple in Bangalore. *Indian J. Hort., 1* (1) : 31-34, Bangalore (263)

JAYARAJ, S., A. ABDUL KAREEM AND P.P. VASUDEVA MENON, 1960 : *Myllocerus viridanus* F. (Curculionidae : Coleoptera) a new pest of papaya in South India. *Curr. Sci., 29* (10) : 413, Bangalore (81)

JOHNSON, J., 1975 : Note on the biology of *Acanthopsyche mimima* Hamp. (Psychidae) a pest of plaintain. *Agric. Res. J Kerala, 12* (2) : 198-199, Vellayani, Trivendrum (45)

JOSEPH, K.V. AND C.N. OOMMEN, 1960 : Notes on some insect pests infesting dry tamarind fruits in Kerala State. *Indian J, Ent., 22* (3) : 172-180, New Delhi (110,120)

JOSHI, R.K., C.L.KAUL AND B.P. SRIVASTAVA, 1969 : Studies on chemical control of white grub *(Holotrichia consanguinea* Bl.). *Indian J. Ent., 31* (2) : 143-148, New Delhi (65)

JOTWANI, M.G. AND PRAKASH SARUP, 1966 : Relative toxicity of different pesticides to the adults of singhara beetle, *Galerucella birmanica* Jacoby (Coleoptera : Chrysomelidae). *Indian J. Ent., 28* (2) : 210-214, New Delhi (108)

KADAM, M.V. AND PATEL, G.A., 1957 : Pests of crucifers, chillies, onions, cucurbits and bhendi. In : *Crop Pests and How to Fight Them* : 91-100. Directorate of Publicity, Government of Bombay, Bombay (246)

KADYAN, A.S., S.N. KAUSHIK AND D.S. GUPTA, 1971 : Phytotoxicity of some insecticides to musk melons. *Indian J. Ent., 33* (4) : 463-465, New Delhi (247)

KAIFER, H.H., 1943 : Eriophid studies - 13. *Calif. Dept. Agric.Bull. 32* (3) : 212, California (189)

KALRA, A.N. AND J.P. KULSHRESHTHA, 1961 : Studies on the biology and control of *Lachnosterna consanguinea* (Blanch.), a pest of sugarcane in Bihar. *Bull. ent. Res., 52* (3) : 577-587, London (177)

KAPOOR, V.C., 1970 : Indian Tephritidae with their recorded hosts. *Oriental Insects, 4* (2) : 207-251, New Delhi (89,297)

KAPOOR, V.C., 1972 : Identification of common fruit flies (Tephritidae : Diptera) of India. *Ent. Rec., 84* : 165-169, London (21,150,227)

KAPUR, A.P. AND M.R. FOTIDAR, 1943 : First record of the sexual form and oviparous reproduction of wooly aphid, *Eriosoma lanigerum* Hausm. from Kashmir, India. *Curr. Sci., 12* (3) : 84-85, Bangalore (263)

KHAN, A.W., 1955 : Studies on stocks immune to wooly aphis of apple (*E. lonigerum* Hausmn.). *Punjab Fr. J., 19 : 28-35*, Lahore (263)

KHAN, K.M. AND GHAI, SWARAJ, 1974 : White grubs and their control in India. *Pesticides, 8* (12) : 19-25, Bombay (242)

KHANNA, S.S., 1952 : Biology of *Deporaus marginatus* Pasc. (Curculionidae Coleoptera). *Proc. nat. Acad. Sci. India, B-22* (1-5) : 72-80, Bangalore (32)

KHATIB, MAHMOOD HASAN, 1934 : The life-history and biology of *Galerucella birmanica* Jac. (Coleoptera, Phytophaga, Chrysomelidae, Galerucinae) and external morphology of larva and pupa. Part - I. *Indian J. agric. Sci., 4* (4) : 715-723, New Delhi (106,107)

KHURANA, A.D. AND O.P. GUPTA, 1972 :Bark-eating caterpillars pose serious threat to fruit trees. *Indian Farmers' Digest, 5* (4) : 51-52, Pantnagar (16, 188)

KODAMA,G., 1931 : Studies on *Aleurocanthis spiniferus* Quaint. (In Japanese). *Kagoshima-Ken, Kyusha*, 38 pp., Japan (133)

KRISHNAMOORTHY, C. AND K. RAMASUBHIAH, 1962 : Termites affecting cultivated crops in Andhra Pradesh and their control. Retrospect and prospect. In : *Termites in Humid Tropics* : 243-245, UNESCO, Paris (228)

KULKARNY, H.L., 1955 : Incidence of mango flower galls in Bombay Karnatak. *J. Bombay nat. Hist. Soc., 53* (1) : 147-148, Bombay (24)

KUNHIKANNAN, K., 1923 : The lime (citrus) tree borer *Chelidonium cinctum* (Guer.). *J. Mysore agric. exp. Un., 5* : 129-131, Bangalore (146)

KUNHIKANNAN, K., 1928 : The large citrus borer of South India, *Chelidonium cinctum* (Guer.). *Bull. Dept. Agric. Mysore, Ent. Ser., No. 8*, 24 pp., Bangalore (146)

KUSHWAHA, K.S., 1960 : A note on infestation of termites (Insecta : Isoptera) around Udaipur (Rajasthan). *Sci. Cult., 26* (1) : 39-40, Calcutta (228)

LAL, K.B., 1952 : Insect pests of fruit trees grown in the plains of the Uttar Pradesh and their control. *Agric. & Ani. Husb., U.P., 3* (1-3) : 54-80, Allahabad (220)

LAL, K.B. AND ZAHID ALI SIDDIQI, 1952 : Biology of peach curling aphid on the Kumaun hills. *Indian J. Ent., 14* (3) : 191-196, New Delhi (287)

LAL, K.B. AND R.N. SINGH, 1947 : Seasonal history and field ecology of the wooly aphis in the Kumaun hills. *Indian J. agric. Sci., 17* (4) : 211-218, New Delhi (262,263)

LALL, B.S. AND S.N. SINHA, 1959 : On the biology of the melonfly, *Dacus cucurbitae* (Coq.) (Diptera : Trypetidae). *Sci. Cult., 25* (2) : 159-161, Calcutta (249)

LASZIO, HENRY DE AND PAUL S. HENSHAW, 1954 : Plant material used by primitive people to affect fertility. *Science, 119* : 625-631, Lancaster, Pa. (32)

LATIF, A., 1949 : The taxonomic status of *Drosicha stebbingi* (Green) and *Drosicha mangiferae* (Green) (Hem· Coccid·). *Bull. ent. Res., 40* (3) : 351-354, London (14)

LATIF, ABDUL AND CH· MUHAMMAD YUNAS, 1951 : Food plants of citrus leaf miner (*Phyllocnistis citrella* Stn.) in the Punjab. *Bull· ent. Res., 42* (2) : 311-316, London (128)

LEEFMAN, S., 1920 : De palmsnuitkever *Rhynehophorns ferrugieus* F· *Inst. plantenziekten, Meded, 43* : 1-90, The Netherlands (206).

LEFROY, H. MAXWELL, 1910 : Life-history of Indian insects. Coleoptera-I· *Mem. Dept. Agric. India, Ent. Ser., 2* (8) : 139-164, Calcutta (106)

LEFROY, H. MAXWELL AND F.W· HOWLETT, 1909 : *Indian Insect Life - Manual of the Insects of the Plains*, 786 pp. Thacker Spink & Co·, Calcutta (37,81,108)

LEONARD, M.D., 1931 : A bibiliography of the banana root weevil, *Cosmopolites sordidus* Germar· *J. Dept. Agric. Puerto Rico, 15* : 147-176, San Juan (41)

LETHIERRY, M.L·, 1889 : Definitions of three new Hymenoptera. *J. Asiatic Soc. Bengal, 58* : 252-253, Calcutta (10)

LINFORD, M.B·, 1932 : Transmission of the pineapple yellow spot virus by *Thrips tabaci*. *Phytopath., 22* : 301-324, Lancaster, Pa. (84)

LINNAEUS, CAROLUS, 1758 : *Systema Naturae, 10 Ed·, 1 :* 346, 464 Laurentil Salvii, Holmige (124, 202)

LUH, NIEN-TSIN, 1936 : A Survey on the falling off of citrus fruit due to insect pests in Hangyuen during 1935. *Ent. Phytopathol., 4* (6) : 102-107, Hangchow (149)

MADHAVA RAO, V.N., 1965 : The Jackfruit in India, *Farm. Bull. (New Series), No. 34,* 18 pp, Indian Council of Agricultural Research, New Delhi (96)

MALDONADA - COPRILES, J., 1964 : Studies on Idiocerinae leaf hoppers : II - The Indian and Philippine species of *Idiocerus* and the genus *Idioscopus* (Homoptera : Cicadellidae). *Proc. ent. Soc. Wash·, 66* (2) : 89-100, Washington,D.C. (10)

MANGAT, B S·, 1966 : Biology and control of citrus psylla, *Diaphorina citri* Kuw. *Pl. Prot. Bull., 18* (3) : 18-20, New Delhi (131)

MARSHALL, ROY E., 1954 : *Cherries and Cherry Products*, 283 pp. Interscience Publishers Inc·, New York (316)

MAULIK, S·, 1919 : *The Fauna of British India -* Coleoptera : Chrysomelidae (Hispinae and Cassidinae), 439 pp. Taylor & Francis, London (178)

MAULIK, S·, 1926 : *The Fauna of British India -* Coleoptera (Chrysomelinae and Halticinae) : 422-423. Taylor & Francis, London (108)

MAULIK, S·, 1936 : *The Fauna of British India -* Coleoptera (Galerucinae) : 32-34 & 214-222· Taylor & Francis, London (106)

MAY, A.W·S., 1946 : Pests of cucurbit crops. *Qd. agric. J., 62* (3) 137-150, Brisbane (254)

MC BRIDE, O.C. AND A.C. MASON, 1934 : The effect of sub freezing temperatures on the mango weevil. *J. econ. Ent.*, *27* : 902, Wiscounsin (22)

MEHTA, P.R. AND B.K. VARMA, 1968 : *Plant Protection*, 587 pp. Directorate of Extention, Ministry of Food, Agriculture, Community Development and Cooperative, New Delhi (57,108)

MENON, K.P.V. AND K.M. PANDALAI, 1958 : *The Coconut Palam — A Monograph* : 251-292. Indian Central Coconut Committee, Erankulum (206)

MEYRICK, EDWARD, 1924 : *Exotic Microlepidoptera, 3* : 107. Taylor & Francis, London (272)

MISHRA S.C. AND N.D. PANDEY, 1965 : Studies on bionomics of *Argyroploce leucaspis* Meyr. (Eucosmidae : Lepidoptera). *Labdev J. Sci. Tech.*, *3* (2) : 126-127, Kanpur (193)

MISRA, C.S., 1912 : Litchi leaf curl. *Agric. J. India, 7* : 286-293, Pusa (Bihar) (189,190)

MISRA, C.S., 1920 : Index to Indian fruit-pests. *Rept. Proc. 3rd Ent. Mtg., Pusa (Bihar), February 1919, 2* : 564-595, Calcutta (87,262)

MISRA, C.S., 1924 : A'list of Coccidae in the Pusa collection. *Rept. Proc. 5th Ent. Mtg., Pusa (Bihar), February 1923* : 345-351, Calcutta (19)

MITRI, TALAAT K. AND LEWIS J. STANNARD (Jr.)., 1962 : *Chaetanaphothrips clarus* (Moulton), a new combination, with notes on its genus (Thysanoptera : Thripidae). *Ann. Ent. Soc. Amer , 55* (1) : 383-386, Colombus, O.(52)

MOOKHERJEE, P.B. AND S.R. WADHI, 1956 : Further studies on the phytotoxicity of modern organic inseeticides. *Indian J. Ent.*, *18* (4) : 332-335, New Delhi (247)

MOUND, L.A., 1968 : A review of R.S. Begnall's Thysanoptera collection. *Bull. Br. Mus. Nat. Hist. Ent. Suppl., 11 :* 161 pp., London (91)

MOUND L.A., 1977 : *Thrips tobaci* Lind. In : *Diseases, Pests and Weeds in Tropical Crops* : 280-282, Verlag Paul Parey, Berlin, (84)

MUKHERJEE, T.D., 1941 : *Arcyophora dentula* Hamps., a new pest of pomegranate. *Indian J. Ent.*, *3* (2) : 337, New Delhi (226)

MURTHY, D.V., 1958 : A short note on the banana leaf thrip *Caliothrips (Heliothrips) kadaliphila* Ramkr. M. *Mysore agric. J.*, *33* (1) : 2-4, Bangalore (51)

MUTHUKRISHNAN, T.S., K.R. NAGARAJA, T.R. SUBRAMANIAN, I.P. JANAKI AND E.V. ABRAHAM, 1958 : Brief notes on a few crop pests noted for the first time in Madras. *Madras agric. J.*, *45* (9) : 363-364, Madras (102)

NAGARAJA RAO, K.R., 1955 : Farmers try new way of controlling chilli thrip. *Indian Fmg., 5 (2)* : 7, New Delhi (116)

N AGARAJA RAO, K.R., 1959 : Studies on phytotoxicity of certain systemic pesticides. *Indian J. Ent.*, *21* (2) : 89-92, New Delhi (247)

NAGARAJA RAO, K.R. AND T.R. SUBRAMANIAM, 1958 : *Glyphodes bivitralis* Guen. as a new pest of jack. *Indian J. agric. Sci.*, **28** (4) : 561-562, New Delhi (102)

NAIR, M.R.G.K., 1975 : *Insects and Mites of Crops in India*, 404 pp. Indian Council of Agricultural Research, New Delhi (38,43,45, 54, 67, 98, 99, 100, 113, 118, 133, 147, 162, 173, 185, 209, 212, 253, 262)

NARAYANAN, E.S., 1953 : The red pumpkin beetle and its control. *Indian Fmg .*, **3** (2) : 8-9, New Delhi (247)

NARAYANAN, E.S., 1954 : Seasonal pests of crops : the anar butterfly, *Virachola isocrates* Fabr. *Indian Fmg.*, **4** (1) : 8-9, New Delhi (220)

NARAYANAN, E.S. AND BATRA, H.N., 1960 : *Fruit flies and Their Control*, 68 pp. Indian Council of Agricultural Research, New Delhi (20,80,234,248)

NAYAR, K.K., T.N. ANANTHAKRISHNAN AND B.V. DAVID, 1976 : *General and Applied Entomology*, 589 pp. Tata Mc Graw-Hill Publishing Co. Ltd., New Delhi (9, 10, 38, 70, 89, 114, 123, 129, 175, 194, 235, 247, 281, 285)

NEHRU, JAWAHARLAL, 1946 : *The Discovery of India*, 498 pp. Signet Press, Calcutta (3)

NIRULA, K.K., 1955 : Investigations on the pests of coconut palm-Part II-*Oryctes rhinoceros* L. *Indian Cocon. J.*, **8** (4) : 161-180, Ernakulam (204)

NIRULA, K.K., J. ANTONY AND K.P.V. MENON, 1951 : Investigations on the pests of the coconut palm. The coconut caterpillar, *Nephantis serinopa* Meyr. control with DDT spray. *Indian Cocon. J.*, **4** : 225-234, Ernakulam (209)

NIRULA, K.K., J. ANTONY AND K.P.V. MENON, 1953 : Some investigations on the control of termites. *Indian Cocon. J.*, **7** (1) : 26-33, Ernakulam (209)

NISHIDA, T. AND G.F. HOLDAWAY, 1955 : The erinose mite of Lychee. *Hawaii Agric. Expt. Sta.*, *Circ. No. 48*, 10 pp., Honolulu (190)

OOMMEN, C.N., 1962 : Studies on *Phycita orthoclina* Meyr. (Lepidoptera : Pyralidae) a new pest of tamarind in Kerala. *Indian J. Ent.*, **24** (3) : 188-190, New Delhi (119)

OVIEDO, G.F. DE, 1520 : *History of the Indies*. Book 7, Chapter 14 (A hand written manuscript in the Huntington Library, No. HM 117), San Marino, California (82)

PAGDEN, H.T., 1931 : Two citrus fruit borers. *Malay States Dept. Sci. Ser.*, **7**, 29 pp., Kuala Lumpur (157)

PALO, M.A., 1932 : Anthracnose and important insect pests of mango in the Philippines with a report on blossom spraying experiments. *Philipp. J. Sci.*, **48** (2):209-235, Manila (37)

PANDEY, N.D. AND Y.D. PANDEY, 1964 : Bionomics of *Phyllocnistis citrella* Stt. (Lepidoptera : Gracillariidae). *Indian J. Ent.*, **26** (4) : 417-422, New Delhi (128)

PANDEY, N.D., T.R. SUKHANI AND R.L. GUPTA, 1966 : Bionomics of *Achaea janata* Linnaeus (Lepidoptera : Noctuidae). *Labdev J. Sci. Tech.*, 5 (2) : 127-128, Kanpur (224)

PARSHAD BALDEV, 1960 : On a new aquatic weevil from India *Bagous trapae* n. sp. (Curculionidae : Coleoptera). *Indian J. Ent.*, 22 (4) : 298, New Delhi (109)

PATEL, R.C., 1977 : *Spodoptera litura* (F.). In : *Diseases, Pests and Weeds in Tropical Crops* : 503-505, Verlag Paul Parey, Berlin (47)

PATEL, R.C. AND H.L. KULKARNY 1956 : Bionomics of the pumpkin caterpillar *Margaronia india* Saund. (Pyralidae : Lepidoptera). *J. Bom. nat. Hist. Soc.*, 54 (1) : 118-127, Bombay (254).

PATEL, V.C., H.K. PATEL AND R.C. PATEL, 1964 : Effect of various insecticides on the bark[borer : (*Indarbela* sp.) of road side trees in Gujarat. *Indian J. Ent.*, 26 (1) : 103-106, New Delhi (16, 58)

POPENOE, PAUL B., 1913 : *Date Growing in the Old and New World*, 316 pp. West India Gardens, Altadena, California (202)

PRADHAN, S. 1969 : Integrated pest control - Philosophy and feasibility. *Pesticides*, 3 (4) :15-24, Bombay (261)

PRADHAN, S., 1969 : *Insect Pests of Crops*, 208 pp. National Book Trust India, New Delhi (281)

PRADHAN, S., M.G. JOTWANI AND B.K. RAI, 1958 : Bioassay of insecticides VIII. Relative toxicity of some insecticides to red pumpkin beetle, *Aulacophora foveicollis* Lucas (Coleoptera : Chrysomelidae). *Indian J. Ent.*, 20 (2) : 104-107, New Delhi (247)

PRADHAN, S., M.G. JOTWANI AND PRAKASH SARUP, 1963 : Failure of BHC and DDT to control singhara beetle, *Galerucella birmanica* Jacoby (Chrysomelidae : Coleoptera). *Indian J. Ent.*, 25 (2) : 176-179, New Delhi (108)

PRADHAN,S., M.G. JOTWANI AND PRAKASH SARUP, 1964 : Save your singhara crop. *Indian Fmg.*, 14 (8) : 33-34, New Delhi (108)

PRADHAN, S. AND S.R. WADHI, 1962 : Quarantine problems facing the introduction of mango. *Bull. nat. Inst. Sci. India*, 19 : 106-118, New Delhi (10)

PRASAD, D., 1957 : On the distribution, bionomics and control of mango shoot gall psyllid, *Apsylla cistellata* Buckton. *Indian J. Ent.*, 19 (2) : 78-83, New Delhi (29)

PRASAD, S.N., 1970 : Immature stages of *Erosomica indica*. *Cecidologia indica*, 6 (1 & 2) : 66-76, Allahabad (24)

PRASAD, S.N., 1972 : On the control of mango midge pests. *Cecidologia indica*, 7 (2) : 51-63, Allahabad (24)

PRASAD, S.N., 1974 : Mango midge pests. *Horticulture in Uttar Pradesh* : 55-60, Souvenir, Directorate of Horticulture & Fruit Utilization, U.P., Lucknow (24)

PRUTHI, HEM SINGH, 1969 : *Textbook of Agricultural Entomology,* 977 pp.. Indian Council of Agricultural Research, New Delhi (169, 220, 235,278,282)

PRUTHI, HEM SINGH AND H.N. BATRA, 1938 : *A Preliminary Annotated List of Fruit Pests of North-West Frontier Province. Bull. No. 19.* 22 pp., Imperial Council of Agricultural Research, New Delhi (291)

PRUTHI, HEM SINGH AND H.N. BATRA, 1939 : The phytophagous chalcidoid *Eurytoma samsonovi* Vas., a serious pest of apricot and some other fruits in India. *Indian J. agric. Sci., 9* (2) : 277-283, New Delhi (314)

PRUTHI, HEM SINGH AND H.N. BATRA, 1960 : *Important Fruit Pests of North-West India. Bull. No. 80,* 113 pp., Indian Council of Agricultural Research, New Delhi (34, 39, 67, 80, 107, 137, 161, 182, 249, 262, 288, 291,329)

PRUTHI, HEM SINGH AND M.S. MANI, 1945 : *Our Knowledge of Insect and Mite Pests of Citrus in India and Their Control. Sci. Mongr., No. 16,* 42 pp., Indian Council of Agricultural Research, New Delhi (73, 123, 128, 133, 139)

PRUTHI, HEM SINGH AND L.N. NIGAM, 1939 : The bionomecs, life-history and control of the grasshopper *Poecilocerus Pictus* (Fab.) a new pest of cultivated crops in North India. *Indian J. agric Sci., 9* (4) : 626-641, New Delhi (76)

PRUTHI, HEM SINGH AND V. PRABHAKAR RAO, 1951 : *San Jose Scale in India Bull., No. 71,* 48 pp., Indian Council of Agricultural Researc, New Delhi (261)

PUREWAL, S.S. AND U.S. KANG, 1956 : You Shoud grow singhara, *Indian Fmg., 6* (5) : 7-9, New Delhi (108, 109)

PUTTARUDRIAH, M. AND G.P. CHANNA BASAVANNA, 1959 : A preliminary account of phytophagous mite of Mysore. *Proc. 1st All India Cong. Zool., Jabalpur, 2* : 530-539, Calcutta (189)

QUAYLE, H.J., 1938 : *Insects of Citrus and Other Sub-tropical Fruits.* Comstock Publishing Co. Inc., Ithaca, New York (137,139,149)

RAGHANATH, T.A.V.S. AND DHAMO K. BUTANI, 1977 : *Creatonotus gangis* (Linnaeus) a pest of pomegranate. *Entomologists' Newsletter,* 7(5) : 21, New Delhi (227)

RAHMAN, KHAN A., 941 : Occurrence of the gypsy moth, *Lymantria obfiuscata* Wlk. in Simla hills. *Indian J.Ent., 31* (2) : 338, New Delhi (275)

RAHMAN, KHAN A. AND ATIQUAR RAHMAN ANSARI, 1941 : Scale insects of the Punjab and North-west Frontier Province usually mistaken for San, Jose scale (with description of two new species). *Indian J. agric. Sci 11* (5) : 816-830, New Delhi (19, 260)

RAHMAN, KHAN A. AND ANSHI LAL BATRA, 1944 : The onion thrip (*Thrips tabaci* Lind. : Thripidae : Terebrantia : Thysanoptera) *Indian J. agric. Sci., 14* (4) : 308-310, New Delhi (84)

RAHMAN, KHAN A. AND NAND KISHORE BHARDWAJ, 1937 : The grapevine thrip (*Rhipiphothrips cruentatus* Hood) (Thripidae : Terebra ntia : Thysanoptera). *Indian J. agric. Sci.,* 7 (4) : 633-651, New Delhi (164)

RAHMAN, KHAN A. AND ASA NAND KALRA, 1940 : Volant animals which act as carriers of San Jose Scale. *Curr. Sci., 9* (5) : 235, Bangalore (261)

RAHMAN, KHAN A. AND ASA NAND KALRA, 1943 : Tent caterpillar (*Malocosoma indica* Wlk) in the Simla hills. *Proc. Indian Acad. Sci., B-18* (2) : 41-44, Bangalore (268, 312)

RAHMAN, KHAN A. AND ABDULWAHID KHAN, 1941 : Biology and control of wooly aphis *Eriosoma lanigerum* Hausm (Aphididae : Rhynchota) in the Punjab. *Indian J. agric Sci., 11* (2) : 265 278, 357 New Delhi (262).

RAHMAN, KHAN A. AND ABDUL WAHID KHAN, 1942 a : Bionomics and control of *Aeolesthes holosericea* F. (Cerambycidae : Coleoptera). *Proc. Indian Acad. sci., B-15* (4) : 181-185, Bangalore (317)

RAHMAN, KHAN A. AND ABDUL WAHID KHAN, 1942 b : A study of the life-history and control of *Batocera horsefieldi* Hope (Lamiidae : Coleaptera) - a borer pest of walnut trees in the Punjab. *Proc. Indian Acad. Sci., B-15* (4): 202-205, Bangalore (327)

RAHMAN, KHAN A. AND M. ABDUL LATIF, 1944 : Destruction, bionomics and control of the giant mealy bug, *Drosicha stebbingi* Green. *Bull ent. Res., 35* : 197-209, London (13, 14)

RAI, B.K., H.C. JOSHI, Y.K. RATHOD, S.M. DUTTA AND V.K.R. SHINDE, 1969: Studies on the bionomics and control of white grubs (*Holotrichia consanguinea* Bl.) in Lalsot dist. Jaipur (Rajasthan,). *Indian J. Ent., 31* (2) : 135-142, New Delhi (65)

RAIZADA, USHA, 1965 : Life-history of *Scirtothrips dorsalis* Hood with detailed external morphology of the immature stages. *Bull. ent. Loyolla Coll., 6* : 30, Madras (115, 164)

RANJHEN, P.L., 1949 : On the morphology of the immature stages of *Dacus (Stumeta) cucurbitae* Coq. (the melon fruit fly) with notes on its biology. *Indian J. Ent., 11* (1) : 83-100, New Delhi (249)

RAO, D. SESHAGIRI, 1972 : *A Handbook of Plant Protetion*, 841 pp. S.V. Rangaswamy & Co. Pvt. Ltd., Bangalore (83)

RAO, S.N., 1956 : Notes on biology of gall midges (Honididae : Diptera) from India : X. *Indian J. Ent., 18* (1) : 79-81, New Delhi (23)

RAO, V. PRABHAKAR AND S.N. CHATERJEE, 1948 : The San Jose scale and other scale insects usually mistaken therefor in India. *Indian J. Ent., 10* (1) : 5-29, New Delhi (262, 292)

RAO, Y. RAMACHANDRA, 1930 : The mango hopper problem in South India. *Agric. J. India., 25* (1) : 17-25, Pusa (Bihar) (11)

RATHORE, V.S., V.K. KAUSHIK AND N.K. SOOD, 1970 : Control of cotton thrips : Insecticide trials. *PANS, 16* (2) : 263-264, London (116)

REDDY, D. BAP., 1968 : *Plant Protection in India*, 454 pp. Allied Publishers, Calcutta (50, 78, 106)

REINKING, A.O., 1921 : Notes on coccides and aleurodides on various hosts in Indo-China and Siam, *Philip. Agriculturist, 9* (6-7) : 185, Manila (134)

RICHARDS, O.W. AND R.G. DAVIES, 1977 : *Imms' General Textbook of Entomology*, 1354 pp. Chapman and Hall, London

358 FRUIT PESTS

ROY, R.S., AND CHANDESHWAR SHARMA, 1952 : Diseases and pests of bananas and their control. *Indian J. Hort.*, *9* (4) : 39-52, New Delhi (44, 52, 54)

SAMUEL, C.K., 1940 : Plant hosts of two important aphids, *Myzus persicae* (Sulzer) and *Aphis maidis* (Fitch) at Pusa. *Indian J. Ent.*, *2* (2) : 244-245, New Delhi (289)

SAN JUAN, J.M., 1924 : *Prays citri* Mill., a rind insect pest of Philippines oranges. *Philipp. Agric.*, *12* : 339-348, Los Banos (147)

SANDHU, S.S., 1973 : Chiku can bring profits to Punjab growers. *Punjab Hort. J.*, *13* (1) : 25-29, Ludhiana (92)

SARAIVA, A. COUPTINHO, 1964 : O gorguiho da bananeira - *Cosmopolites sordidus* (Germar) — no arquipelago de Cabo Verde (The banana borer, *Cosmopolites sordidus* in the Cape Verde Islands). *Garcia de Orta, 12* (2) : 241-249, Lisben (42)

SARMA, A.K. AND P.N. SAIKIA, 1967 : The banana beetle can be checked. *Indian Fmg.*, *17* (2) : 18-19, New Delhi (44)

SATHIANANDAN, V.K., R. BETTAI GOWDER AND T. SANTHANARAMAN, 1972 : Control of mango hoppers, *Idiocerus* spp. *Proc. 3rd int. Symp. Subtrop. & Trop. Hort. (Abstracts)*: 83-84, Bangalore (12)

SAUNDERS, W.W., 1850-51 : On insects injurious to cotton plant. *Trans. Ent. Soc.*, *1* : 158-166, London (253)

SCHMUTTERER, H., 1977 : Isoptera. In : *Diseases, Pests and Weeds in Tropical Crops* : 283-285, Verlag Paul Parey, Berlin (25)

SCOTT, P., 1957 : Das brauen von bananbier in Uganda in British Ost-Afrika. *Brau - u - Malz.*, *10* (18) : 32-36, Wien (40)

SEKHAR, P.S. AND SUSEELA SEKHAR, 1964 : Investigations on thrips occurring on arabica coffea, *Coffea arabica* Linn. — I. Bionomics. *Indian Coffee, 28* (8) : 173-179, Bangalore (143)

SEN, A.C. AND D. PRASAD, 1953 : Pests of banana in Bihar. *Indian J. Ent.*, *15* (3) : 240-256, New Delhi (41,44)

SEN, P.C., 1923 : The mango weevil. *Bengal agric. J.*, *3* : 66, Calcutta (23)

SENGUPTA, GOKUL CHANDRA AND BASANTA KUMAR BEHURA, 1955 : Some new records of crop pests from India. *Indian J. Ent.*, *17* (2) : 283-285, New Delhi (39)

SENGUPTA, GOKUL CHANDRA AND BASANTA KUMAR BEHURA, 1957 : Annotated list of crop pests in the State of Orissa. *Mem. Ent. Soc. India, 5* : 44 pp., New Delhi (23)

SESHADRI, A.R. AND T.N. ANANTHAKRISHNAN, 1954 : Some new Indian Thysanoptera, I. *Indian J. Ent.*, *16* (3) : 210-226, New Delhi (164, 214)

SETHI, S.L. 1967 : Insecticides for the control of citrus psylla in India. *J. econ. Ent.*, *60* : 270-271, Wiscounsin (131)

SHAH, M.I., H.N. BATRA AND P.L. RANJHEN, 1948 : Notes on biology of *Dacus (Strumeta) ferrugineus* Fabr. and other fruit flies in North-West Province. *Indian J. Ent.*, *10* (2) : 249 266, New Delhi (20)

SHAH, R., 1946 : An effective and inexpensive method for the control of stem-borers in fruit trees with special reference to *santra* trees in C.P. and Berar. *Curr. Sci., 15* (5) : 135, Bangalore (188)

SHARMA, P.L., S.C. AGARWALA AND B.S. ATTRI, 1971 : Beetles eating pome and stone fruits in Himachal Pradesh and their control. *Himachal J. agric. Res., 1* (1) : 57-60, Solan (271, 293)

SHARMA, P.L. AND B.S. ATTRI, 1969 : Control of stem bores, *Aeolesthes holoserica* Fabricius (Cerambycidae : Coleoptera). *Indian J. Ent., 31* (1) : 79-80, New Delhi (318)

SHARMA, P.L., B.S. ATTRI, R.L. SHANDIL AND O.P. BHALLA, 1968 : Biology and control of peach leaf curl aphid, *Brachycaudus* (*anuraphis*) *helichrysi* Kalt. *Indian J. Ent., 30* (4) : 289-294, New Delhi (287)

SHARMA, P.L. AND O.P. BHALLA, 1962 : Preliminary studies on the control of San Jose scale, *Q. perniciosus* (Comstock) with miscible oil as compared to diesel oil emulsion. *Himachal Hort., 3* : 235-237, Simla (261)

SHARMA, P.L. AND O.P. BHALLA, 1963 : Occurrence of the thrips damaging apple blossoms in Himachal Pradesh. *Indian J. Ent., 25* (1) : 85-86, New Delhi (276)

SHARMA, P.L. AND O.P. BHALLA 1965 : Studies on the control of San Jose scale, *Quadraspidiotus perniciosus* (Comstock). *Indian J. Ent., 27* (3) : 323-330, New Delhi (261)

SHARMA, P.L., A.C. DAVID AND O.P. BHALLA. 1963 : Flat headed borer, *Sphenoptera lafertei* Obb. and its control in Himachal Pradesh. *Himachal Hort., 4* (1) : 1-5, Simla (296)

SHARMA, P.L., S. SRIVASTAVA AND H.L. DHALIWAL, 1973 : Chemical control of the peach fruit flies. *Pesticides, 7* (10) : 20-21, Bombay (297, 298)

SHUKLA, G.S. AND K. KUMAR, 1969 : A new record of *Odioporus longicollis* (Ol.) (Coleoptera, Curculionidae) from Uttar Pradesh. *Sci., Cult., 35* (9) : 481-482, Calcutta (43)

SIMMONDS, N.W., 1966 : *Bananas,* 512 pp. Longmans, Green & Co. Ltd., London (40, 43, 46, 49, 53)

SIMWAT, G.S. AND A.S. SIDHU, 1974 : Biology of the fig leaf roller, *Phycodes minor* Moore (Glyphipterygidea : Lepidoptera). *Indian J. Ent., 36* (2) : 149-150, New Delhi (238)

SINGH, A.B., 1972 : Studies on the transmission of papaya mosaic virus by *Aphis gossypii* Glover. *Indian J. Ent., 34* (3) : 240-245, New Delhi (78)

SINGH, CHHOTEY, 1964 : Temperate fruit pests. In : *Entomology in India* : 213-276. Silver Jubilee Number. Entomological Society of India, New Delhi (261, 269, 277, 292, 297, 300)

SINGH, DALJIT, 1967 : Cultivation of some less known fruits in India. *Indian Hort., 11* (4) : 51-54, New Delhi (240)

SINGH, GAJENDRA, ASHOK KUMAR AND P.K. PATHAK, 1974 : Studies on the efficiency of some insecticides against mango hopper, *Amritodus atkinsoni* (Lethierry). *Indian J. Hort., 31* (3) : 283-285, Bangalore (12)

SINGH, J. P., 1970 : *Elements of Vegetable Pests,* 275 pp. Vora and Co. Pvt. Ltd., Bombay (255)

SINGH, KARAM LAMBA, 1931 : A contribution towards our knowledge of Aleyrodidae (whiteflies) of India. *Mem. Dept. Agric. India, Ent. Ser., 12,* 98 pp. Imperial Council of Agricultural Research, New Delhi (213)

SINGH, LAL BIHARI, 1960: *The Mango,* 438 pp. Leonard Hill (Book) Limited, London (9, 19, 21, 84)

SINGH, LAL BIHARI AND UDAY PRATAP SINGH, 1954 : Insect pests and diseases. In : *Litchi* : 23-26, Superintendent Printing and Stationary, Lucknow, U. P., Lucknow (187,194)

SINGH, M.P. AND R.B. SHARMA, 1961 : Observations on the life-history and biology of litchi bug, *Chrysocoris stolii* (Wolff.) (Heteroptera : Pentatomidae). *Indian J. Ent., 23* (3): 214-219, New Delhi (191)

SINGH, R.N., 1946 : Factors affecting the development and time of hatching of eggs of *Drosicha stebbingi* Green. *Indian J. Ent., 8* (1) : 132, New Delhi (14)

SINGH, RANJIT, 1958 : *The Litchi in India.* *Farm Bull. No. 44,* 24, pp. Indian Council of Agricultural Research, New Delhi (187)

SINGH, RANJIT, 1969 : *Fruits,* 213 pp. National book Trust Ltd., New Delhi (9, 40, 75, 123)

SINGH, S.M., 1654 a : A note on serious damage of mango crop by *Lymantria mathura* Moore in Doon valley. *Indian J. Hort., 11* (4) : 150, New Delhi (16, 188)

SINGH, S.M., 1954 b : Studies on *Apsylla cistellata* Buckton causing mango galls in India. *J. econ. Ent., 47* (4) : 563-564, Wiscounsin (29)

SINGH, S.M., 1945 c : *Rhynchaenus mangiferae,* a serious mango pest in Uttar Pradesh. *Curr. Sci., 23* (8) : 270-271, Bangalore (33)

SINGH, S.M., 1957 : A serious damage to mango shoots by the borer *Chlumetia transversa* Walker in Uttar Pradesh. *Indian J. Hort., 14* (4) : 236-238, New Delhi (37, 38)

SINGH, S.P. AND N.S. RAO, 1976 : Evaluation of some insecticides against black citrus aphid *Toxoptera aurantii* (B. de F.). *Pesticides, 10* (11) : 21-22, Bombay (141)

SINGH, S.P. AND N.S. RAO : 1977 : Effectiveness of different contact insecticides against soft green scale, *Coccus viridis* Green (Coccidae : Homoptera) on citrus. *Pesticides, 11* (3) : 33-36, Bombay (198)

SINGH, SHAM, S. KRISHNAMURTHI AND S. L. KATYAL, 1963 : *Fruit Culture in India,* 451 pp. Indian Council of Agricultural Research, New Delhi (202, 219)

SINGH, SHUMSHER, 1942 : A contribution to our knowledge of Indian Thysanoptera. *Indian J. Ent , 4* (2) : 111-135, New Delhi (105)

SINHA, M·M· AND S.S. SINHA, 1961 : *Spectrotropha sordidalis* Hamps· (Pyralidae: Lepidoptera) a new pest of mango in Bihar. *Indian J. Ent.*, **23** (1) : 72, New Delhi (31)

SIVAGAMI, R., K.R. NAGARAJA RAO AND S. VIJAYARAGHAVAN, 1964 : Record of a few insect pests of tamarind at Coimbatore. *Madras agric. J.*, **15** (5) : 221-222, Coimbatore (116, 118)

SMITH, R·I·, 1977 : *Cosmopolites sordidus* Germ· In : *Diseases, Pests and Weeds in Tropical Crops* : 406-410, Verlag Paul Parey, Berlin (42)

SOHI, B.S. AND O. S. BINDRA, 1974 : Comparative effectiveness of different insecticides for the control of the grapevine leaf roller, *Sylepta lunalis* Guen· (Lepidoptera : Pyralidae) in the Punjab, India. *Indian J. Hort.*, **31** (4) : 389-392, Bangalore (168)

SRIVASTAVA, A.S·, 1956 a : Damage caused by tortoises to singhara crop in Uttar Pradesh· *Bull. Pl. Prot. India*, **8** (1) : 260, New Delhi (106)

SRIVASTAVA, A.S·, 1956 b : Bionomics and control of singhara beetle (*Galerucella birmanica* Jacoby - Chrysomelidae) in Uttar Pradesh· *Agric. & Ani. Husb.*, *U.P.*, **6** (7) : 5-10, Allahabad (106, 107, 108)

SRIVASTAVA, A.S., 1964 : Pests of fruit trees. In : Entomological Research during the last ten years. Section of Entomologist to Government of Uttar Pradesh· *Res. Memoir, III* : 66-67, Allahabad (16)

SRIVASTAVA, A·S·, B.P. GUPTA, G·P· AVASTHI, B.P· PANDE AND H·P· SAXENA, 1969 : Bionomics, nature of damage and control of *Galerucella birmanica* Jacoby - a pest of water nut· *Beitr. Ent.*, **19** : 607-616, Berlin (107,108)

SRIVASTAVA, A· S·, Y.P. SINGH, R.C. PANDEY AND B.K. AVASTHY, 1973 : Bionomics and control of mango mealy bugs· *World Crops.*, **25** (2) : 87-88, London (14)

SRIVASTAVA, B·K·, 1958 : On the external morphlogy of *Idiocerus clypealis*. *Beitr. Ent.*, **8** (5-6) : 732-744, Berlin (12)

SRIVASTAVA, B.K. AND R.M· KHAN, 1963 : Bionomics and control of *Holotrichia insularis*· *Indian J· Ent.*, **25** (4) : 347-354, New Delhi (65, 114)·

SRIVASTAVA, D.N., 1957 : *The Loquat in India. Farm Bulletin No. 38*, 20 pp. Indian Council of Agricultural Research, New Delhi (196)

SRIVASTAVA, P.D·, 1976 : *Economic Zoology*, 300 pp. Commercial Publication Bureau, New Delhi (106)

SRIVASTAVA, R.P· AND DHAMO K· BUTANI, 1972 : A method to prevent the mango mealy bug (*Drosicha mangiferae* Green) damage. *Entomologists' Newsletter*, **2** (5) : 35, New Delhi (14, 101, 111)

SRIVASTAVA, U·S· AND S·B. AGARWAL, 1966 : Observations on the life-history and biomomics of *Anjeerodiplosis peshawarensis* Mani (Diptera, Cecidomyiidae)· *Bull. ent. Res.*, **57** (1) : 121-137, London (237)

STANNARD, L.J. (Jr.), 1956 : Three new species and two new records of thrips (Thysanaptera) in Illinois. *Trans. Illinois Acad. Sci.*, **49** : 172-176, Springfield (52)

STANNARD, LEWIS J. AND TALAAT K. MITRI, 1962 : Preliminary studies on the *Tryphacothrips* complex in which *Anisopilothrips, Mesostenothrips* and *Elixothrips* are erected as new genera (Thripidae : Heliothripinae). *Trans. Amer. ent. Soc.,* **87** : 183-224, Philadelphia (51)

SUBRAMANYAM, C.K., 1926 : A note on the life history of *Cryptorrhynchus mangiferae* Fabricius. *Madras agric. Dept. Yrbk.,* 1925 : 29-36, Government of Madras, Madras (22)

SUNDRA BABU, P.C., 1969 : Studies on the varietal incidenee of the mango nut weevil, *Sterochetus mangiferae* Fabricius. *South India Hort.,* **17** (1&2): 34-40, Combatore (21)

SUNDRA BABU, P.C. AND B. VASANTHARAJ DAVID, 1973 : A note on unprecedent occurrence of fruit-piercing moths on grapevines. *South India Hort.,* **21** (4) : 134-136, Coimbatore (170)

TALHOUK, ABDUL MON'IM S., 1969 : Insects and mites injurious to crops in middle eastern countries. *Monographien zur angew Entomologie No. 21,* 239 pp. Verlag Paul Parey, Berlin (135,138)

TALGERI, G.M., 1967 : Entomology section in retrospect. *Poona agric. Coll. Mag.,* **57** (4) : 84-100, Poona (21, 57, 90)

TANDON, P.L., K. KRISHNAIAH AND V.G. PRASAD, 1975 : Chemical control of mango shoot borer, *Chlumetia transversa* Walker (Noctuidae : Lepidoptera). *Indian J. Hort.,* **32** (3 & 4) : 154 - 155, Bangalore (38)

TANDON, P.L., A.C. MATHUR AND K. KRISHNAIAH, 1974 : Chemical control of mango fruit fly, *Dacus dorsalis* Hendel. *Prog. Hort.,* **6** (1) : 11-13, Ranikhet (21)

TAYLOR, T.H.C., 1935 : The campaign against *Aspidiotus destructor* Sign. in Fiji. *Bull. ent. Res.,* **26** : 1-102, London (19)

TEOTIA, T.P.S. AND S. CHAUDHRI, 1966 : Some observations on life history of *Euproctis fraterna* Moore (Lepidoptera : Lymantriidae) on castor. *Labdev J. Sci. Tech.,* **4** (1) : 45-47, Kanpur (241)

TREHAN, K.N., P.V. WAGLE, S.R. BAGAL AND G.M. TALGERI, 1947 : Biology and control of *Scelodonata strigicollis* Mots (Chrysomelidae : Coleoptera). *Indian J. Ent.,* **9** (1) : 93-107, New Delhi (161)

UPPAL, B.N. AND P.V. WAGLE, 1944 : Control of mango hoppers in Bombay Province. *Indian Fmg.,* **5** (9) : 401-403, New Delhi (10)

USMAN, S. AND M. PUTTARUDRIAH, 1955 : A list of the insects of Mysore including the mites. *Mysore Dept. Agric. Bull. (Ent. Ser.) No.* **16,** 194 pp. Bangalore (119)

VARMA, G.C. AND O.S. BINDRA, 1972 : New record of defoliating beetles on falsa, *Grewia asiatica* Linn. *Indian J. Ent.,* **34** (3) :254, New Delhi (242)

VARMA, GANPATI SINGH, 1955 : *Miracles of Fruits,* 338 pp. G.S. Ayurvedic Researeh Products, Delhi (110)

VEERESH, G.K., 1974 : Root grub control campaign in Karnataka. *White grub Newslet.,* 1 (1) :4-7, Bangalore (113)

VEERESH, G.K. AND D. RAJAGOPAL, 1977 : Occurrence of bagworm, *Clania crameri* Westwood (Lepidoptera : Psychidae) as a serious pest of grapevine in Karnataka. *Curr. Res.,* 6 : 171-172, Bangalore (170)

VEERESH, G.K. AND M. PUTTARUDRAIAH, 1968 : A study of the bionomics of *Lymantria ampla* Wlk. (Lymantriidae : Lepidoptera). *Mysore J. agric. Sci.,* 11 (2) : 61-69, Bangalore (276)

VENUGOPAL, S., 1958 : *Oenospila (Thalassodes) flavifusata* Wlk. a geometrid pest of *Eugenia jambolana* in South India. *Madras agric. J.,* 45 (3) : 120-121 Madras (216)

VERMA, A.N., J.S. BALAN AND S.M. SINGHVI, 1970 : Relative efficacy of different insecticides for the control of *Ak* grasshopper *Poecilocerus pictus* F. on castor. *Telhan Patrika,* 2 (2) : 1-5, Hyderabad (76)

VERMA, A.N. AND A.D. KHURANA, 1974 : Further new host records of *Indarbela* sp. (Lepidoptera : Metarbelidae). *HAU. J. Res.,* 4 (3) : 253-254, Hissar (188)

VERMA, A.N., G.S. SANDHU AND P.U. SARAMMA, 1967 : Singhara beetle and its control. *Prog. Farming, PAU.,* 3 (5) : 21-22, Ludhiana (108)

VERMA, K.K., 1969 : Functional and developmental anatomy of reproductive organs of the male of *Galerucella birmanica* Jac. (Coleoptera : Phytophaga, Chrysomelidae). *Annls. Sci. nat.,* 11 : 129-234, Paris (107)

VEVAI, E.J., 1969 a : Know your crop, its pest problems and control - 9 : Grapevine. *Pesticides,* 3 (6) : 31-37, Bombay (160)

VEVAI, E.J., 1969 b : Know your crop, its pest problems and control - 11 : Citrus. *Pesticides,* 3 (8) : 31-37, Bombay

VEVA, E.J, 1969 c : Know your crop, its pest problems and control - 15 : Mango. *Pesticides,* 3 (12) : 21-31, Bombay (9, 21, 39)

VEVAI, E.J., 1971 a : Know your crop, its pest problems and control - 33 : Banana - Plantain. *Pesticides,* 5 (6) : 37-56, Bombay (47, 43, 49)

VEVAI, E.J., 1971 b : Know your crop, its pest problems and control - minor tropical fruits. *Pesticides,* 5 (11) : 31-54, Bombay (69, 173, 187, 197)

VISWANATH, B.N. AND B.L.V. GOWDA, 1974 : A preliminary study of the bionomics of *Trabala vishnu* Lefevere (Lepidoptera : Lasiocampidae). *Curr. Sci.,* 31 (1) : 34-35, Bangalore (218)

VOUTE, A.D., 1935 a : *Crytorrhynchus gravis* F. und die Ursachen seiner Massenvermehrung in Java. *Arch. Neerl. Zool.,* 2 : 112, (23, 27)

VOUTE, A.D., 1935 b : Twee beschadigers van jouge manggaloten : I - De manggalot-boorder (*Chlumetia transversa* Wlk). II - De plome manggarups (*Bombotelia jacosatrix* Gn.). *Landbouw.,* 10 (7) : 255-271, Wageningen (37)

WADHI, S.R., 1964 : A basis for the separation of stone weevil infested mango fruits from uninfested ones. *Indian J. Ent.*, **26** (3) : 362-363, New Delhi (10)

WADHI, S.R., 1972 : Investigations to develope quarantine disinfestation procedure for mango stone weevil, *Sternochetus mangiferae* (Fab.) and the oriental fruit fly, *Dacus dorsalis* Hendel. *Entomologists' Newsletter*, **2** (3) : 14, New Dehli (22)

WADHI, S R., 1973 : Codling moth in India. *Entomologists' Newsletter*, **3** (11) : 67, New Delhi (278)

WADHI, S.R. AND H.N, BATRA, 1964 : Pests of tropical and subtropical fruit trees. In : *Entomology in India* : 227-260, Silver Jubilee Number. Entomological Society of India, New Delhi (123, 187, 192, 231)

WADHI, S.R. AND G.R. SETHI, 1975 : Eradication of codling moth - a suggestion. *J. Nuclear agric. biol.*, **4** (1) : 18-19, New Delhi (279)

WADHI, S.R. AND D.N. SHARMA, 1972 : Bromide residues in pulp of mango fruits subjected to hot aqueous ethylene dibromide dips *Acta Horticulture*, **24** : 239, The Hague (22)

WALKER, FRANCIS, 1859 : *Catalogue Lepidoptera, British Museum*, **17** : 499, London (95)

WANG, PING-YUAN, 1963 : Notes on Chinese pyralid moths of the genus *Diaphorina* Hubner. *Acta Ent. Sin.*, **12** (3) : 358-365, Peking (95, 253)

WATT, GEORGE, 1908 : *The Commercial Products of India*, 1189 pp. John Murray, London (111)

YAWALKAR, K.S. AND A.F. CHITRIV, 1969 : Singhara beetle (*Galerucella birmanica* Jacoby) and its control. *Pesticides*, **3** (10) : 26-28, Bombay (108)

7

APPENDICES

INSECT PESTS
(Classified list)

ORTHOPTERA Latreille, 1793

ACRIDIDAE :

Atractomorpha crenulata Fabricius – Peach, plum
Aularches miliaris Linnaeus – Mango, coconut
Chrotogonus species – Peach, plum
Patanga cuccincta (Linnaeus) – Citrus
Poekilocerus pictus (Fabricius) – Banana, *ber*, citrus, fig, grapevine, guava, melons, papaya, peach

GRYLLIDAE :

Brachytrypes portentosus (Lichtenstein) – Citrus, mango
Platygryllus melanocephalus (Serville) – Mango
Teratodus monticollis Grey – Grapevine

ISOPTERA Brulle, 1832

KALOTERMITIDAE :

Neotermes bosei Snyder - Mango
N. mangiferae Roonwal and Sen-Sarma – Mango

TERMOPSIDAE :

Stylotermes fletcheri Holmgren – Mango

RHINOTERMITIDAE :

Coptotermes heimi Wasmann – Mango, mulberry
Heterotermes indicola Wasmann – Grapevine, mango

TERMITIDAE :

Microtermes mycophagus Desneux – Peach
M. obesi (Holmgren) – Grapevine, mango, mulberry
Odontotermes assmuthi Holmgren – Mango
O. feae (Wasmann) – Mango
O. obesus (Rambur) – Apple, citrus, grapevine, guava, sapota, pomegranate.
O. wallonensis Wasmann – Mango
Trinervitermes biformis (Wasmann) – Mango
T. rubidus (Hegen) – Mango

HEMIPTERA Linnaeus, 1758

FLATIDAE :

Flaka ocellata Fabricius – Sapota

EURYBRACHIDAE :

Eurybrachys tomentosa Fabricius – *Ber*

CERCOPIDAE :

Clovia lineaticollis (Motschulsky) – Jack-fruit
Cosmoscarta funeralis Butler – Citrus
C.niteara Distant – Fig
C.relata Distant – Jack-fruit
Phymatostetha deschampes Linnaeus – Banana
Ptyelus species – Jack-fruit

MACHAEROTIDAE :

Machaerota planitiae Distant – *Ber*

CICADIDAE :

Pycna repanda (Linnaeus) – Apple

MEMBRACIDAE :

Gargara mixta (Buckton) – *Falsa*
G. varicolor Stal – Litchi
Leptocentrus obliquis Walker – *Ber*, tamarind
L. taurus (Fabricius) – *Ber*, citrus, *falsa*
Otinotus oneratus (Walker) – *Ber*, custard apple, mango, tamarind
Oxyrhachis tarandus (Fabricius) – Apple, citrus, grapevine, tamarind
Tricentrus bicolor Distant – *Ber*, citrus

CICADELLIDAE :

Amrasca biguttula biguttula (Ishida) – *Ber*
A. lybica (de Berg) – *Ber*
A. splendens Ghauri – Mango
Amritodus atkinsoni (Lethierry) – Mango
Empoasca punjabensis Pruthi – Guava
E minor Pruthi – Grapevine
Erythroneura species – Fig
Erythroneuropsis species – Grapevine
Idioscopus clypealis (Lethierry) – Mango
I. nigroclypeatus (Melichar) – Mango
I. niveosparsus (Lethierry) – Mango
Kolla diaphana Distant – Fig
Nirvana species – Fig
Tettigoniella illustris Distant – Citrus
Typhlocyba species – Grapevine

PSYLLIDAE :

Apsylla cistellata (Buckton) – Mango
Arytania fasciata Laing – Walnut
A. obscura Crawford – Mango
Diaphorina citri Kuwayama – Citrus
D. communis Mathur – Citrus
Dynopsylla grandis Crawford – Fig
Euphyllura caudata Mathur – *Jamun*
E. concolor Mathur – *Jamun*
Leuronota minuta (Crawford) – Citrus
Megatrioza vitiensis (Kirkaldy) – *Jamun*

Paurocephala menoni Mathur – *Falsa*
Pauropsylla brevicornis Crawford – Mango
P. depressa Crawford – Fig
Psylla pyricola Foerster – Apple, pear
Trioza jambolanae Crawford – Mango

ALEYRODIDAE :

Acaudaley·odes rachipora (Singh) – Jack-fruit, tamarind
Aleurocanthus citriperdus Quaintance and Baker – Citrus
A. husaini Corbett – Citrus, litchi, peach, pear, plum
A. mangifarae Quaintance and Baker – Mango
A. rugosa Singh – *Jamun*
A. spiniferus (Quaintance) – Citrus, grapevine, pear
A. woglumi (Ashby) – Avocado, citrus, grapevine, guava, mango, pear, plum, pomegranate, quince, sapota
Aleurocybotus setiferus Quaintance and Baker – Citrus
Aleurolobus marlatti (Quaintance) – Citrus, mulberry
Aleurothrixus flocculus (Maskell) – Citrus
Aleurotrachelus caerulescens Singh – Jack-fruit
Aleurotuberculatus citrifolii (Corbett) – Citrus
A. psidii (Singh) – Guava
A. murrayae (Singh) – Citrus
Bemisia giffardi (Kotinsky) – Citrus
B. tabaci (Gennadius) – *Falsa*, papaya
Dialeurodes citri (Ashmead) – Citrus, jamun
D. citrifolii (Morgan) – Citrus
D. vulgaris Singh – *Jamun*
Dialeurolonga elongata Dozier – Citrus, litchi
Dialeuropora decempuncta (Quaintance and Baker) – Custard apple
Pealius schimae Takahashi – Jack-fruit
Rhachisphora trilobitoides (Quaintance and Baker) – *Jamun*
Singhiella bicolor (Singh) – *Jamun*
Siphoninus phillyreae (Haliday) – Peach, pear, pomegranate
Trialeurodes ricini (Misra) – Sapota

LACHNIDAE :

Cinara krishni (George) – Peach, pear
Pterochlorus persicae Cholodkovsky – Almond, apricot, cherry, peach, plum

APHIDIDAE :

Aphis epillabina Kulkarny – Mango
A. gossypii Glover – Apple, citrus, guava, grapevine, papaya, melons
A. medicaginis Koch (auct. *A. craccivora* Koch) – Papaya
A. pomi de Geer – Apple, citrus, peach, pear, walnut
A. punicae – Pomegranate
A. rumicis Linnaeus (auct. *A. fabae* Scopoli) – Pear
Brachycaudus helichrysi (Kaltenbach) – Almond, apricot, citrus, peach, plum
Chromaphis jugladicola (Kaltenbach) – Walnut
Greenidia artocarpi (Westwood) – Jack-fruit
G. formosana Maki – Guava
G. mangiferae Takahashi – Mango

Hyalopterus pruni (Geoffroy) – Appicot, grapevine, peach, plum
Macrosiphum euphorbiae (Thomas) – Mango
M. sonchi Linnaeus – Papaya
Myzus cerasi (Fabricius) – Cherry, peach
M. persicae (Sulzer) – Almond, apricot, apple, cherry, citrus, papaya, peach,
                                               pear, plum
Panaphis juglandes (Geoz) – Walnut
Pentanolia nigronervosa Coquillett – Banana
Rhopalosiphum nymphae (Linnaeus) – Water nut
Taxoptera aurantii (Boyer de Fonscolombe) – Citrus, custard apple, jack-
                            fruit, *jamun*, litchi, loquat, sapota, tamarind
T. citricidus (Kirkaldy) – Citrus
T. odinae van der Goot – Citrus, mango

PEMPHIGIDAE (ERIOSOMATIDAE) :
   Eriosoma lanigerum (Hausmann) – Apple, crab apple, pear, quince

MARGARODIDAE :
   Crypticerya jaihind Rao – Guava
   Drosicha contrahens (Walker) – Apple, apricot, citrus, mulberry, peach, plum
   D. dalbergiae (Green) – Citrus, guava, *jaman*, litchi, mango, papaya,
                                   pineapple, sapota
   D. mangiferae (Green) – Apple, apricot, ˜ber, cherry, citrus, *falsa*, fig,
                     grapevine, guava, jack-fruit, *jamun*. litchi, mango,
                     mulberry, papaya, peach, pear, plum, pomegrnate
   D. stebbingi (Green) – Fig, mango, tamarind
   Hemiaspidoproctus cinerae (Green) – Citrus, guava, pomegranate, sapota
   Icerya aegyptiaca (Douglas) – Are canut, citrus, custard apple, date palm,
                           guava, grapevine, jack-fruit, pear
   I. minor Green – Apple, citrus, guava, mango
   I. pulcher (Leonardi) – Mango
   I. purchasi Maskell – Almond, apple, apricot, citrus, fig, grapevine, guava,
                       mango, peach, pomegranate, walnut
   I. seychellarum (Westwood) – Citrus, guava, mango, palmae, pear
   Labinoproctus poleii (Green) – Citrus, guava
   Perissopneumon tamarindus (Green) – Apple, banana, *ber*, citrus, *falsa*,
                          fig, mulberry, tamarind
   Steatococcus assamensis Rao – Citrus

ORTHEZIIDAE :
   Orthezia insignis Browne – Citrus

LECCIFERIDAE (KERRIIDAE) :
   Kerria albizziae (Green) – Litchi
   K. communis (Mahdihassan) – Ber, custard apple, grapevine
   K. fici (Green) – *Ber*, *Ficus* spp.
   K. indicola (Kapur) – *Ber*
   K. lacca (Kerremann) – Citrus, *ber*, grapevine, mango, tamarind
   K. jhansiensis Misra – *Ber*
   K. mysorensis (Mahdihassan) – *Ber*
   K. pusana (Misra) – *Ber*

Tachardia minuta Morrison – Pomegranate
Tachardina lobata (Green) – Apple, citrus, custard apple, mango

PSEUDOCOCCIDAE :

Dysmicoccus brevipes (Cockerell) – Banana, mulberry, pineapple
Ferrisia virgata (Cockerell) – *Aonla,* banana, citrus, grapevine, guava, jack-fruit
Maconellicoccus hirsutus (Green) - Citrus, grapevine, mulberry
Nipaecoccus viridis (Newstead) – *Ber,* citrus, fig, grapevine, guava, jack-fruit, mango, mulberry, tamarind,
Planococcus citri (Risso) – Citrus, fig, pineaple, sapota
P. lilacinns (Cockerell) – Citrus, custard apple, fig, guava, pomegranate, sapota, tamarind
Pseudococcus adonidum (Linnaeus) – Citrus, fig
P. bromiliae Bouche – Pineapple
P. citriculus Green – Citrus
P. comstocki (Kuwana) – Banana, citrus
Rastrococcus iceryoides (Green) – Citrus, mango, sapota
R. mangiferae (Green) – Mango

COCCIDAE :

Ceroplastes actiniformis Green – Fig, grapevine, mango, pear
C. ceriferus (Anderson) – Citrus, guava, mango, mulberry
C. floridensis Comstock – Apple, cashewnut, citrus, custard apple, fig, guava, mango
C. pseudoceriferus Green – Mango
C- rubens Maskell – Citrus, fig, jack-fruit, mango, pear
Ceroplastodes cajani (Maskell) – *Ber,* fig, guava
Chloropulvinaria polygonata (Cockerell) – Apple, grapevine, *jamun,* mango, mulberry, peach
C. psidii (Maskell) – Citrus, fig, guava, jack-fruit, *jamun,* litchi, loquat, mango, sapota
Coccus acutissimus (Green) – Banana, *ber,* jack-fruit, mango
C. adersi (Newstcad) – Mango
C. bicruciatus (Green) – Citrus, mango
C. discrepans (Green) – Banana, *ber,* citrus, fig, grapevine, *jamun,* mango mulberry
C. elongatus (Signoret) – Grapevine, sapota
C. hesperidum Linnaeus – Banana, citrus, guava, mango
C. longulum (Douglas) – Sapota
C. mangiferae (Green) – Mango
C. viridis (Green) – *Bael,* citrus, guava, loquat, mango
Eriochiton theae Green – Peach, pear, plum
Eulecanium tiliae (Linnaeus) – Almond, apple, apricot, cherry, loquat, peach, pear, plum
Lecanium latioperculatum Green – Mango
L. ramakrishnae Ramakrishna – Fig, pear
Parasaissetia nigra (Nietner) – Citrus, grapevine, guava, litchi
Parthenolecanium corni (Bouche) – Apple, plum
P. persicae (Fabricius) – mulberry
Pulvinaria burkilli Green – *Ber*

P. cellulosa Green – Citrus, mango
P. ixorae Green – Mango
P. maxima Green – *Ber*, citrus, grapevine, mulberry
Saissetia coffee (Walker) – Citrus, guava, loquat, litchi, mango
S. depressum Cockerell – Banana
S. oleae (Bernard) – Citrus, fig, guava, olive, sapota, tamarind
Vinsonia stellifera (Westwood) – Citrus, *jamun*, mango

DIASPIDIDAE:

Abgrallaspis cyanophylli (Signoret) – Banana
Andaspis hawaiiensis (Maskell) – Pomegranate
A. leucophleae Rao – Pomegranate
Aonidia zizyphi Rahman and Ansari – *Ber*
Aonidiella aurantii (Maskell) – Banana, *ber,* citrus, fig, guava, *jamun,*
      mulberry, peach, pear
A. citrina (Coquillett) – Citrus, custard apple, mango, mulberry, palmae
A. comperei Mc Kenzie – Citrus
A. orientalis (Newstead) – *Bael*, banana, *ber,* citrus, guava, *jamun,* mulberry,
      peach, pomegranate, tamarind
Aspidiotus destructor Signoret – Avocado, banana, *ber,* citrus, date palm, fig,
      grapevine, guava, *jamun,* mango, papaya, peach, sapota, tamarind
A. exisus Green – Citrus
A. tamarindi Green – Citrus, tamarind
Aulacaspis rosae (Bouche) – Mango
A. tubercularis Newstead – Mango
Cardiococcus castilleae (Green) – Tamarind
Chionaspis pusa Rao – Citrus
Chrysomphalus ficus Ashmead – Almond, apple, avocado, banana, *ber,*
      citrus, date palm, grapevine, mango, mulberry
C. dictyospermi (Morgan) – Apple, citrus, mango
Cornuaspis beckii (Newman) – Citrus
Dentachionaspis centripetalis Rao – Apple
Duplaspidiotus tesseratus (Charmoy) – Apple, grapevine, *jamun,* pomegranate
Fiorinia nephelii Maskell – Litchi
F. proboscidaria Green – Citrus
F. theae Green – Citrus, olive
Genaparlatoria pseudaspidiotus (Lindinges) – Mango
Hemiberlesia lataniae (Signoret) – Apple, banana, *ber,* citrus, fig, grapevine,
      guava, jack-fruit, tamarind
H. palmae (Cockerell) – Citrus
H. punicae (Signoret) – Pomegranate
H. rapax (Comstock) – Apple, citrus
Howardia biclavis (Comstock) – Apple, *ber,* citrus, custard apple, peach,
      tamarind
Insulaspis citrina Borchsenius – Citrus
I. gloverii (Packard) – Citrus, guava, mango
I. lasianthi (Green) – Citrus
I. pallidula Williams – Citrus, guava, mango
Lepidosaphes conchiformis Shimer (non Gamelin) – Fig
L. tapleyi Williams – Mango

Leucaspis riecae Targioni – Fig

Lindingaspis floridana Ferris – Mango

L. greeni (Brain and Kelly) – Guava, mango, pomegranate

L. rossi (Maskell) – Citrus, guava, mango, pomegranate

Parlaspis papillosa (Green) – Jack-fruit

Parlatoria blanchardi (Targioni) – Date palm

P. boycei Mc Kenzie – Pear

P. camelliae Comstock – *Bael*, citrus, grapevine, mango, olive

P. cinerea Hadden – Apple, apricot. *bael*, citrus, *falsa*, *jamun*, litchi, loquat, mango, peach, plum

P. cristifera Green – Citrus

P. crypta Mc Kenzie – Mango

P. marginalis Mc Kenzie – *Jamun*

P. oleae (Colvee) – Apple, apricot, *ber*, citrus, fig, grapevine, loquat, mango, mulberry, olive, peach, pear, plum, pomegranate

P. pergandii Comstock – *Bael*, citrus, guava, mango, olive, palmae

P. pseudopyri Kuwana – *Ber*, *jamun*, litchi, loquat

P. zizyphus (Lucas) – Citrus, date palm

Phenacaspis cockerelli (Cooley) – Mango

P. megaloba (Green) – *Ber*

P. vitis (Green) – Grapevine, mango

Pinnaspis aspidistrae (Signoret) – Citrus, fig, jack-fruit, mango

P. minor (Maskell) – Citrus, tamarind

P. strachani (Cooley) – Citrus, tamarind

P. temporarin Ferris – Tamarind

P. theae (Maskell) – Pomegranate

Pseudaonidia trilobitiformis (Green) – Guava, mango

Pseudaulacaspis barberi (Green) – Mango

P. pentagona (Targioni) – Almond, apple, apricot, cherry, peach, pear, plum

Pseudoparlatoria ostreata Cockerell – Papaya

Quadraspidiotus perniciosus (Comstock) – Almond, apple, apricot, cherry, chestnut, citrus, crab apple, mulberry, peach, pear, plum, raspberry, strawberry, walnut

Radionaspis indica (Marlatt) – Mango

Selenaspidus articulatus (Morgan) – Citrus, fig

Semelaspidus artocarpi (Green) – Jack-fruit

Unaspis atricolor (Green) – Citrus, tamarind

TINGIDAE :

Monosteira edeia Drake and Livingstone – *Ber*

M. minutula Montandon – *Ber*

Stephanitis cahriesis Drake and Mahanasundarun – Jack-fruit

S. typicus Distant – Banana

Urentius ziziphifalius Menon and Hakk – *Ber*

MIRIDAE (CAPSIDAE):

Creontiades pallidifer Walker – Melons

Gymnaspis manglferae Green and Parlatoria cristifera Green mentioned by Ayyar (1930) under manuscript names, are yet nomen nudum species.

Helopeltis antonii Signoret – Grapevine, guava
H. febriculosa Bergroth – *Ber*, guava
Prodromus subviridis Distant – Banana

LYGAEIDAE :

Spilostethus hospes (Fabricius) – Peach, pear, plum
S. pandurus (Scopoli) – Apple, citrus, grapevine, *amun*, litchi, mango,
                             peach, pear, plum

PYRRHCORIDAE :

Dysdercus koenigii (Fabricius) (auct. *D. cingulatus* Fabricius) – Mango

COREIDAE :

Anoplocnemis phasiana (Fabricius) – Citrus, grapevine
Dasynus antennatus (Kirby) – Citrus
D. fuscescens (Distant) – Banana
Leptoglossus australis (Fabricius) – Banana, citrus, melons, pomegranate
Riptortus linearis Fabricius – Fig

PENTATOMIDAE :

Agenoscetis nubila (Fabricius) – *Ber*
Antestictsis cruciata (Fabricius) – *Ber*, citrus, mango, peach, plum
Bagrada cruciferarum Kirkaldy – Mango
Cantoa ocellatus Thunberg – Grapevine
Cappaea taprobanensis (Dallas) – Citrus
Chrysocoris grandis (Thunberg) – Citrus
C. particius (Fabricius) – Mango
C. stolli (Wolff) – Litchi
Coptosoma nazirae Atkinson – Mango
Dalpada species – Cherry
Erthesima fulle (Thunberg) – Apple
Eurostus grossipes Dallas – Citrus
Halyomorpha picus (Fabricius) – Pomegranate
Halys dentatus (Fabricius) – *Jamun*, litchi, mango, mulberry
Jurtina indica Dallas – Pomegranate
Nezara viridula (Linnaeus) – Citrus
Rhynchocoris humeralis (Thunberg) – Citrus
Scutellera nobilis (Fabricius) – *Aonla*, *falsa*, grapevine
S. perplexa (Westwood) – Citrus
Tessaratoma javanica (Thunberg) – *Ber*, litchi
T. quadrata Distant – Apple, litchi
Vitellus orientalis Distant – Citrus

THYSANOPTERA Haliday, 1836

AEOLOTHRIPIDAE :

Aeolothrips collaris Priesner – Mango

THRIPIDAE (including HELIOTHRIPIDAE) :

Anaphothrips oligochaetus Karny – Pomegranate
A. sudanensis Trybom – Banana

Astrothrips parvilimbus Stannard and Mitri – Banana
Caliothrips indicus (Bagnall) – Mango
Chaetanaphothrips orchidii (Moulton) – Banana
Frankliniella dampfi Priesner – Apple, citrus, grapevine, melons, sapota
Gigantothrips elegans Zimmermann – Fig
Helionothrips kadaliphilus (Ramakrishna and Margabandhu) – Banana
Heliothrips haemorrhoidalis (Bouche) – Banana, citrus
Heliothrips species – Loquat
Hercinothrips bicinctus Bagnall – Banana
Megalurothrips distalis (Karny) – Litchi
M. usitatus (Bagnall) – Litchi
Panchaetothrips indicus Bagnall – Banana
Pseudodendrothrips dwivarna (Ramkarishna and Margabandhu) – Jack-fruit
Ramaswamiahiella   subnudula Karny – Citrus,   mango,   pomegranate,
tamarind
Retithrips syriacus (Mayet) – Custard apple, grapevine, pomegranate
Rhipiphorothrips cruentatus Hood – Country almond,   custard apple,
grapevine, *jamun,* mango, pomegranate
Rhopalandrothrips nilgiriensis (Ramakrishna) – Citrus
Scirtothrips aurantii Faure – Citrus
S. citri (Moulton) – Citrus
S. dorsalis Hood – *Ber,* citrus, grapevine, mango, pomegranate, tamarind
Scolothrips asura Ramakrishna and Margabandhu – *Ber,* banana
Selenothrips rubrocinctus (Giard) – Avocado, guava, mango, pear
Taeniothrips rhopalantennalis Shumsher – Apple
T. traegardhi (Trybom) – Grapevine
Taeniothrips species – Apricot, pear, plum
Thrips flavus Schrank – Apple, citrus
T. florum Schmut – Apple, banana, citrus, grapevine
T. pandu Ramakrishna – Citrus
T. tabaci Lindemann – Fig, pineapple

PHLOEOTHRIPIDAE :
Dolichothrips indicus (Hood) – *Ber, jamun,* litchi
Haplothrips ceylonicus Schmutz – Tamarind
H. ganglbaueri Schmutz – Mango
Haplothrips species – Loquat
Leeuwenia ramakrishnae Ananthakrishnan – *Jamun*
Mallothrips indica Ramakrishna – *Jamun*
Neoheegetia mangiferae (Priesner) – Mango
N. zizyphi Bagnall – *Ber*
Teuchothrips eugeniae Seshadri and Ananthakrishnan – *Jamun*

COLEOPTERA Linnaeus, 1758

CARBIDAE :
Neocollyris bonelli Guerin-Meneville – *Ber*

PASSALIDAE :
Tiberioides kuwerti Arrow – Rotten wood of walnut

LUCANIDAE :
Lucanus lunifer Hope – Apple, peach, plum

SCARABAEIDAE* :
Catharcius molossus Linnaeus – Apple, peach, plum
Popillia complanata Newman – Apple, peach
P. cupricollis Hope – Apple
P. cyanea Hope – Apple, plum, peach
P. sulcata Redtenbacher – Apple
Clinteria chloronota Blanchard – Citrus
C. spilota Arrow – Apple, peach
Epicometis squalida Linnaeus – Peach
Macronota quadrilineata Hope –Apple, peach
Mimela fulgidividata Blanchard – Apricot
M. passerini Arrow – Apple, cherry
M. pectoralis Blanchard – Apple, apricot
M. pusilla Hope – Cherry, walnut
Oxycetonia albopunctata (Fabricius) – Citrus, *falsa*
O. histrio (Olivier) – Citrus
O. jucunda (Faldermann) – Apple, citrus
O. versicolor (Fabricius) – *Falsa*
Protaetia impavida Janson – Cherry, peach, pear, plum
P. neglecta (Hope) – Apple, cherry, peach, pear, plum
Rhomborrhina glabervima Westwood – Apple
Stalagmosoma albella (Pallas) – Cherry
Torynorrhina opalina (Hope) – Apple, apricot, citrus
Apogonia ferruginae (Fabricius) – *Ber*
A. uniformis Blanchard – *Falsa*
Apogonia species – Grapevine
Eupatorus hardwisked Hope – Apple
Oryctes rhinoceros (Linnaeus) – Banana, date palm, pineapple
Xylotrupes gideon (Linnaeus) – Apple, peach, plum
Autoserica insanabilis Brenske – Citrus
A. nilgirensis (Sharp) – Citrus
Autoserica species – Apple
Brahmina coriacea (Hope) – Apple, apricot, fig, grapevine, peach, pear, plum, walnut
Ectinophoplia nitidiventris Arrow – Citrus
Hilyotrogus holosericeus Redtenbacher – Apple, peach, pear, plum
Holotrichia consanguinea (Blanchard) – Citrus, *falsa*, fig, grapevine, guava, *jamun*, mango, pear, plum
H. crinicollis Redtenbacher – Apple, apricot
H. insularis Brenske – *Ber*, *falsa*, guava, *jamun*, mango, tamarind
H. longipennis (Blanchard) – Apple, apricot, cherry, peach, pear, plum, strawberry, walnut
H. repetita Sharp – Citrus
H. rufoflava Brenske – Citrus
H. serrata (Fabricius) – Guava
Hoplosternus furicaudus (Ancy) – Apple, apricot, cherry, peach

¹This includes Cetoniinae, Dynastinae, Melolonthinae and Rutelinae (Richards and Davies, 1977).

H. indica Hope – Apple
H. nepalensis Hope – Apple
Melolontha indica Hope – Apple
Microtrichia cotesi Brenske – Apple, peach
Schizonycha ruficollis (Fabricius) – Citrus, guava, tamarind
Schizonycha species – *Falsa*
Serica clypeata Brenske – Cherry
S. maculosa Brenske – Cherry
S. marginella Hope – Cherry
Serica species – Apricot
Adoretosoma citricola Ohaus – Citrus
Adoretus bengalensis Brenske – Grapevine
A. bicaudatus Arrow – Mango
A. bimarginatus Ohaus – Apple, cherry
A. brachypygus Burmeister – Grapevine
A. duvauceli Blanchard – Apple, apricot, grapevine, guava, fig, loquat, peach, pear, plum
A. epipleuralis Arrow – Apple, apricot, peach, plum
A. horticola Arrow – Apple, fig, grapevine, guava, loquat, peach, pear, plum
A. lasiopygus Burmeister – Apple, fig, graphvine, guava, loquat, mango, peach, pear, plum
A. nitidus Arrow – Almond, *ber*
A. pallens Arrow – *Ber*
A. simplex Sharp – Apple, apricot, cherry, peach, pear, plum
A. versutus Harold – Apple, apricot, *ber*, fig, grapevine, guava, loquat, peach, pear, plum, walnut
Anomala aurora Arrow – Peach
A. bengalensis Blanchard – Ber, guava, *jamun*, *falsa*, grapevine
A. decorata – Peach
A. dimidiata Hope – Apple, apricot, *ber*, grapevine, peach, pear, plum, pomegranate
A. dorsalis (Fabricius) – *Ber*, grapevine
A. dussumieri (Blanchard) – Mango
A. flavipes Arrow – Apple, apricot, cherry, peach, pear, plum
A. lineatopennis Blanchard – Apple, apricot, peach, pear, plum
A. polita Blanchard – Apple, apricot, plum
A. ruficapilla Burmeister – *Ber*, grapevine
A. rufiventris Redtenbacher – Apple, apricot, peach, pear, plum
A. rugosa Arrow – Apple, apricot, walnut
A. transversa Burmeister – Cherry
A. varicolor Gyllenhal – Mango, peach, pear, plum
A. variivestis Arrow – Apple
Pachyrhinadoretus frontatus (Burmeister) – Grapevine
Singhala tenella Blanchard – Citrus

BUPRESTIDAE :

Agrilus grisator Kerremans – Citrus
A. mediocris Kerremans – Citrus
Belionota prasina Thunberg – Citrus, guava, mango
Capnodis carbonaria Klug – Almond, apple

Pailoptera cupreosplendens Saunders – *Ber*
Sphenoptera lafertei Thomson – Apple, apricot

LYCIDAE :

Calochromus darjeelingensis Bourgeois – Rotten wood and stumps of walnut
C. kaschmirensis Kleine – Rotten wood and stamps of walnut
C. tarsalis Waterhouse – Rotton wood and stumps of walnut

BOSTRYCHIDAE :

Dinoderus distinctus Lesne – Mango
Heterobostrychus aequalis Waterhouse – Mango, mulberry
H. hamatipennis Lesne – Mango
H. pileatus Lesne – Mango
Lyctoxylon convixtor Lesne – Mango
Micrapate simplicipennis Lesne – Mango, mulberry
Minthea sugicollis Walker – Mango
Parabostrychus elongatus Lesne – Mango
Rhizopertha dominica Fabricius – Grapevine, tamarind
Schistocerus ambioides Waterhouse – Guava, mango
Sinoxylon anale Lesne – *Ber*, grapevine, *gular, jamun,* mango, mulberry
S. conigerun Gerstacher – Mango
S. crassum Lesne – Citrus, mango
S. c. dekhanense Lesne – *Ber*, mango
S. indicum Lesne – *Ber*, mango
S. oleare Lesne – Mango
S. sudanicum Lesne – Mango
Trogoxylon spinifrons Lesne – Mango
Xylocis tortilicornis Lesne – *Ber*
Xylodectes ornatus Lesne – *Ber, gular,* mango
Xylopsocus capucinus Fabricius – Litchi, mango, mulberry
Xylothrips flavipes Illiger – *Gular,* mango

LYCTIDAE :

Lyctus africanus Lesne – *Gular*, mango
L. malayanus Lesne – Mango

NITIDULIDAE :

Amphicrossus species – Guava, peach, pear, plum, pomegranate
Carpophilus dimidiatus Fabricius – Grapevine, guava, mango, peach, pear, plum, pomegranate
Soronia species – Citrus

CUCUJIDAE :

Laemophloeus species – Apricot, grapevine
Oryzaephilus mercator Fauvel – Apricot, walnut
O. suronamensis (Linnaeus) – Almond, apricot, apple, date palm, fig, grapevine, melons, peach, pear, plum, walnut

TENEBRIONIDAE :

Alphitobius laevigatus (Fabrius) – Tamarind
Dasus depressum (Fabricius) – Grapevine
D. hoffmannseggi Steven – Grapevine

Opatrum species – Grapevine
Tribolium castaneum (Herbest) – Apricot, almond, apple, grapevine, peach,
pear, plum
Uloma species – Tamarind
MELOIDAE :
Epicauta horticornis (Hagg-Rutenberg) – Citrus
Mylabric macilenta Marseul – Apple, pear, plum
M. phalerata (Pullas) – Melons, plum
M. puslulata (Thunberg) – Melons
CERAMBYCIDAE :
Acanthophorus serraticarnis Olivier – Mango
Aeolesthes holosericea Fabricius – Apple, apricot, cherry, guava, mulberry,
peach, pear, plum, walnut
A. sarta Solsky – Almond, apple, apricot, cherry, peach, quince, walnut
Aesopida malasiaca Thomson – Citrus
Ag laophis humerosus Chevrolat – Walnut
Anoplophora versteegii (Ritsema) – Mango, citrus
Apriona cinerea Chevrolat – Apple, fig, mulberry, peach
A. rugicollis Chevrolat – Fig, jack-fruit, mulberry
Batocera horsfieldi Hope – Walnut
B. namitor (Newmann) – Mango
B. roylei (Hope) – Mango
B. rubus (Linnaeus) – Mango, jack-fruit
B. rufomaculata (de Geer) – Apple, fig, guava, *gular*, jack-fruit, mango,
pomegranate, wulnut
B. titana (Thomson) – Mango
Blepephaeus succinctor (Chevrolat) – Citrus
Coelosterna (auct. *celosterna*) *scabrator* Fabricius – Apple, *ber*, grapevine
C. spinator Fabricius – Apple, *ber*, pomegranate
Chelidonium cinctum (Guerin-Meneville) – Citrus
Chloridolum alcmene Thomson – Citrus
Coloborhombus fulvus Bates – *Ber*
C. hemipterus Olivier – Guava, *jamun*
Derolus volvulus Fabricius – *Bael, ber*
Dialeges undulatus Gahan – Pear
Demonax balyi Pascoe – Citrus
Dorysthenes Lophosternus hugeli Redtenbacher – Apple, apricot, peach,
pear, plum, walnut
Epepeotes luscos (Fabricius) – Fig, jack-fruit, mango
E. ficicola – Mango
E. parrotiae – Walnut
Glenea beesoni – Walnut
G. belli Gahan – Jack-fruit
G. multiguttata Guerin-Meneville – Mango
G. quatuordecimmaculata (14-maculata) Hope – Cherry
Gnatholea eburifera (Thomson) – Citrus
Linda nigroscutata (Fairmaire) – Apple
Macrotoma crenata Fabricius – Mango
M. plagiata Waterhouse – Apple, walnut

Morimopsis lachrymans Thomson – Walnut
Oberea mangalornsis Foerster – Citrus
Olenecamptus bilobus Fabricius – Fig, *gular*, jack-fruit, mango, mulberry,
                                                                        pomegranate
Peltotrachelus pubes Foerster – Citrus
Perissus laetus Lameere – Pear
Pharsatia proxima Gahan – Mango
Plocaederus obesus Gahan – Mango
P. pedestris White – Mango
Prionus corpulentus Bates – Walnut
Rhaphuma horsfieldi White – Walnut
Rhytidodera bowringi White – Mango
Sthenias grisator Fabricius – Almond, apple, grapevine, citrus, jack-fruit
                                                                        mango, mulberry
Stromatium barbatum (Fabricius) – Citrus, mango
Xoanodera trigona Pascoe – Citrus
Xylotrechus contortus Gahan – Walnut
X. smei Laporte and Gory – *Jamun*, mango, mulberry

CHRYSOMELIDAE :
Altica caerulescens Baly – Apple, cherry, peach, pear, plum, strawberry,
                                                                        walnut
A. cyanea Weber – Water nut
Aphthona species – Apple
Aplosonyx trifasciatus Jacoby – Apple
Aspidomorpha santaecrucis (Fabricius) – Citrus
Caryoborus gonagra (Fabricius) – Tamarind
Cerogria nepalensis Hope – Apple, *ber*, peach, plum, walnut
Colasposoma coeruleatum Baly – Citrus
C. semicostatum Jacoby – Citrus
C. splendidum (Fabricius) – Citrus
Cryptocephalus insubidus Suffrain – Litchi
Dactylispa lohita Maulik – Cherry, walnut
Diapromorpha melanopus Lacordaire – Litchi
D. quadripunctata Jacoby – Litchi
Galerucida rutilans Hope – Apple
Gynadrophthalma species – Mango
Hispa dama Chapuis – Cherry, walnut
Hoplasoma sexmaculata Hope – Peach
Luperus species – Apple
Merista sexmaculata Hope – Apple
M. trifasciata Hope – Apple
Mimastra cyanura Hope – Almond, apple, apricot, chestnut, fig, *falsa*,
                                        grapevine, mulberry, peach, pear plum, pomegranate
Monolepta erthrycephala Baly – Apple, grapevine, peach, pear, plum,
                                                                        walnut
M. fulvifrons Jacoby – Apple
M. khasiensis Weise – *Ber*
M. signata Olivier – Grapevine, mango
Oides scutellata (Hope) – Grapevine

Oocassida cruenta (Fabricius) – *Ber*
O. obscura (Fabricius) – *Ber*
O. pudibunda (Boheman) – *Ber*
Pachnephorus bistriatus Mulsant – Cherry
Phyllotreta chotanica Duvivier – Apple, cherry, walnut
Platypria andrewesi (Weise) – *Ber*
P. echidna Guerin-Meneville – *Ber*
P. erinaceus Fabricius – *Ber*
P. hystrix Fabricius – *Ber*, walnut
Pyrrhalta birmanica (Jacoby) – Water nut
P. rugosa (Jacoby) – Water nut
P. rutilans Hope – Apple
Rhaphidopalpa cincta (Fabricius) – Melons
R. foveicollis (Lucas) – Mango, melons
R. intermedia (Jacoby) – Melons
Rhytidodera bowringi Gahan – Mango
R. simulans White – Mango
Rhytidodera species – Fig
Scelodonta strigicollis Motschulsky – Grapevine, mango
S. dillwyni – Grapevine
Throscoryssa citri Maulik – Citrus
Triclina species – Walnut
Wallaceana dactyliferae – Date palm

EUMOLPHIDAE :

Nodostoma fulvicorne Jacoby – Litchi
N. plagiosum Baly – Walnut
N. subcastatum Jacoby – Banana, grapevine
N. viridipennis Motschulsky – Banana, grapevine

ANTHRIBIDAE :

Araecerus fasciculatus de Geer – Arecanut, citrus, custard apple, dry stems of papaya
A. suturalis Boheman – Dry fruits of custard apple, dry mango buds, dry stems of papaya
Autotropis modesta Jordan – Dry stems of tamarind
Basitropis hamata Jordan – *Ber*
B. nitidicutis Jek. – *Jamun*, mango
Dendrotogus angustipennis Jordan – *Jamun*

CURCULIONIDAE :

Acythopius citrulli Marshall – Water melon
Alcidodes frenatus (Faust) – Mango
A. mali Maskell – Apple
A. porrectirostris Marshall – Walnut
Aclees cribratus Gyllenhal – Fig
Amblyrrhinus poricollis Boheman – Almond, apple, apricot, *bael, ber*, citrus, litchi, mango, peach, plum
Apoderus bistriolatus Faust – Litchi
A. blandus Faust – Walnut
A. tranquebaricus (Fabricius) – Country almond, mango
Atmetonychus peregrinus Olivier – *Ber*, mango, pear, plum

Camptorrhinus mangiferae (Marshall) – Mango
Cionus hortulana Geoffroy – Apple
Cosmopolites sordidus (Germar) – Banana
Deiradoleus species – Apple, pear
Deporaus marginatus Pascoe – Mango
Dereodus pollinosus Redtenbacher – Apple, *ber*
Desmidophorus hebes Fabricius – Mango
Dyscerus clathratus (Pascoe) – Apple
D. fletcheri Marshall – Apple
D. malignus Marshall – Apple
Emperorrhinus defoliator Marshall – Apple, apricot, cherry, peach, pear
Hypera variabilis Herbest – Apple, pear
Hypolixus pica Fabricius – *Ber*
Hypomeces squamosus (Fabricius) – Citrus
Myllocerus blandus Faust – Guava
M. delecatulus Boheman – Litchi
M. dentifer (Fabricius) – Citrus
M. discolor Boheman – Almond, apple, apricot, *bael, ber*, citrus, grapevine,
                         guava, loquat, mango, papaya
M. dorsatus (Fabricius) – Litchi
M. evasus Marshall – Citrus
M. laetivirens Marshall – Almond, apple, apricot, *ber*, citrus, mango,
                         peach, pear, plum, pomegranate, strawberry
M. lefroyi Marshall – Cherry
M. sabulosus Marshall – Apple, *ber*, guava, mango, peach, pear, plum
M. spinicollis Marshall – Apple
M. subfasciatus Guerin-Meneville – Apple
M. transmarinus (Herbest) – *Ber*
M. undecimpustulatus maculosus Desbrochers – Almond, apple, apricot, *ber*,
           guava, litchi, mango, peach, pear, plum, pomegranate, strawberry
M. viridanus (Fubricius) – Guava, walnut
Ochyromera artocarpi Marshall – Jack-fruit
Odioporus lorgicollis Olivier – Banana
Oncideres repandator Faust – Mango
Onychocnemis careyae Marshall – Jack-fruit
Peltotrachelus pubes Faust – *Ber*, citrus, mango
Phytoscaphus indicus Boheman – Citrus
P. triangularis Olivier – Apple, *ber*, pear
Platymycterus sjostedti Marshall – *Ber*, mango
Polytus mullberborgi Boheman – Banana
Rectosternum poriolle (Faust) – Mango
Rhynchaenus mangiferae Marshall – Mango
Rhynchophorus ferrugineus (Olivier) – Date palm
Sitones crinitus Olivier – *Ber*
Sternochetus frigidus (Fabricius) – Mango
S. mangiferae (Fabricius) – Mango
S. decipiens Marshall – Citrus
Tanymecus circumdatus Wiedemann – Apple
T. hispidus Marshall – *Ber*

Teluropus ballardi Marshall – Jack-fruit
Xanthochelus faunus Olivier – *Ber*
X. superciliosis Gyllennal – *Ber*

SCOLYTIDAE (IPIDAE) :

Cryphalus javanus (Eggers) – Tamarind
Hypocryphalus mangiferae Stebbing – Mango
Hypothenemus eupolyphagus Beeson – Mango, tamarind
Pityogenes species – Apple
Scolytoplaytatypus raja Blandford – Apple
Scolytus juglandis Sampson – Apple, walnut
Xyleborus affinis Eichhoff – Guava
X. andrewesi Blandford – *Jamun*, mango
X. bicolor Blandford – *Jamun*
X. discolor Blandford – Walnut
X. fornicatus Eichhoff – Guava
X. kraatzi Eichhoff – Mango
X. laticollis Blandford – *Jamun*
X. semigranosus Blandford – *Jamun*, mango
X. semiopacus Eichhoff – Grapevine
X. testaceus (Walker) – Citrus

PLATYPODIDAE :

Crossotarsus fairmairei Chapuis – Walnut
C. saundersi Chapuis – *Ber, gular, jamun*, mango
Platypus cupulatus Chapuis – *Jamun*
P. indicus Strohmeyer – Jack-fruit
P. solidus Walker – *Gular*, mango

DIPTERA Linnaeus, 1758

CECIDOMYIDAE :

Allassomyia tenuispatha Keiffer – Mango
Amradiplosis allahabadensis Grover – Mango
A. echinogalliperda Mani – Mango
Anjeerodiplosis peshawarensis Mani – Fig
Dasyneura amaramanjarae Grover – Mango
D. citri Grover – Citrus, mango
Erosomyia indica Grover – Mango
Procontarinia mangiferae (Grover) – Mango
P. matteiana Kieffer and Cecconi – Mango
Procystiphora indica Grover – Mango
P. mangiferae (Felt) – Mango
Rhabdophaga mangiferae Mani – Mango
Stictodiplosis moringai Mani – Mango
Udumbarie nainiensis Grover – Fig

TRYPETIDAE :

Carpomyia vesuviana Costa – *Ber*
Dacus ciliatus (Loew) – Apple, melons

D. correctus (Bezzi) – *Bael, ber,* citrus, mango, peach, sapota

D. cucurbitae (Coquillett) – *Bael,* citrus, grapevine, guava, date palm melon, papaya, peach

D. diversus (Coquillett) – Banana, citrus, guava, *jamun,* mango, melons, papaya

D. dorsalis (Hendel) – Apple, apricot, *bael,* banana, *ber,* citrus, fig, guava, jack-fruit, loquat, mango, peach, pear, plum, persimmon, pomegranate, quince, sapota

D. duplicatus (Bezzi) – Peach, pear

D. incisus Walker – Guava, mango

D. maculipennis (Doleschall) – Peach

D. pedestris Bezzi – Papaya

D. scutellaris (Bezzi) – Citrus

D. tau (Walker) – Citrus, mango, mulberry, sapota

D. zonatus (Saunders) – Apple, *bael, ber,* cherry, citrus, custard apple, fig, guava, mango, melons, peach, pear, pomegranate, sapota

Myiapardalis pardalina Bigot – Melons

Toxotrypana curvicauda Gerstacker – Papaya

LEPIDOPTERA Linnaeus, 1758

HELIOZELIDAE :

Antispila anna Fletcher – *Jamun*

A. argosioma Meyrick – Grapevine

TISCHERIIDAE :

Tischeria ptarmica Meyrick – *Ber*

TINEIDAE :

Cerostoma rugosella Stainton – Papaya

Coconympha irriarcha Meyrick – Coconut, palmae

Tinea species – Mango

YPONOMEUTIDAE :

Prays citri Milliere – Citrus

P. endocarpa Meyrick – *Bael,* citrus

PSYCHIDAE :

Acanthopsyche minima Hampson – Banana

A. plagiophleps Hampson – Tamarind

Acanthopsyche species – Mango, pomegranate

Arytania obscura Crawford – Mango

Chalioides vitrea Hampson – Tamarind

Eumeta crameri (Westwood) – Citrus, grapevine, pomegranate, tamarind

Cremastopaychae pendula de Joannis – Plum

Kophene cuprea Moore – Banana

GRACILLARIIDAE :

Acrocercops cramerella Snellen – Litchi

A. gemoniella (Stainton) – Sapota

A. heirocosma Meyrick – Litchi

A. loxias Meyrick – *Jamun*

A. phaeospora Meyrick – *Jamun*
A. syngramma Meyrick – Mango, *jamun*
A. telestis Meyrick – *Jamun*
A. zygonoma Meyrick – Mango
Gracillaria zachrysa Meyrick – Apple
Norosa propolia Hampson – Citrus
Oecadarchis species – Tamarind
Phyllonorycter epichares Meyrick – Apple
P. ganodes (Meyrick) – Apple
P. hapalotoxa Meyrick – Apple
P. iochrysis Meyrick – *Ber*

PHYLLOCNISTIDAE :
Phyllocnistis citrella Stainton – Citrus
P. toparcha Meyrick – Grapevine

GLYPHITERYGIDAE :
Phycodes minor Moore – Fig
P. radiata (Ochsenheimer) – Fig

HELIODINIDAE :
Stathmopoda sycastis Meyrick – Fig
S. theoris Meyrick – Mango
S. trissorrhiza Meyrick – Grapevine
Stathmopoda species – Peach

OECOPHORIDAE :
Psorosticha zizyphi (Stainton) – *Aonla*, citrus

COSMOPTERYGIDAE :
Sathrobrota simplex (Walsingham) – Mango, peach, pomegranate
Microcolona leucosticta Meyrick – Guava
M. technographa Meyrick – Guava

XYLORYCTIDAE :
Nephantis serinopa Meyrick – Date palm
Procometis spoliatrix (Meyrick) – Litchi

GELECHIIDAE :
Anarsia lineatella Zeller – Apricot, mango, peach, plum
A. melanoplecta Meyrick – Mango
A. sagittaria Meyrick – *Ber*
Anarsia species – Sapota
Chelaria rhicnota Meyrick – Mango
C. spathota Meyrick – Mango
Chelaria species – Sapota
Holcocera pulverea (Meyrick) – Tamarind
Hypatima haligramma Meyrick – Mango
Idiophantis acnthopa Meyrick – *Jamun*
Sitotroga cerealella (Olivier) – Apricot, fig, grapevine

CARPOSINIDAE :
Meridarchis reprobata Meyrick – *Ber, jamun*, olive
M. scyrodes Meyrick – *Ber, jamun*, olive

COSSIDAE :

Zeuzera coffeae N ietner – Citrus, custard apple, guava, loquat, litchi, pomegranate

Zeuzera Species – Apple, cherry, plum

METARBELIDAE :

Indarbela dea (Swinhoe) – Mango

I. quadrinotata (Walker) – *Ber,* citrus, guava, *jamun,* litchi, loquat, mango, mulberry, pomegranate

I. tetraonis (Moore) – *Aonla, ber,* citrus, *falsa,* guava, jack-fruit, *jamun,* litchi, mango

I. theivora (Hampson) – Mango

LIMACODIDAE (COCHLIDIDIIDAE) :

Belippa laleana Moore – Apple, banana, mango, pear

Contheyla rotunda Hampson – Banana

Latoia lepida (Cramer) – Banana, citrus, country almond, fig, mango, pomegranate

L. repanda (Walker) – Almond, apricot

Microlimax species – Banana

Miresa decedena Walker – Banana

Nemata lohor (Moore) – Banana

Phocoderma velutina (Kollar) – Mango

Thosea aperiens (Walker) – Tamarind

ZYGAENIDAE :

Aglaope ahyalina (Kollar) – Apple

Eterusia pulchella Kollar – Apple

TORTICIDAE (EUCOSMIDAE) :

Ancylis aromatias Meyrick – *Ber*

A. lutescens Meyrick – *Ber*

Argyroploce mormopa Meyrick – *Jamun*

Cryptophlebia illepida (Butler) – *Bael, ber,* citrus, litchi, tamarind

Cydia palamedes (Meyrick) – Tamarind

Dudua aprobola (Meyrick) – *Jamun,* litchi, mango

Enarmonia anticipans walker – Mango

Eucosma ocellana-Schiffer Muller – Apple, apricot, pear, plum, quince

Gateseclarkeana erotias (Meyrick) – Mango

Hedyia cellifera (Meyrick) – *Jamun*

Laspeyresis pomonella (Linnaeus) – Apple

L. pseudonectis Meyrick – Banana

Olethreutes leucaspis (Meyrick) – Litchi

Strepsicrates rhothia (Meyrick) – Guava, *jamun,* mango

Acleris extensana (Walker) – Apple

Archips micaceana (Walker) – Apple, apricot, citrus, guava, *jamun,* litchi, peach, pear

A. subsidiaria (Meyrick) – Apple

A. termias (Meyrick) – Apple, apricot, cherry, pear

Argyrotaenia franciscana (Walsingham) – Apple

Homona coffearia (Nietner) – *Jamun*

Rhopobota raevana Hubner – Apple, pear
Ulodemis trigrapha Meyrick – Citrus
PYRALIDAE* :
    Anonaepestis bengalella (Regonot) – Custard apple
    Aphomia gularis Zeller – Tamrind
    Assara albicostalis Walker – Tamarind
    Cadra, cautella (Walker)–Almond, fig, grapevine, peach, pear, plum, tamarind
    Euzophera punicaella (Moore) – Apple, pear, pomegranate
    Hyapsila leuconeurella Ragonot – Mango
    Nephopteryx eugraphella (Ragonot) – Mango, sapota
    Phycita orthoclina Meyrick – Tamarind
    P. umbrayelis Hampson – Mango
    Plodia interpunctella (Hubner) – Apricot, grapevine, walnut
    Lamida moncusalis (Walker) – Mango
    L. sordidalis (Hampson) – Mango
    Orthaga euadrusalis Walker – Mango
    O. exvinacea Hampson – Mango
    O. mangifarae Misra – Mango
    Orthaga species – *Jamun*
    Stericta species – Mango
    Trioza jambolanae Crawford – *Jamun*
    Ctenomereistia ebriola Meyrick – Mango
    Diaphania bivitralis (Guenee) – Jack-fruit
    D. caesalis (Walker) – Jack-fruit, melons
    D. indica (Saunders) – Melons
    D. itysalis (Walker) – Fig
    D. pyloâlis (Walker) – Fig
    D. stolalis (Guenee) – Fig
    Dichocrocis punctiferalis (Guenee) – Apricot, citrus, guava, jack-fruit,
        mango, mulberry, peach, pear, plum, pomegranate, tamarind
    Herculia species – Mango
    Lygropia quaternalis (Zeller) – *Falsa*
    Noorda albizonalis Hampson – Mango
    Sylepta lunalis Guenee – Grapevine
    Synclera univocalis (Walker) – *Ber*

PTEROPHORIDAE :
    Exelastis atmosa (Walsingham) – Grapevine, *jamun*
    Oxyptilus rugulus Meyrick – Grapevine

NYMPHALIDAE :
    Euthalia garuda (Moore) – Mango
    Kallima inachns Boisduval – Peach

LYCAENIDAE :
    Baspa melampus (Stoll) – Mango
    Chilades laius (Cramer) – Citrus
    Deudorix epijarbas (Moore) – Litchi, pomegranate
    Freyeria putli (Kollar) – Citrus

---

*This has been split into five families, the Phycitidae, Crambidae, Galleriidae, Pyralididae and Pyraustidae – but these have been treated as sub-families. (Richards and Davies 1977)

Jamides bochus Stoll – Citrus
Rapala nissa (Kollar) – Apple
R. varuna (Horsfield) – Guava
Rapala species – Litchi
Tarucus theophastus (Fabricius) – *Ber*, citrus
Virachola isocrates (Fabricius) – *Aonla*, apple, citrus, guava, litchi, loquat, peach, pear, plum, pomegranate, sapota, tamarind

PAPILIONIDAE :

Papilio demoleus Linnaeus – *Bael, ber*, citrus
P. helenus Linnaeus – Citrus
P. h. daksha Moore – Citrus
P. machaon Linnaeus – Citrus
P. m. asiatica Menestries – Citrus
P. memnon Linnaeus – Citrus
P. polyctor Boisduval – Citrus
P. polymnestor Cramer – Citrus
P. polytes Linnaeus – Citrus
P. porinda Moore – Citrus
P. protenor Cramer – Citrus
P. rumanzovia Eschscholz – Citrus

HESPERIIDAE :

Erionota thrax (Linnaeus) – Banana
Gangara thyrsis (Fabricius) – Coconut, palmae
Suastus gremius (Fabricius) – Date palm

GEOMETRIDAE :

Chloroclystis species – Mango
Comostola laesaria (Walker) – Mango
Delinia nedardaria Herrich-Schaffer – *Ber*
Gymnoscelis imparatilis (Walker) – Mango
Hyposidra successaria (Walker) – *Jamun*
Oenospila flavifusata (Walker) – *Jamun*
O. quadraria (Guenee) – Litchi, mango, tamarind
O. veraria (Guenee) – Litchi, mango

LASIOCAMPIDAE :

Malacosoma indica (Walker) – Almond, apricot, cherry, pear, walnut
Metanastria hyrtaca (Cramer) – Country almond, *jamun*, sapota
Paralebeda plagifera (Walker) – Citrus
Streblote siva (Lefebvre) – *Ber*, guava
Suana concolor (Walker) – Citrus
Trabala vishnou (Lefebvre) – Country almond, *jamun*

BOMBYCIDAE (including EUPTEROTIDAE) :

Eupterote geminata Walker – Banana
Ocinara varians Walker – Fig, jack-fruit

SATURNIIDAE :

Actias selene (Hubner) – Apple, cherry, pear, walnut
Antheraea pernyi Guerin-Meneville – Apple, pear, walnut
A. paphia (Linnaeus) – *Ber*

Caligula simla Westwood – Apple, pear, walnut
Cricula trifenestrata (Helfer) – *Ber*, mango

SPHINGIDAE

Acherontia styx (Westwood) – Mango, grapevine
Agnosia microta (Hampson) – *Falsa*
Agrius convolvuli (Linnaeus) – Grapevine
Hippotion boerhaviae (Fabricius) – Grapevine
H. celerio (Linnaeus) – Grapevine
Langia zeuzeroides Moore – Apple, apricot, cherry, pear
Macroglossum belis (Linnaeus) – Citrus
Oxyambulyx sericeipennis Butler – Walnut
Rhyncholoba acteus Cramer – Grapevine
Thereta alecto Linnaeus Grapevine
T. gnoma (Fabricius) – Grapevine
T. oldenlandiae (Fabricius) – Grapevine
T. pallicosta (Walker) – Grapevine

NOTODONTIDAE :

Neostauropus alternus (Walker) – Mango, tamarind
Thiacidas postica Walker – *Ber*

ARCTIIDAE :

Asura ila Moore – Banana
A. ruptofascia Hampson – Mango
Celama analis William and Westwood – Mango
C. fasciata (Walker) – Mango
Creatonotus gangis (Linnaeus) – Banana, pomegranate
C. transiens (Walker) – Citrus, *gular*
Diacrisia obliqua (Walker) – Banana, papaya
Estigmene perrottetic Guenee – Citrus
Pericallia ricini (Fabricius) – Banana

NOCTUIDAE :

Achaea janata (Linnaeus) – *Ber*, citrus, grapevine, guava, mango, pomegranate
Adris sikkimensis Butler – Citrus
Agrotis ipsilon (Hufnagel) – Almond, apples, apricot, peach, pear, plum, strawberry
A. segetum (Schiffer-Muller) – Apricot, peach, strawberry
Amyna octo Guenee – Apple
Anomis flava Fabricius – Grapevine
A. fulvida Guenee – Citrus, guava
Anua mejanesi Guenee – Citrus
A. triphaenoides Walker – *Jamun*
Arcyophora dentula Ledrev – Pomegranate
Bombotelia delatrix Guenee – *Jamun*
B. jocosstrix Guenee – Mango
Calpe emarginata (Fabricius) – Citrus, guava, mango
C. fasciata Moore – Citrus
C. ophideroides Guenee – Citrus

Calyptra bicolor (Moore) – Citrus, *jamun*
Carea chlorostigma Hampson – Jamun
C. subtilis Walker – Jamun
Chlumetia transversa (Walker) – Litchi, mango
C. alternans Moore – Mango
Ercheia cyllaria Cramer – Citrus
Erebus heiroglyphica (Drury) – Citrus
Eublemma abrupta (Walker) – Mango
E. angulifera Moore – Mango, tamarind
E. brachygonia Hampson – Mango
E. silicula Swinhoe – Mango
E. versicolor (Walker) – Mango
Giaura sceptica Swinhoe – *Falsa*
Grammodes geometrica Fabricius – Citrus, guava
G. stolida Fabricius – Grapevine
Heliothis armigera (Hubner) – Apple, citrus, fig
H. assulata Guenee – Fig
Lagoptera dotata (Fabricius) – Citrus, grapevine
L. honesta (Hubner) – Citrus
L. submira (Walker) – Citrus
Maruca testulalis Geyer – Citrus, guava
Mocis frugalis (Fabricius) – Citrus, grapevine
Nanaguna breviuscula Walker – Mango
Ophiusa coronata (Fabricius) – Citrus, grapevine
O. tirhaca (Cramer) – Citrus, pomegranate
Othreis ancilla (Cramer) – Citrus
O. aurantia Moore – Citrus
O. cajeta (Cramer) – Citrus, grapevine
O. cocalus Cramer – Citrus
O. discrepans Walker – Citrus
O. fullonia (Clerck) – Citrus, grapevine, guava, mango, peach, pear, plum, pomegranate
O. hypermnestra Cramer – Citrus
O. materna (Linnaeus) – Citrus, grapevine. mango, pear
O. slaminia (Fabricius) – Citrus
O. tyrannus (Guenee) – Citrus
Pandesma quenavadi Guence – Citrus
Parallelia algira (Linnaeus) – Ber, citrus, grapevine
P. palumba Guenee – Citrus
Pelochyta astrea Drury – Citrus
Pericyma glaucinans (Guenee) – Citrus, grapevine
Perigea capensis Guenee – Grapevine
Phyllodes consobrina Westwood – Citrus
Polydesma umbricola Boisduval – Grapevine
Remigia archesia Cramer – Grapevine
Selepa celtis Moore – *Aonla*, fig, litchi, mango
Serrodes inara Cramer – Citrus, grapevine
Sphingomorpha chlorea Cramer – Citrus, grapevine
Spodoptera litura (Fabricius) – Apple, banana, grapevine

Sympis rufibasis Guenee – Litchi
Tiracola plagiata (Walker) – Apple, banana

LYMENTRIIDAE :

Beara dichromella Walker – *Ber*
Dasychira mendosa (Hubner) – Apple, banana, citrus, guava, mango, mulberry, peach, pear
D. m. fusiformis (Walker) – Citrus
Euproctis flava (Bremer) – Apple, *ber*, grapevine, mango, pear, plum
E. fraterna (Moore) – Apple, apricot, banana, *ber*, citrus, *falsa*, grapevine, jamun, mango, peach, plum, pomegranate, quince, strawberry
E. lunata (Walker) – Apple, *ber*, grapevine, mango, mulberry, plum, pomegranate
E. signata Blanchard – Apple, plum
E. subnotata (Walker) – Grapevine, guava
E. virguncula Walker – Grapevine
E. xanthosticha (Hampson) – Mango
Lymantria ampla Walker – Apple, guava, mango, mulberry, pomegranate
L. beatrix Stoll – Mango
L. concolor Walker – Apple
L. dispar (Linnaeus) – Apple
L. mathura Moore – Mango
L. obfuscata Walker – Apple, apricot, peach
Perina nuda (Fabrtcius) – Fig, jack-fruit, mango
Porthesia scintillans (Walker) – Apple, mango, pomegranate

HYPSIDAE :

Aganais ficus (Fabricius) – Fig, grapevine
Argina cribraria (Clerck) – Citrus, guava
Asota alciphron (Cramer) – Fig

HYMENOPTERA Linnaeus, 1758

CHALCIDIDAE (EURYTOMIDAE) :

Eurytoma samsonowi Vasiljev – Apricot
Lipothymus bakeri Joseph – Fig
Myrmicaria brunnea Saunders – Citrus

FORMICIDAE :

Camponotus species – Citrus
Dorylus orientalis Westwood – Mango
Oecophylla smaragdina (Fabricius) – Citrus, *jamun*, jack-fruit, litchi, mango, sapota

VESPIDAE :

Polistes olivaceus (Fabricius) – Citrus, grapevine
Vespa auraria Smith – Apple
V. cincta Fabricius – Sapota
V. orientalis Linnaeus – Citrus, grapevine, peaeh, pear, plum, pomegranate

MEGACHILIDAE :

Megachile anthracina Smith – Loquat

7.2
MITES
(Attacking fruit trees)

ACARIDAE, Ewing Nesbitt
 Tyrophagus putrescentiae (Schrank) – Mango

ERYTHRAEIDAE, Oudemans
 Abrolophus ripicola (Womersley) – Orchard
 Leptus giganticus Khot – Citrus
 L. poonaensis Khot – Fig, pear

TARSONEMIDAE, Kramer
 Polyphagotarsonemus patus (Banks) – Citrus

CHEYLETIDAE, Leach
 Cheletogenes ornatus (Canestrini and Fangago) – Fig, mango
 Cheletomorpha lepidopterorum (Shaw) – Orange

STIGMAEIDAE, Oudemans
 Agistemus fleschneri Summers

TENUIPALPIDAE, Berlese
 Brevipalpus californicus (Banks) – Citrus
 B. deleoni Pritchard and Baker – Citrus, guava
 B. oboratus (Donnadieu) – Citrus
 B. phoenicis (Geijskes) – Citrus, grapevine, guava, papaya, pomegranate
 Lervacarus transitans (Ewing) – *Ber*
 Raoiella indica Hirst – Banana, palmae
 Tenuipalpus punicae Pritchard and Baker – Pear, pomegranate

TETRANYCHIDAE, Donnadieu
 Bryobia praetiosa Koch – Pear
 Eotetranychus hirsti Pritchard and Baker – Fig
 E. kankitus Ehara – Citrus
 E. mandensis Manson – Citrus
 E. pamelae Manson – Citrus
 E. suginamensis (Yokoyama) – Mulberry
 Eutetranychus truncatus Estebanes and Baker – Grapevine
 E. orientalis (Klein) – Almond, *ber*, citrus, peach, pear
 E. africanus (Tucker) – Orange
 F. coffeae (Nietner) – Citrus
 Oligonychus indicus (Hirst) – Banana
 O. biharensis (Hirst) – Arecanut, grapevine, litchi
 O. mangiferus (Rahman and Sapra) – Grapevine, *jamun*, mango
 O. punicae (Hirst) – Grapevine, pomegranate

Panonychus citri (McGregor) – Citrus
P. ulmi (Koch) – *Ber*, peach, pear
Petrobia latens (Muller) – Citrus
Petrobia species – Pear
Schizotetranychns baltazarae Rimando – Citrus
S. hindustanicus (Hirst) – Citrus
Tetranychus cinnabarinus (Boisduval) – Apple, almond, cherry, citrus, grapevine, guava, melons, papaya, peach, pear, strawberry
T. fijiensis Hirst – Citrus
T. neocaledonicus Andre – Citrus, fig, grapevine, mango, papaya
T. sexmaculatus Riley – Citrus

TYDEIDAE, Kramer

Pronematus elongatus Baker – Grapevine
P. fleschderi Baker – Citrus

ERIOPHYOIDAE, Naplepa

Aceria ficus (Cotte) – Fig
A. granati (Canestrini and Massalongo) – Pomegranate
A. litchi (Keifer) – Litchi
A. mangiferae Sayed – Mango
A. mori (Keifer) – Mulberry
A. sheldoni (Ewing) – Citrus
Diptilomiopus assamica Keifer – Citrus
Eriophyes cernnus Massee – *Ber*
E. ficivorus Channa Basavanna – Fig
E. pyri (Pagenstecher) – Pear
E. rubifolii Channa Basavanna – Raspberry, strawberry
E. vitis (Pagenstecher) – Grapevine
Floracarus fleshneri Keifer – Mandarin orange
Metaculus mangiferae (Attiah) – Mango
Neocalacarus mangiferae Channa Basavanna – Mango
*Paratetra murraya*ᵉ Channa Basavanna – Tamarind
Phyllocoptruta oleivorus (Ashmead) – Citrus
Rhyncaphytoptus ficifoliae Keifer – Fig

AMEROSEIIDAE, Berlese

Neocypholaelaps pradhani Gupta – Apple, pear

ASCIDAE, Voigts and Oudemans

Asca biswasi Bhattacharya – Soil under banana
A. Imnuta Bhattacharya – Soil under banana
A. pseudospicata Bhattacharya – Soil under banana and citrus
Lasioseius quadrisetosus Chant – Citrus
Melichares fici Narayanan and Ghai – Fig

EVIPHIDIDAE, Berlese

Eviphis convergens Berlese – Soil under banana

PHYTOSEIIDAE, Berlese

Amblyseius aerialis (Muma) – Citrus
A. asiaticus (Evans) – Mango

A. assamensis Chant – Citrus
A. coccosocius Ghai and Menon – Citrus, mango
A. cucumeris (Oudemans) – Citrus
A. delhiensis (Narayanan and Kaur) – Citrus, grapevine
A. finlandicus (Oudemans) – Guava, pear
A. hima Pritchard and Baker – Citrus
A. hibisci (Chant) – Citrus
A. kalimpongensis Gupta – Citrus
A. largoensis (Muma) – Citrus
A. mangiferae Ghai and Menon – Mango
A. mcmurtryi Muma – Citrus
A. ovalis (Evans) – Banana, fig
A. para aeriallis Muma – Citrus
A. rykei Pritchard and Baker – Fig
A. salebrosus Chant – Citrus
A. sojaesnis Ehara – Almond, peach
Indodromus meerutensis Ghai and Menon – Mango
Phytoseius coheni Swirski and Schechter – Guava
P. intermedius Evans and Macfarlane – Fig, mango
P. macropilis (Banks) – Fig, mulberry
Typhlodromus bakeri (Garman) – Grapevine, litchi
T. caudiglans Schuster – Mango
T. fleschaeri Chant – Orange
T. insularis Ehara – *Jamun*
T. nesbitti Womersley – Mango
T. pruni Gupta – Peach
T. rhenanus (Oudemans) – Citrus, mango
T. rickeri Chant – Citrus
T. roshanlali Narayanan and Ghai – Mango

NEMATODES

(Associated with fruit trees

Aphelenchoides besseyi – Strawberry
Criconema species – Pomegranate
Dolichodorus pulvinus – Palmae
Discocriconemella recensi – Palmae
Helicotylenchus dihystera – Banana
Helicotylenchus erythrinae – Mango
H. multicinctus – Banana, pineapple
Hemicriconemoides mangiferae – Mango
Hoplolaimus indicus – Mango
Longidorus africanus – Fig, grapevine
L. saginus – Palmae
Meloidogyne arenaria – Banana, grapevine
M. exigua – Banana, citrus
M. hapla – Strawberry
M. incognita – Grapevine, peach, plum
M. javanica – Fruits
Paranlogidorus flexus – Palmae
Pratylenchus brachyurus – Citrus, peach, pineapple
P. coffeae – Banana, citrus
P. vulnus – Grapevine, olive, peach, walnut
Radopholus similis – Banana, citrus
Rotylenchus reniformis – Banana, papaya, pineapple
Scutellonema bradys – Banana, palmae, pineapple
Tylenchulus semipenenetrans – Citrus, grapevine, olive
Xiphinema bakeri – Raspberry
X. index – Grapevine

What's the use of their having names
if they won't answer to them ?
No use to them – but it is useful
to the people that name them

7.4

FRUITS

(Grown in India)

PALMACEAE :
    Areca catechu Linnaeus – Àrecanut, *Sopari*
    Cocos nucifera Linnaeus – Coconut palm, *Nariyal*
    Phoenix dactylifera Linnaeus – Date palm, *Khajoor*
    P. sylvestris Roxburghii – Toddy palm, *Taari*

BROMELIACEAE :
    Ananas comosus Merril – Pineapple. *Ananas*

MUSACEAE :
    Musa paradisiaca Linnaeus – Plantain, *Sabzi kela*
    M. sapientum Linnaeus – Banana, *Kela*

TRAPACEAE :
    Trapa bispinosa Roxburghii – Waternut, *Singhara*
    T. natans Linnaeus – Water chestnut, *Paniphal*

JUGLANDACEAE :
    *Juglans regia Linnaeus – Walnut, *Akhrot*

BETULACEAE :
    Corylus avellana Filbert – Hazelnut, *Bhotia badam*

FAGACEAE :
    Castanea sativa Miller – Chestnut

MORACEAE :
    Artocarpus altilis Fosberg – Bread-fruit
    A. lakoocha Roxburghii – Monkey-jack, *Barahal*
    A. heterophyllus Lamark – Jack-fruit, *Katahal*
    Ficus carica Linnaeus – Fig, *Anjeer*
    F. glomerata Roxburghii – Cluster fig, *Gular*
    F. hispida Linnaeus – Bokeda, *Jhangli anjeer*
    Morus alba Linnaeus – White mulberry, *Shahtoot*
    M. nigra Linnaeus – Black mulberry, *Toot*
    M. rubra Linnaeus – Red mulberry

ANNONACEAE :
    Annona atemoya Hort – Atemoya, *Lakshmanphal*
    A. cherimolia Miller – Cherimoya, *Hanumanphal*
    A. muricata Linnaeus – Soursop, *Mamphal*
    A. reticulata Linnaeus – Bullock's heart, *Ramphal*
    A. squamosa Linnaeus – Custard apple, *Sitaphal*, *Sharifa*

LAURACEAE :
    Persea americana Miller – Avocado, *Avocado*

ROSACEAE :
    Amygdalis communis Linnaeus – Almond, *Badam*
    Armeniaca vulgaris Lamark – Apricot, *Khurmani*

*Juglans nigra (black walnut) and J. cinerea (butter nut or white walnut) are grown mainly in Eastern U.S.A. and Canada.

Cydonia oblonga Miller – Quince, *Amritphal*
Eriobotrya japonica Lindley – Loquat, *Lokat*
Fragaria vesca Linnaeus – Strawberry, *Straberi*
Malus baccata (Linnaeus) – Crab apple, *Jhungli sev*
M. sylvestris Miller – Apple, *Sev*
Prunus avium Linnaeus – Sweet cherry, *Gilas*
P. cerasus Linnaeus – Sour cherry, *Aelche*
P. serotina Ehrhart – Black cherry, *Katsol*
P. domestica Linnaeus – Plum, *Alubukhara*
P. persica Sieborp and Zuccarini – Peach, *Aroo*
P. salicina Lindley – Japanese plum
Pyrus communis Linnaeus – Pear, *Naspati*
P. serotina culta Rehder – Sand pear, *Bagugosha, Nak*
Rubus alleghaniensis Porter – Blackberry
R. ellipticus Smith – Yellow raspberry, *Heer, Hinsar*
R. fruticosus Linnaeus – Blackberry, *Chhanch*
R. idaeus Linnaeus – Raspberry, *Rasberi*
R. lasiocarpus Smith – Black raspberry, *Kala anchu*

LEGUMINOSAE :
   Pithecolobium dulce Benth – Manila tamarind, *Vilaiti Imli*
   Tamarindus indica Linnaeus – Tamarind, *Imli*
OXALIDACEAE :
   Averrhoa bilimbi Linnaeus – Bilimbi
   A. carambola Linnaeus – Carambola, *Kamrakh*
RUTACEAE :
   Aegle marmelos Correa – Bengal quince, *Bael*
   Citrus aurantifolia Swingle – Sour lime, *Kagzi nimbu*
   C. aurantium Linnaeus – Sour orange, *Khatta*
   C. grandis Osbeck – Pomelo, *Chakotra*
   C. jambhiri Lush – Rough lemon, *Jambori*
   C. limettioides Tanaka – Sweet lime, *Mitha nimbu*
   C. limon Burmann – Lemon, *Nimbu*
   C. limonia Osbeck – Mandarin lime, *Bara nimbu*
   C. medica Linnaeus – Citron, *Gulgul*
   C. paradisi Macfadyen – Grapefruit, *Vilaiti chakotra*
   C. reticulata Blanco – Mandarin orange, *Santra*
   C. sinensis Osbeck – Sweet orange, Malta, *Mosambi*
   Feronia limonia (Linnaeus) – Wood apple, *Kaith*
   Fortunella japonica Swingle – Kumquat

EUPHORBIACEAE :
   Emblica officinalis Gaerth (= *Phyllanthus emblica*) – Aonla, *Amla*
   Phyllanthus acidus Skeels – Star gooseberry, *Hariphal*
   P. indica – Gooseberry, *Rasbheri*
ANACARDIACEAE :
   Anacardium accidentale Linnaeus – Cashewnut, *Kaju*
   Buchanania lanzan Sprengel – Almondette, *Chivoli, Chironji*
   Mangifera indica Linnaeus – Mango, *Aam*
   Pistacia vera Linnaeus – Pistachio, *Pista*

SAPINDACEAE :
  Litchi chinensis Sonnerat – Litchi, *Lychee*
  Sapindus laurifolius Vahlenberg – Soapnut, *Aritha*
RHAMNACEAE :
  Zizyphus mauritiana Lamarck – Jujube, *Ber*
VITAGEAE :
  Vitis vinifera Linnaeus – Grapevine, *Angoor*
TILIACEAE :
  Grewia asiatica Linnaeus – Phalsa, *Falsa*
STERCULIACEAE :
  Sterculia foetida Linnaeus – *Pinari*
GUTTIFERAE :
  Garcinia indica Choisy – *Kookam*
  G. maugostana Linnaeus – Mangosteen
PASSIFLORACEAE :
  Passiflora edulis Sims – Passion fruit, *Possion phal*
CARICACEAE :
  Carica papaya Linnaeus – Papaya, *Papita*
PUNICACEAE :
  Prunica granatum Linnaeus – Pomegranate, *Anar*
COMBRETACEAE :
  Terminalia catappa Linnaeus – Country almond, *Desi badam*
MYRTACEAE :
  Psidium guajava Linnaeus – Guava, *Amrood*
  Syzygium cuminii Skeels – Java plum, *Jamun*
  S. jambos Alston – Rose apple, *Gulab jamun*
SAPOTACEAE :
  Archas (Malicarna) zapota Linnaeus – Sapota, *Chikoo*
  Chrysophyllum cainito Linnaeus – Star apple
  Manilkara hexandra Dubard – Indian plum, *Khirnee*
EBENACEAE :
  Diospyros virginiana Linnaeus – Persimmon, *Tendu*
  D. kaki Linnaeus – Japanese persimmon, *Halwa tendu, Japani phal*
  D. lotus Linnaeus – Dateplum persimmon, *Amlot, Chini phal*
APOCYNACEAE :
  Carissa congesta Wt. (=*C. carandas* Linnaeus) – *Karonda*
BORAGINACEAE :
  Cordia dichotoma (Forster) (= *C. myxa* Linnaeus) – Sebasten, *Lesura*
OLEACEAE :
  Olea europaea Linnaeus – Olive, *Zetoon*
SOLANACEAE :
  Cyphomandra betacea Sendtner – Tree tomato, *Tamatar*
  Physalis peruviana Linnaeus – Cape goosebery, *Tepari*
CUCURBITACEAE :
  Citrullus vulgaris Schrader – Watermelon, *Tarbuz*
  Cucumis melo Linnaeus – Musk melon, *Kharbuz*
  C. melo momordica Duthie and Fuller – Snap melon, *Phoot*
  C. melo utilissimus Duthie and Fuller – Long melon, *Kakri*
  C. sativus Linnaeus – Cucumber, *Kheera*

8
INDEX

SPECIES INDEX

Insulaspis gloverii (Packard)

Coccus hesperidum Linnaeus

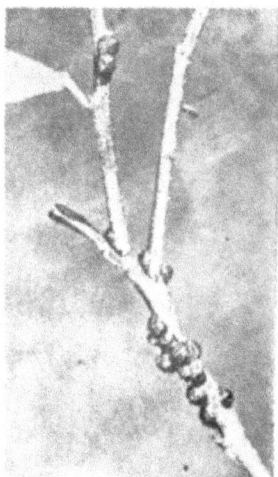

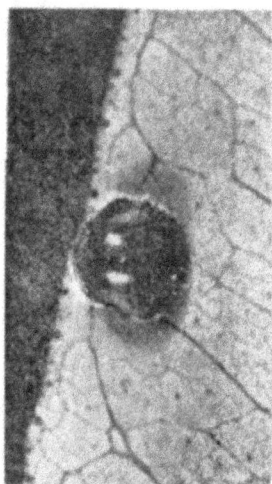

Saissetia coffeae (Walker)

(Courtesy : Gerold Heim)

LEMON BUTTERFLY (*PAPILIO DEMOLEUS* LINNAEUS)

Eggs & Caterpillars

Butterflies ♂♀

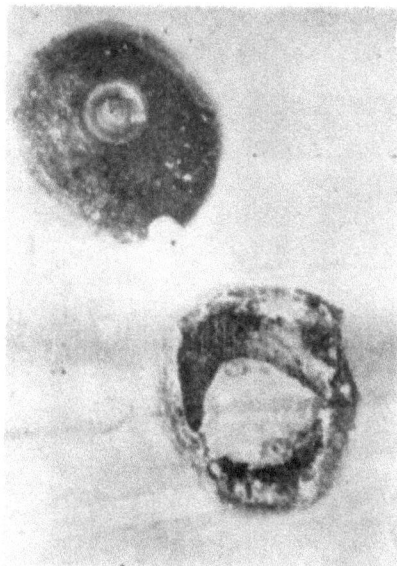

Chrysomphalus ficus Ashmead

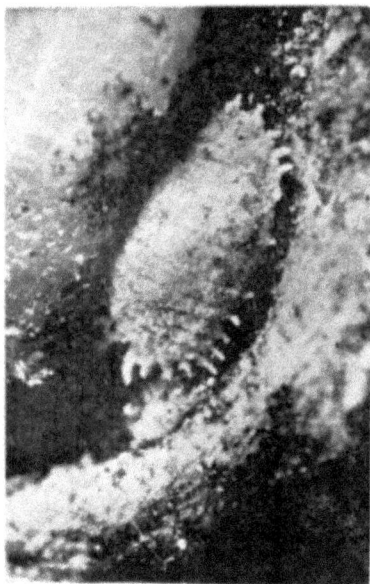

Planococcus citri (Risso)

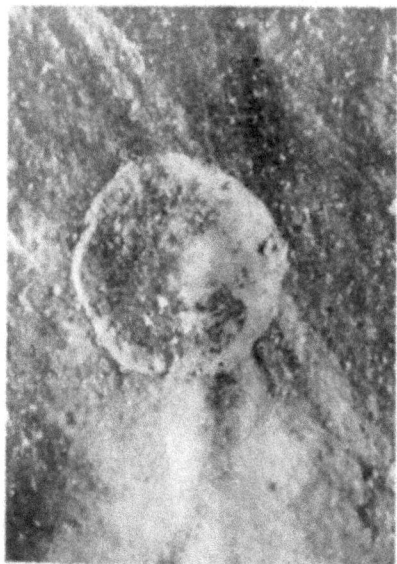

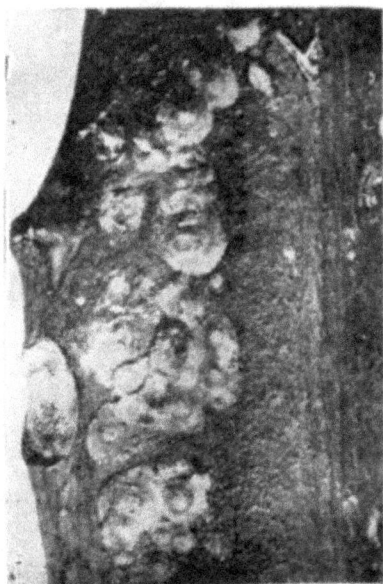

Aonidiella aurantii (Maskell)

(Courtesy : Gerold Heim)

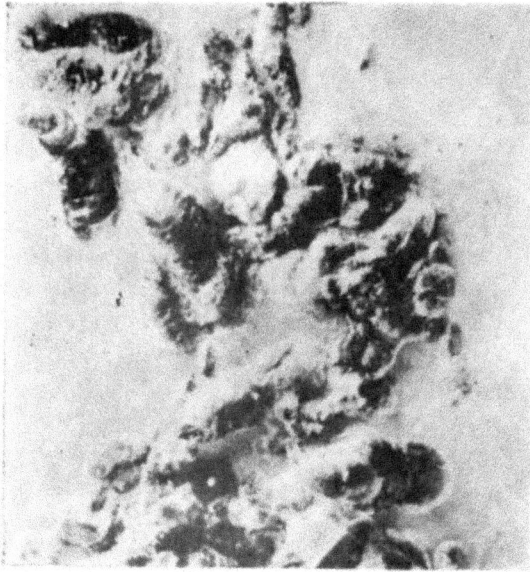

Cornuaspis beckii (Newman)

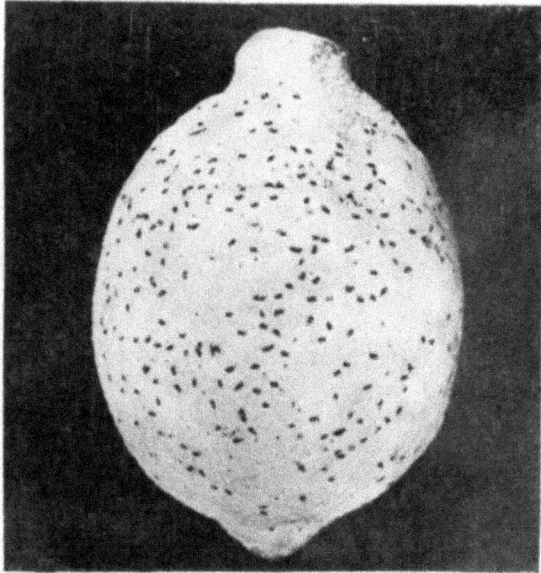

Parlatoria zizyphus Lucas on lemon

(Courtesy : Gerold Heim)

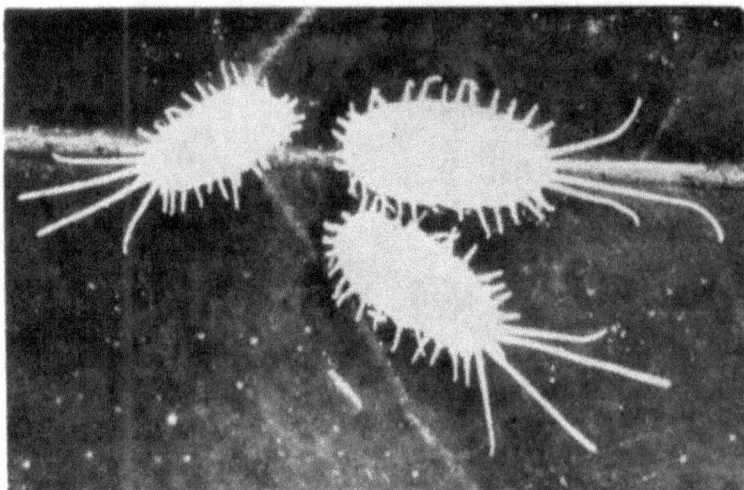

Pseudococcus adonidum (Linnaeus)

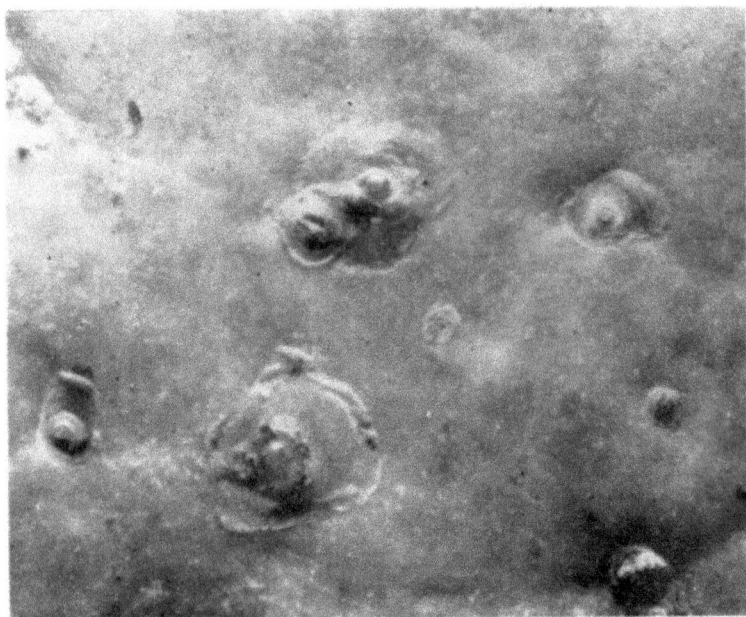

Chrysomphalus dictyospermi (Morgan)
(Courtesy : Gerold Hiem)

(Courtesy : D.O.Garg)

Fruit flies ovipositing on citrus fruit

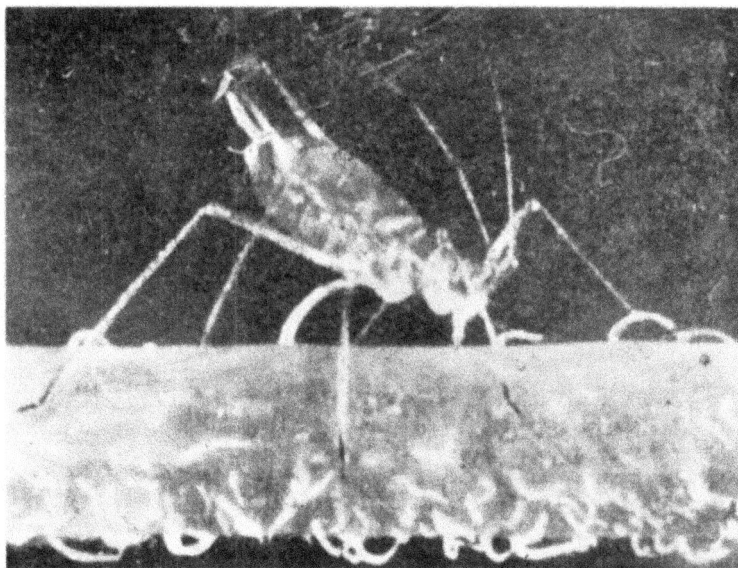

Aphid feeding on citrus leaf petiole

COCKCHAFER BEETLE

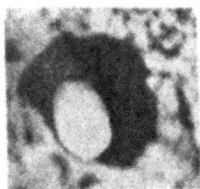

Egg

Grubs

Pupa

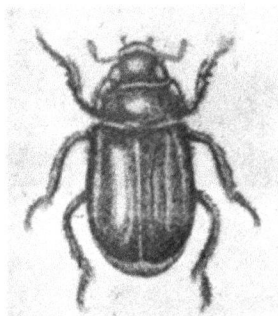

Adult

Damaged grapevine leaves

(Courtesy : M. G. Jotwani)
Pyrrhalta birmanica (Jacoby) – beetles feeding on *singhara* leaves

(Courtesy : G. M. Talati)
Nephantis serinopa Meyrick – larvae feeding on banana leaf

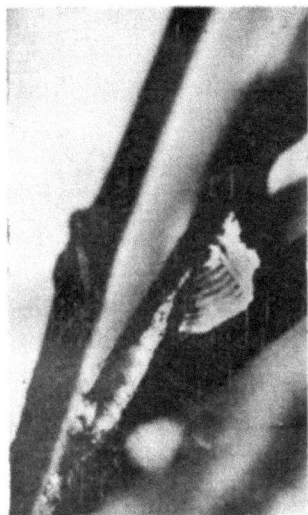

Icerya purchasi Maskell

Parlatoria oleae (Colvee) on peach leaf

Healthy peach fruits

Fruits damaged by
Dacus zonatus (Saunders)

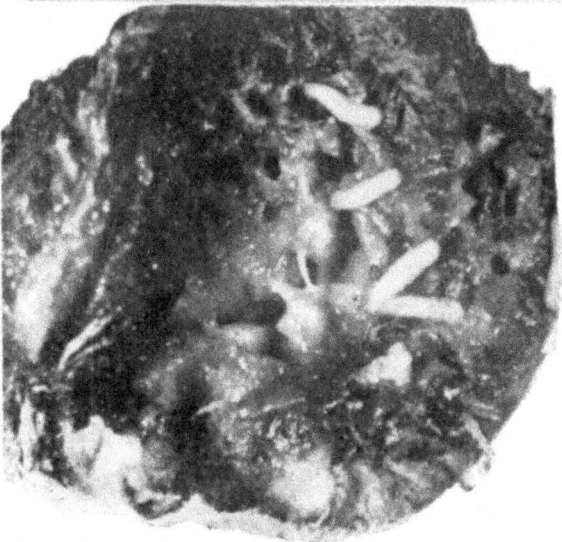

Damaged fruit cut open
to show maggots of fruit fly